Cultural Landscapes Preservation and Social–Ecological Sustainability

Cultural Landscapes Preservation and Social–Ecological Sustainability

Editors

María Fe Schmitz
Cristina Herrero-Jáuregui

MDPI • Basel • Beijing • Wuhan • Barcelona • Belgrade • Manchester • Tokyo • Cluj • Tianjin

Editors

María Fe Schmitz
Biodiversity, Ecology and
Evolution
Complutense University of
Madrid
Madrid
Spain

Cristina Herrero-Jáuregui
Biodiversity, Ecology and
Evolution
Complutense University of
Madrid
Madrid
Spain

Editorial Office
MDPI
St. Alban-Anlage 66
4052 Basel, Switzerland

This is a reprint of articles from the Special Issue published online in the open access journal *Sustainability* (ISSN 2071-1050) (available at: www.mdpi.com/journal/sustainability/special_issues/ cultural_landscapes_preservation).

For citation purposes, cite each article independently as indicated on the article page online and as indicated below:

LastName, A.A.; LastName, B.B.; LastName, C.C. Article Title. *Journal Name* **Year**, *Volume Number*, Page Range.

ISBN 978-3-0365-2571-6 (Hbk)
ISBN 978-3-0365-2570-9 (PDF)

Contents

About the Editors . . . . **vii**

Preface to "Cultural Landscapes Preservation and Social–Ecological Sustainability" . . . . . . **ix**

María Fe Schmitz and Cristina Herrero-Jáuregui
Cultural Landscape Preservation and Social–Ecological Sustainability
Reprinted from: *Sustainability* **2021**, *13*, 2593, doi:10.3390/su13052593 . . . . **1**

Nicolas Marine, Cecilia Arnaiz-Schmitz, Cristina Herrero-Jáuregui, Manuel Rodrigo de la O Cabrera, David Escudero and María F. Schmitz
Protected Landscapes in Spain: Reasons for Protection and Sustainability of Conservation Management
Reprinted from: *Sustainability* **2020**, *12*, 6913, doi:10.3390/su12176913 . . . . **7**

Xavier Martín, Anna Martínez and Isabela de Rentería
The Integration of Campsites in Cultural Landscapes: Architectural Actions on the Catalan Coast, Spain
Reprinted from: *Sustainability* **2020**, *12*, 6499, doi:10.3390/su12166499 . . . . **25**

Ge Chen, Jiaying Shi, Yiping Xia and Katsunori Furuya
The Sustainable Development of Urban Cultural Heritage Gardens Based on Tourists' Perception: A Case Study of Tokyo's Cultural Heritage Gardens
Reprinted from: *Sustainability* **2020**, *12*, 6315, doi:10.3390/su12166315 . . . . **53**

Peter Chrastina, Pavel Hronček, Bohuslava Gregorová and Michaela Žoncová
Land-Use Changes of Historical Rural Landscape—Heritage, Protection, and Sustainable Ecotourism: Case Study of Slovak Exclave Čív (Piliscsév) in Komárom-Esztergom County (Hungary)
Reprinted from: *Sustainability* **2020**, *12*, 6048, doi:10.3390/su12156048 . . . . **67**

Pedro Molina Holgado, Lara Jendrzyczkowski Rieth, Ana-Belén Berrocal Menárguez and Fernando Allende Álvarez
The Analysis of Urban Fluvial Landscapes in the Centre of Spain, Their Characterization, Values and Interventions
Reprinted from: *Sustainability* **2020**, *12*, 4661, doi:10.3390/su12114661 . . . . **93**

Amy Hudson and Kelly Vodden
Decolonizing Pathways to Sustainability: Lessons Learned from Three Inuit Communities in NunatuKavut, Canada
Reprinted from: *Sustainability* **2020**, *12*, 4419, doi:10.3390/su12114419 . . . . **123**

Fabio Pollice, Antonella Rinella, Federica Epifani and Patrizia Miggiano
Placetelling® as a Strategic Tool for Promoting Niche Tourism to Islands: The Case of Cape Verde
Reprinted from: *Sustainability* **2020**, *12*, 4333, doi:10.3390/su12104333 . . . . **143**

Carl Österlin, Peter Schlyter and Ingrid Stjernquist
Different Worldviews as Impediments to Integrated Nature and Cultural Heritage Conservation Management: Experiences from Protected Areas in Northern Sweden
Reprinted from: *Sustainability* **2020**, *12*, 3533, doi:10.3390/su12093533 . . . . **159**

Chika Udeaja, Claudia Trillo, Kwasi G.B. Awuah, Busisiwe C.N. Makore, D. A. Patel, Lukman E. Mansuri and Kumar N. Jha
Urban Heritage Conservation and Rapid Urbanization: Insights from Surat, India
Reprinted from: *Sustainability* **2020**, *12*, 2172, doi:10.3390/su12062172 . . . . **175**

Tsele T. Nthane, Fred Saunders, Gloria L. Gallardo Fernández and Serge Raemaekers
Toward Sustainability of South African Small-Scale Fisheries Leveraging ICT Transformation
Pathways
Reprinted from: *Sustainability* **2020**, *12*, 743, doi:10.3390/su12020743 . . . . **201**

Yoshinori Tokuoka, Fukuhiro Yamasaki, Kenichiro Kimura, Kiyokazu Hashigoe and Mitsunori Oka
Spatial Distribution Patterns and Ethnobotanical Knowledge of Farmland Demarcation Tree
Species: A Case Study in the Niyodo River Area, Japan
Reprinted from: *Sustainability* **2020**, *12*, 348, doi:10.3390/su12010348 . . . . **223**

Haiyun Xu, Tobias Plieninger, Guohan Zhao and Jørgen Primdahl
What Difference Does Public Participation Make? An Alternative Futures Assessment Based on
the Development Preferences for Cultural Landscape Corridor Planning in the Silk Roads Area,
China
Reprinted from: *Sustainability* **2019**, *11*, 6525, doi:10.3390/su11226525 . . . . **233**

Qiushan Li, Kabilijiang Wumaier and Mikiko Ishikawa
The Spatial Analysis and Sustainability of Rural Cultural Landscapes: Linpan Settlements in
China's Chengdu Plain
Reprinted from: *Sustainability* **2019**, *11*, 4431, doi:10.3390/su11164431 . . . . **257**

About the Editors

María Fe Schmitz

María Fe Schmitz is a Professor of Ecology at the Complutense University of Madrid (Department of Biodiversity, Ecology and Evolution). Her research interests focus on the theoretical-academic and applied aspects of landscape ecology and social-ecological systems and global change. She has extensive experience in the spatial analysis of cultural landscapes, the implications of traditional rural systems (agrarian and silvo-pastoral systems) in the maintenance and conservation of ecological processes (such as biodiversity and ecological connectivity), land planning, management effectiveness of protected areas and sustainable development. She has developed models of relationships between the rural-cultural landscape and the socio-economy of the local populations, with proven usefulness for land planning, management and tourism.

Cristina Herrero-Jáuregui

Cristina Herrero is an Assistant Professor at the Complutense University of Madrid. She has carried out her research on tropical (Brazilian Amazon), subtropical (Argentine Chaco) and Mediterranean (Spain) ecosystems, focusing on the structural and functional responses of ecosystems and their management by local populations. Among the factors analyzed, human management and environmental gradients stand out. For this, she has used both analytical and demonstrative experiments, as well as observational and descriptive studies, analysis tools typical of both the experimental and social sciences, and contemplating very different scales of analysis. All this shows the interdisciplinary nature of her career, which is necessary to plan the management of natural resources and seek solutions to environmental problems.

Preface to "Cultural Landscapes Preservation and Social–Ecological Sustainability"

Cultural landscapes are the result of social-ecological processes that have co-evolved throughout history, shaping high-value sustainable systems. The current processes of global change, such as agricultural intensification, rural abandonment, urban sprawl, and socio-economic dynamics, are threatening cultural landscapes worldwide. Whereas this loss is often unstoppable due to rapid and irreversible social-ecological changes, there are also examples where rationale protection measures can preserve cultural landscapes while promoting the sustainability of social-ecological systems. However, not all conservation policy-making processes consider the value of cultural landscapes, which makes their preservation even more difficult. Indeed, conservation policies focused on the wilderness paradigm are often counterproductive to conserving highly valuable cultural landscapes. The chapters in this book cover a wide spectrum of topics related to the preservation and sustainability of cultural landscapes, using different methodological approaches and involving regions from all over the world. This book can be useful for both researchers and professionals interested in using the socio-ecological framework in their scientific and applied work. The editors thank all those scholars who have participated by sharing their cutting-edge ideas in this book. The constant support of all managing editors who handled the submission, review and publication processes has been indispensable. Additionally, special thanks are given to the reviewers, who with their time and valuable suggestions contributed to the improvement of the contents of this book.

María Fe Schmitz, Cristina Herrero-Jáuregui
Editors

sustainability

MDPI

Editorial

Cultural Landscape Preservation and Social–Ecological Sustainability

María Fe Schmitz and Cristina Herrero-Jáuregui *

Department of Biodiversity, Ecology and Evolution, Complutense University of Madrid, 28040 Madrid, Spain; ma296@bio.ucm.es
* Correspondence: crherrero@bio.ucm.es

1. Introduction

Cultural landscapes are the result of social–ecological processes that have co-evolved throughout history, shaping high-value sustainable systems. They are an interface between nature and culture, characterized by the conservation and protection of ecological processes, natural resources, landscapes, and cultural biodiversity [1]. The adaptation to the environment and the social–ecological resilience of cultural landscapes depends, to a great extent, on the transmission of culture associated with the so-called traditional ecological knowledge (TEK) of recognized importance in the sustainable use of natural resources and the conservation of ecological processes and biodiversity [2]. Therefore, the conservation of naturalness and culturalness must be considered together within the socioecological framework of biocultural heritage, which requires adequate protection and management [3,4].

The current processes of global change, such as agricultural intensification, rural abandonment, urban sprawl, and socioeconomic dynamics, are threatening cultural landscapes worldwide. Even though this loss is often unstoppable because of rapid and irreversible social–ecological changes, there are also examples where rational protection measures can preserve cultural landscapes while promoting the sustainability of social–ecological systems. In Europe, significant efforts have been made in recent decades to preserve TEK and cultural landscapes [5]. However, not all conservation policymaking processes consider the value of cultural landscapes, which makes their preservation even more difficult.

Indeed, protected areas (PAs) designed to safeguard remaining habitats and species represent the cornerstone of conservation efforts [6], and their effectiveness can range from areas with inclusive and adaptive programs for sustainable management to areas with no active management, known as "paper parks" [7,8]. Land conservation policies have frequently been defensive, and management plans have often neglected or even restricted traditional rural activities, forgetting the local population, which has contributed to the high conservation values recognized in cultural landscapes [9]. These nature conservation efforts based on wilderness and naturalness have resulted in the decline of functional species composition and plant diversity of pasture systems [10], loss of natural and biocultural diversity and, ultimately, in the abandonment of the rural landscape and the reduction or disappearance of traditional knowledge [9]. Additionally, management plans of PAs are too often dependent on administrative boundaries and political legislation, and not on social–ecological relationships, biophysical processes, and ecosystem services fluxes, which reduces their effectiveness in protecting landscapes based on social–ecological interactions [1,11].

Therefore, there is a growing need of developing innovative theoretical and methodological approaches that are focused on management strategies for preserving cultural landscapes and natural heritage.

check for
updates

Citation: Schmitz, M.F.; Herrero-Jáuregui, C. Cultural Landscape Preservation and Social–Ecological Sustainability. *Sustainability* **2021**, *13*, 2593. https://doi.org/10.3390/su13052593

Received: 18 February 2021
Accepted: 22 February 2021
Published: 1 March 2021

Publisher's Note: MDPI stays neutral with regard to jurisdictional claims in published maps and institutional affiliations.

2. Focus of This Special Issue

With the aim of opening up the debate on how cultural landscapes are protected around the world, in this Special Issue, we publish state-of-the-art research concerning management of cultural landscapes and natural heritage. We included thirteen papers in this Special Issue, accepted out of all the submitted works and after the review process. The accepted papers can be divided into four main groups depending on their focus. While, as expected, most of papers focus on cultural landscapes, heritage, and sustainability (Figure 1), five of them specially emphasize the need for local participation, three of them focus on urban development and cultural heritage, two talk about tourism, and two are descriptive.

Figure 1. Word cloud generated from the publications of the Special Issue on Cultural Landscape Preservation and Social–Ecological Sustainability.

3. Overview of the Papers

In the first group, authors highlight the need for local participation in decision making in order to manage cultural landscapes around the world. Indeed, they demonstrate in the different contexts of China, Sweden, Spain, Canada, and South Africa that management strategies and conservation policies based exclusively on decision-makers' criteria are counterproductive for the conservation of cultural landscapes, particularly if they are influenced by the wilderness paradigm. Specifically, Xu et al. use a Chinese case of landscape corridor planning to analyze how citizen involvement may enrich sustainable spatial planning in respect to ideas considered and solutions developed. The authors demonstrate concrete differences between planning solutions developed with and without public participation, showing that collaborative processes can minimize spatial conflicts and demonstrating that public participation does indeed contribute to innovations that could enrich the corridor plan that had been produced exclusively by the decision-makers. On the other hand, Österlin et al. identify that the dominance of a wilderness discourse influences both the objectives and management of the protected areas in Sweden. They demonstrate how this wilderness discourse functions as a barrier against including cultural heritage conservation aspects and local stakeholders in management, as wilderness-influenced objectives are defining protected areas as environments "untouched" by humans. Moreover, Marine et al. assess the management effectiveness of several cultural landscapes by quantifying the evolution of the spatial pattern inside and outside protected landscapes in Spain. They conclude that the land protection approach adopted is not useful for the protection of cultural landscapes, particularly of the most rural ones and that the concepts of uniqueness and naturalness are not appropriate to preserve cultural landscapes. They recommend that different protection measures focused on the needs and desires of the rural population are taken into account in order to protect cultural landscapes that are shaped by traditional rural activities. In fourth place, Hudson and Vodden demonstrate that Inuit-led planning efforts can strengthen community sustainability planning interests and potential in Canada. They suggest that decolonizing efforts must be understood and updated within an Indigenous-led research and sustainability planning paradigm that facilitates autonomous place-based decision making. Finally, Nthane et al. explore how information and communication technologies (ICTs) in South African small-scale fisheries are leveraged towards value chain upgrading, collective action, and institutional sustainability—key issues that influence small-scale fishery contributions to marine resource sustainability. Authors demonstrate that Abalobi's ICT platform has the potential to facilitate deeper meanings of democracy that incorporate socioeconomic reform, collective action, and institutional sustainability in South Africa's small-scale fisheries.

The second group of papers focuses on cultural heritage around cities and how urban development is affecting its conservation and thus, its potential as tourist attraction. Authors offer innovative strategies and solutions in order to achieve a win–win balance where heritage is preserved in the context of urban renewal, contributing to the sustainable development of cultural heritage landscape and urbanization. Chen et al. explore the coexistence between the protection and management of cultural heritage landscapes in Japan and urban development in cities from a novel perspective. They propose an indicator of landscape morphology, namely sky view factor, that predicts the perception of tourists in heritage gardens in an urban context. Authors find that tourists' attitudes towards the high-rise buildings outside the traditional gardens are increasingly diversified, and the impact of this phenomenon is not necessarily negative. Meanwhile, Udeaja et al. recommend a thoughtful integration of sustainable heritage urban conservation into local urban development frameworks and the establishment of approaches that recognize the plurality of heritage values in India. Their paper reveals a myriad of challenges such as inadequacy of urban conservation management policies and processes focused on heritage, absence of skills, training, and resources amongst decision makers and persistent conflict and competition between heritage conservation needs and developers' interests. Furthermore, they denounce that values and significance of Surat's tangible and intangible heritage are not fully recognized by its citizens and heritage stakeholders. In addition, Molina et al. emphasize the need of specially considering urban and peri-urban riverbanks as landscapes in expansion due to the continuous growth of built-up spaces. Their paper characterizes four urban Mediterranean riverbanks describing the richness and composition of bird species and examines the interventions and urban planning criteria applied. Authors infer the need to reassess urban planning in river areas to ensure its compatibility with their operation, values, and possible uses of these systems.

In the third group of papers, authors focus on tourism development and offer innovative solutions for the sustainable preservation of cultural landscapes. On the one hand, Martin et al. foster cultural identity preservation and responsible communal living in nature by presenting a set of architectural actions for the integration of campsites in cultural landscapes along the Catalan coast. The paper highlights the capacity of these settlements to preserve the identity of the place and its culture, organizing communities based on itinerancy and temporality with a high degree of respect for the environment. On the other hand, Pollice et al. show the first results of what they define as a "maieutic reworking of local heritage" in Cape Verde through the sharing of narrative and symbolic artifacts. They highlight the use of Placetelling® as a particular type of storytelling of places that promotes local development and helps to develop a sense of identity and belonging among the members of the community. They defend Placetelling®for supporting local communities to become directly engaged in the preservation of their common legacy in order to transmit it to coming generations and promote it as a particularly useful tool for sustainable tourism development.

The last group of papers is descriptive. This group characterizes and describes cultural landscapes and TEK in different contexts (China, Hungary, and Japan) as a first step to understanding social–ecological dynamics and, thus, design appropriate land management policies to protect them and promote sustainable tourism. Li et al. explore changes in the relationship between humans and land in the farming area of the unique farming settlements in the Chengdu plain (China). Their study introduces the concept of "demand" and "restriction" in sustainable development to explore a future strategy for maintaining the cultural landscape, which is expected to provide a basis for future policy formulation to protect the traditional rural landscape. Chrastina et al. characterize a landscape colonized by Slovaks at the beginning of the 18th century as part of the cultural heritage of the Slovaks in Hungary, showing that a long-term stable cultural landscape has a similar potential for the development of ecotourism as a landscape protected by its wilderness. Finally, Tokuoka et al. examine the spatial distribution, uses, and folk nomenclature of farmland demarcation trees planted in the Niyodo River area in Japan. Authors highlight both the commonalities and uniqueness of demarcation tree culture in different regions of Japan, deepening our understanding of this agricultural heritage.

4. Conclusions

The presented papers cover a wide spectrum of topics from community management to urban heritage preservation and TEK with different methodological approaches and involving regions from all over the world. Therefore, in our opinion, this Special Issue can be interesting for both researchers and practitioners. The presented papers also suggest management policies that we believe can be of interest for further increasing sustainability in cultural landscapes.

Author Contributions: All authors have contributed equally. All authors have read and agreed to the published version of the manuscript.

Funding: This work was funded by the project LABPA-CM: Contemporary Criteria, Methods and Techniques for Landscape Knowledge and Conservation (H2019/HUM-5692), funded by the European Social Fund and the Madrid Regional Government.

Acknowledgments: The authors, as Guest Editors, acknowledge the constant support from all managing editors that handled submissions, review, and publishing processes. Also, special thanks go to the reviewers, who with their time and valuable suggestions contributed to the enhancement of those papers accepted in this Special Issue. The authors want to thank all those scholars who have participated by sharing their cutting-edge ideas in this Special Issue.

Conflicts of Interest: The authors declare no conflict of interest.

References

1. Sarmiento-Mateos, P.; Arnaiz-Schmitz, C.; Herrero-Jáuregui, C.; Pineda, F.D.; Schmitz, M.F. Designing protected areas for social–ecological sustainability: Effectiveness of management guidelines for preserving cultural landscapes. *Sustainability* **2019**, *11*, 2871. [CrossRef]
2. Berkes, F.; Folke, C.; Gadgil, M. Traditional Ecological Knowledge, Biodiversity, Resilience and Sustainability. In *Biodiversity Conservation*; Perrings, C.A., Mäler, K.G., Folke, C., Holling, C.S., Jansson, B.O., Eds.; Springer: Dordrecht, The Netherlands, 2011; pp. 281–299.
3. Gavin, M.C.; McCarter, J.; Mead, A.; Berkes, F.; Stepp, J.R.; Peterson, D.; Tang, R. Defining biocultural approaches to conservation. *Trends Ecol. Evol.* **2015**, *30*, 140–145. [CrossRef] [PubMed]
4. Vlami, V.; Kokkoris, I.P.; Zogaris, S.; Cartalis, C.; Kehayias, G.; Dimopoulos, P. Cultural landscapes and attributes of "culturalness" in protected areas: An exploratory assessment in Greece. *Sci. Total Environ.* **2017**, *595*, 229–243. [CrossRef] [PubMed]
5. Plieninger, T.; Höchtl, F.; Spek, T. Traditional land-use and nature conservation in European rural landscapes. *Environ. Sci. Policy* **2006**, *9*, 317–321. [CrossRef]
6. DeFries, R.; Hansen, A.; Newton, A.C.; Hansen, M.C. Increasing isolation of protected areas in tropical forests over the past twenty years. *Ecol. Appl.* **2005**, *15*, 19–26. [CrossRef]
7. Nagendra, H.; Lucas, R.; Honrado, J.P.; Jongman, R.H.; Tarantino, C.; Adamo, M.; Mairota, P. Remote sensing for conservation-monitoring: Assessing protected areas, habitat extent, habitat condition, species diversity, and threats. *Ecol. Indic.* **2013**, *33*, 45–59. [CrossRef]
8. Arnaiz-Schmitz, C.; Herrero-Jáuregui, C.; Schmitz, M.F. Losing a heritage hedgerow landscape. Biocultural diversity conservation in a changing social-ecological Mediterranean system. *Sci. Total Environ.* **2018**, *637*, 374–384. [CrossRef] [PubMed]
9. Schmitz, M.F.; Matos, D.G.G.; De Aranzabal, I.; Ruiz-Labourdette, D.; Pineda, F.D. Effects of a protected area on land-use dynamics and socioeconomic development of local populations. *Biol. Conserv.* **2012**, *149*, 122–135. [CrossRef]

10. Peco, B.; Sánchez, A.M.; Azcárate, F.M. Abandonment in grazing systems: Consequences for vegetation and soil. *Agric. Ecosyst. Environ.* **2006**, *113*, 284–294. [CrossRef]
11. Schmitz, M.F.; Herrero-Jáuregui, C.; Arnaiz-Schmitz, C.; Sánchez, I.A.; Rescia, A.J.; Pineda, F.D. Evaluating the Role of a Protected Area on Hedgerow Conservation: The Case of a Spanish Cultural Landscape. *Land Degrad. Dev.* **2017**, *28*, 833–842. [CrossRef]

 *sustainability*

Article

Protected Landscapes in Spain: Reasons for Protection and Sustainability of Conservation Management

Nicolas Marine [1,*], Cecilia Arnaiz-Schmitz [2,*], Cristina Herrero-Jáuregui [2], Manuel Rodrigo de la O Cabrera [1], David Escudero [1] and María F. Schmitz [2]

[1]   Department of Architectural Composition, Escuela Tecnica Superior de Arquitectura de Madrid, Universidad Politécnica de Madrid, 28040 Madrid, Spain; rodrigo.delao@upm.es (M.R.d.l.O.C.); david.escudero@upm.es (D.E.)

[2]   Department of Biodiversity, Ecology and Evolution, Complutense University of Madrid, 28040 Madrid, Spain; crherrero@bio.ucm.es (C.H.-J.); ma296@ucm.es (M.F.S.)

*   Correspondence: nicolas.marine@upm.es (N.M.); caschmitz@ucm.es (C.A.-S.)

Received: 10 July 2020; Accepted: 23 August 2020; Published: 25 August 2020

 check for updates

Abstract: Landscape conservation efforts in many European countries focus on cultural landscapes, which are part of the cultural identity of people, have a great heritage significance, improve the living standards of local populations and provide valuable cultural biodiversity. However, despite a wide arrange of protective measures, the management of preserved areas is seldom effective for the protection of cultural landscapes. Through a multi-approach analysis, we characterise the main heritage attributes of 17 Protected Landscapes in Spain and assess their management effectiveness by quantifying the evolution of the spatial pattern inside and outside protected landscapes. Our method has proven useful to quantitatively describe the spatial-temporal patterns of change of the protected and unprotected landscapes studied. We highlight the following results: (i) the concepts of uniqueness and naturalness are not appropriate to preserve cultural landscapes; (ii) the land protection approach currently adopted is not useful for the protection of cultural landscapes, particularly of the most rural ones; (iii) the landscapes studied with greater rural features can be considered as "paper parks". We recommend that different protection measures focused on the needs and desires of the rural population are taken into account in order to protect cultural landscapes that are shaped by traditional rural activities.

Keywords: inside and outside protected areas; intensity of change; IUCN's Category V; landscape structure; management effectiveness; rurality loss; spatial heterogeneity; spatial-temporal patterns

1. Introduction

As a consequence of relevant international initiatives concerning landscape protection, at the end of the 20th century [1–3] the effectiveness of already implemented legal mechanisms for land management has been questioned [4,5]. The assumption that landscape conservation policies are a multi-sector concern essential for sustainable land development calls for an adjustment of existing legal frameworks to more contemporary criteria. Thus, several authors have analysed the capability of existing legislation for the protection of landscape values and for granting collaborative and integrated land management [6–9]. Conservation efforts in many European countries have therefore focused on landscapes depending on human intervention. These cultural landscapes, which are part of the cultural identity of people, have a great heritage significance, improve the living standards of local populations and provide valuable cultural biodiversity [10–13].

The International Union for Conservation of Nature (IUCN) [14] has recognized the value of working landscapes as protected areas, naming them "Protected Landscapes ", the only one of the six IUCN protected area management categories based on the interaction between people and nature (Category V of IUCN Protected Areas). The exceptional natural and cultural values of these landscapes have encouraged measures for their protection. These types of landscape are a relevant reference for the implementation of a sustainable lifestyle [15]. Due to their origin in the long-term coevolution of natural and anthropogenic factors, Protected Landscapes have a strong correspondence with the UNESCO definition of Cultural Landscapes [16,17]. Furthermore, IUCN's Category V set a clear precedent [18] for the protection of Cultural Landscapes and many studies equate both categories for a possible joint implementation [19–21]. However, the disparity between landscapes that are part of Category V is remarkable and Protected Landscapes throughout Europe show many differences not only their natural, cultural, and social characteristics but also in the legislation and the reasons considered for protection.

In Spain, the legal figure of Protected Landscape was incorporated in 1989, as part of a comprehensive natural conservation law. The inclusion of protected landscapes in the Spanish territorial regulation involved a legal novelty of some importance, since, among the different protection categories established, Protected Landscapes were the only ones that lacked some precedents. The Law of 1989 defined Protected Landscapes as "those specific places of the natural environments that deserve special protection because of their cultural and aesthetic values" (art. 17). Both in definition and aim, this figure has a clear precedent in IUCN's Category V, whose definition precedes that of Spanish Protected Landscapes by one year [21,22]. For the first time in Spain, cultural and perceptual values were recognized as key components of landscapes and protected within an environmental legal framework [23]. Furthermore, the recognition of this figure anticipated concepts that would later be brought to the Spanish debate on landscape protection and management policies under the influence of UNESCO and the Council of Europe [24,25] (see Table 1 for a comparison of the definition of Spanish protected landscapes with that of other international protection categories of similar scope). Thus, Spanish Protected Landscapes are connected in concept with several international proposals for sustainable management focused on the relationship between humans and their environment, which has led to important initiatives, such as the Spanish National Plan for Cultural Landscapes (PNPC) [26]. The ratification in 2007 of the European Landscape Convention promoted the promulgation of a revised law that adjusted Spanish Protected Landscapes to the definition given by the Council of Europe [27]. As a consequence, this legal protection figure has not only embraced and adapted the ideas and concepts from various international organisms but also has enough longevity to verify its long term effectiveness.

Table 1. Spanish Protected Landscape compared with similar international categories of protection. Signalled in bold are the laws that legislate Spanish Protected Landscapes.

Year	Text	Protection Category	Status	Definition
1985	United Nation list of national parks and protected areas (IUCN Category V)	Protected Landscape	International guidelines	(1) Landscapes that possess special aesthetic qualities which are a result of the interaction of man and land; (2) landscapes that are primarily natural areas managed intensively by man for recreational and tourism uses.
1989	**Law of conservation of natural areas and wild flora and fauna.**	**Protected Landscape**	**Spanish law**	**Those specific places of the natural environments that deserve special protection because of their cultural and aesthetic values.**
1992	Operational Guidelines for the Implementation of the World Heritage Convention (UNESCO)	Cultural Landscape	International guidelines	The combined works of nature and of man […] illustrative of the evolution of human society and settlement over time, under the influence of the physical constraints and/or opportunities presented by their natural environment and of successive social, economic and cultural forces, both external and internal.
2000	European Landscape Convention (Council of Europe)	Geographic continuum	International guidelines	An area, as perceived by people, whose character is the result of the action and interaction of natural and/or human factors
2007 (Updated in 2015)	**Natural Heritage and Biodiversity Law**	**Protected Landscape**	**Spanish law**	**Part of the territory considered by the competent administrations, through the applicable regulations, as deserving of special protection due to its natural, aesthetic and cultural values. All in accordance with the European Landscape Convention.**
2012	National Plan for Cultural Landscapes	Cultural Landscape	Spanish guidelines	The result of people interacting over time with the natural medium, whose expression is a territory perceived and valued for its cultural qualities, the result of a process and the bedrock of a community's identity.

In spite of all these protection measures, and although they supply considerable amounts of ecosystem services [28,29], several authors have demonstrated that too often, the management of protected areas is not effective for the protection of cultural landscapes [30,31]. To the ubiquitous tendency of rural-urban migration and rural abandonment, strict legal requirements and the lack of encouragement of rural activities do not favour the maintenance of the societal structure that gave rise to the landscapes that are to be preserved. Thus, very often those cultural landscapes subject of conservation, even if protected and in many cases because of mistaken protection measures, evolve to other types of landscapes, dominated by unmanaged forests [32,33]. Following on our previous works, we hypothesise that in Spain the Protected Landscapes category does not have effective management for the protection of cultural rural landscapes.

On this basis, we focus here on empirically identifying the main characteristics of Protected Landscapes in Spain, as well as analysing their management effectiveness. The present paper is proposed with the following objectives: (i) to quantify the evolution of the spatial pattern inside and outside protected landscapes and detect the main indicators of landscape change; (ii) to identify types of protected landscapes according to their land-use composition, spatial structure and intensity of change; (iii) to characterise the different types of protected landscapes according to their heritage attributes; iv) to evaluate the management effectiveness and sustainability of the Protected Landscape category to preserve cultural landscapes.

2. Material and Methods

2.1. Sample Study. Selection of Protected Landscapes

For the declaration of a territory as a protected landscape, the Spanish legislation requires compliance with three basic aspects: (1) the conservation of unique values; (2) the preservation of the harmonious interaction between culture and nature, and (3) the defence of traditional practices. The conceptual differences in the application of this figure are manifested in a wide variety of legal protection instruments. Since 1989 and to date, different Spanish regions have declared a total of 58 Protected Landscapes. Of them, 31 are in the Iberian Peninsula and 27 in the Canary Islands.

Generally, these protected landscapes are representative of historical relationships between societies and their environment, although certain regions have opted for the sole conservation of fauna and flora. Notably, there has also been a tendency to protect special river sections [23]. The extent of protected areas is a less consistent factor and varies greatly from one landscape to another. Thus, an average range between 300 and 6000 ha can be determined, although several landscapes have an area greater than 10,000 ha. Another factor of difference between Spanish regions is that some (such as Cataluña, Castilla y León or Madrid) do not have any protected area under this category.

The current list of Protected Landscapes constitutes a representative sample of the multiple types of cultural landscapes in Spain. The criteria followed by the Spanish Administration for the protection of these landscapes are based on their typological representativeness and geographic diversity. In accordance with these criteria, for this study, we selected a sample of 17 peninsular landscapes illustrative of this great variety. The selected places are located in different regions of Spain (Figure 1) and therefore have different natural and cultural characteristics (Table 2). These areas have been declared protected in different periods of time (Table 2). The selected landscapes also comply with the requirement of having a continuous protection area. The Spanish legislation is unclear in this regard and some landscapes are composed of a collection of discontinuous areas. In this case, the only selected landscape with a divided area included in the sample is the "Paisaje protegido de las Fozes de Fago y Biniés" [34]. This landscape comprises two close areas very similar in size and shape. Considering management schemes, the Spanish legislation regulates the declaration of protected landscapes and the different regions develop the appropriate management plans to achieve their conservation goals. None of the selected landscapes is under other types of national or international protection.

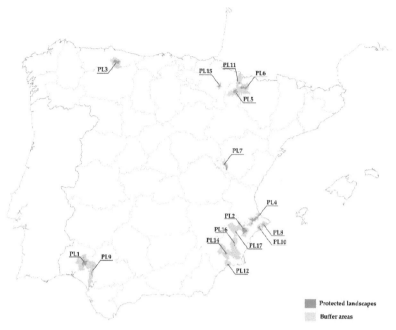

Figure 1. Location of the Spanish Protected Landscapes selected as sample study.

Table 2. List of Protected Landscapes studied. Name, location, extension, declaration dates and main landscape characteristics are indicated (see Figure 1).

Code	Official Names	Administrative Region	Area (ha)	Declaration Date	Main Cultural Features
PL-1	Rio Tinto	Andalucía	16.956	2005	Historic open mining landscape, crossed by a reddish river
PL-2	Serra del Maigmó and Serra del Sit	Comunidad Valenciana	15.842	2007	Mountain landscape with forest resources and cultural heritage
PL-3	Cuencas Mineras	Principado de Asturias	13.225	2002	Mountain landscape with several villages that historically have had a great industrial and mining activity
PL-4	Serpis	Comunidad Valenciana	10.000	2007	Fluvial and agricultural landscape heavily populated
PL-5	Sierra de Santo Domingo	Aragón	9.639	2012	Mountain range with a remarkable geomorphology
PL-6	San Juan de la Peña y Monte Oroel	Aragón	9.514	2007	Medieval mountain landscape with several monasteries and shrines
PL-7	Pinares de Rodeno	Aragón	6.829	1995	Landscape of eroded sandstone and ancient pine forest
PL-8	Sierra de Bernia y Ferrer	Comunidad Valenciana	2.843	2006	Rugged mountain range with remarkable architectural heritage and great scenic value
PL-9	Corredor Verde del Guadiamar	Andalucía	2.706	2003	Agricultural and natural landscape linked to a river of cultural and ecological significance
PL-10	Puigcampana y el Ponotx	Comunidad Valenciana	2.485	2006	Mountain system with high ecological value and with a long history of human occupation dating back to prehistoric times
PL-11 (a, b)	Fozes de Fago y Biniés	Aragón	2.440	2010	Deep ravine with different forest types
PL-12	Sierra de las Moreras	Región de Murcia	2.398	1992	Mountain range with highly particular eroded sandstone formations
PL-13	Ombria del Benicadell	Comunidad Valenciana	2.103	2006	Hydrogeological system with numerous natural springs
PL-14	Barrancos de Gébar	Región de Murcia	1.875	1995	Large ravine system of geomorphological interest
PL-15	Montes de Valdorba	Navarra	1.690	2004	Mosaic landscape alternating natural vegetation and crops
PL-16	Humedal del Ajauque y Rambla Salada	Región de Murcia	1.632	1992	Wetlands formed by the convergence of several watercourses
PL-17	Sierra de Salinas	Región de Murcia	1.332	2002	Mountain range with a heavy presence of human activities

2.2. Methods

We focused on spatial-temporal changes in the landscapes studied and considered the dynamics of their structural characteristics four times over 22 years (1990, 2000, 2006, 2012). A general outline of the steps followed in the methodological approach is shown in Figure 2.

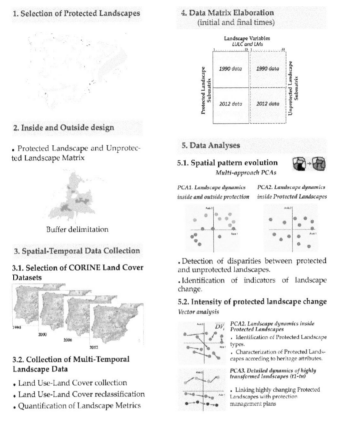

Figure 2. Outline of the steps followed in the methodological approach.

2.2.1. Data Collection and Analyses

Landscape structure was quantified by means of landscape and patch metrics (LMs), whose values are effective indicators of spatial patterns [35,36]. To calculate them: (i) we used CORINE Land Cover Maps of the years 1990, 2000, 2006 and 2012, considering seven land use-land cover types (LULCs) based on the reclassification of the Corine land cover classes into more meaningful and representative categories. The LULCs used were: forest systems, shrublands, *dehesas* (open woodlands used as pastures), agricultural systems, urban systems, rocky areas and wetlands/water bodies; (ii) we selected sixteen spatially explicit and non-redundant LMs, characterized by their easy interpretation and their ability to quantify landscape patterns [30,37]; (iii) we used Fragstats 4.2 [35] for the calculation of the selected LMs [38] (see the method of calculation and a brief description of each LMS in Appendix A): Shannon's diversity index (SHDI) quantifies landscape diversity and it is a good indicator of landscape heterogeneity; Shannon's evenness index, Simpson's evenness index and Modified Simpson's evenness index (SHEI, SIEI and MSIEI, respectively) measure the distribution of areas among patch types. As such, evenness is contrary to dominance; Patch richness density (PRD) measures the number of patch types present; Number of patches (NP) is a simple measure of the extent of subdivision of the patch type; Total edge and Edge density (TE and ED, respectively) inform about the amount of edge created by the patches present in the landscape. NP and edge metrics, together with Landscape Division index (DIVISION) and Splitting index (SPLIT), measure the degree of landscape fragmentation; Core Area index (CAI) is a relative index that quantifies the percentage of core area in a patch. This index could serve as an effective fragmentation index for a particular patch type. Landscape Contagion index (CONTAG) measures both the spatial dispersion of a patch type and the intermixing of units of different patch types at landscape level; Contiguity index (CONTIG) assesses spatial patch contiguity or connectedness; Patch Cohesion index (COHESION) calculates the physical connectedness of the corresponding patch types in an area and Euclidean nearest neighbour distance (ENN) describes the degree of spatial isolation of patches and, therefore, the degree of landscape connectivity; Largest patch index (LPI), measures the size of patches and the amount of edge created by these patches and represents an indirect measure of landscape homogeneity. We generated raster maps of the set of LMs by means of a round moving window with a radius of 100 m and then extracted a mean value of each metric for each of the 17 protected cultural landscapes selected.

We used the same procedure to characterise a buffer around each protected landscape whose area corresponds to that of the administrative divisions of the municipalities included in the respective studied landscape. This design allowed us to quantify inside and outside processes between protected and unprotected landscapes. To quantify the structure and dynamics of cultural landscapes, we elaborated a quantitative data matrix consisting of two sub-matrices describing, respectively, the characteristics of the 17 protected landscapes studied and their surrounding territories in the time span studied, using the 23 selected variables (7 LULCs and 16 LMs). This data matrix was analysed using successive applications of Principal Component Analyses (PCAs), in a multi-approach considering the inside and outside design.

2.2.2. Quantifying the Evolution of Protected and Unprotected Cultural Landscapes

With the collected data, we analysed landscape changes over time taking into account the initial and final periods registered (1990 and 2012): (i) inside and outside the protected cultural landscapes, and (ii) only within protected landscapes. We performed two PCAs at two spatial-temporal scales, respectively. The first PCA was calculated on the whole data matrix and the second one only on the protected landscape submatrix. These two PCAs allowed us to project the distribution of the protected landscape-unprotected matrix systems and inside protected landscapes on their respective ordination planes. Their dimensions represent, according to the factor loadings of the descriptive variables, the two main tendencies of variation of the landscapes and their changes over time.

2.2.3. Calculating the Intensity of Change inside Protected Areas for Landscape Characterisation

We analysed the trajectories of the changes occurred inside the protected landscapes, as well as their intensity, by calculating the displacement vectors, $\vec{DV_i}$ of the coordinates of each landscape on the PCA plane from time t_1 (1990) to time t_2 (2012) (1). The direction of $\vec{DV_i}$ on the ordination plane indicates the tendency of landscape change and their module values the intensity of temporal change.

$$\|\vec{DV_i}\| = \sqrt{\left(x_{t2_i} - x_{t1_i}\right)^2 + \left(y_{t2_i} - y_{t1_i}\right)^2 + \ldots \left(n_{t2_i} - n_{t1_i}\right)^2} \qquad (1)$$

We identified three types of landscapes according to the magnitude of their change, based on the value of the modules of their displacement vectors, by means of the equal intervals method: highly dynamic landscapes (high intensity of change), landscapes of moderate change (medium intensity of change) and landscapes without significant changes (low intensity of change).

We subsequently proceeded to confront the landscape types obtained with the heritage values detected in each of the landscapes. For this purpose, we adapted a list of attributes already tested as a viable method to characterise Spanish cultural landscapes [25]. As a basis for the characterisation, we used the original declaration documents which describe the main historical and geographical attributes of each of the landscapes. The final result is a list divided into two categories: human activities and historical remains. In turn, these are formed by several features described in Table 3. It is significant to note that the list complies with the requirements for cultural characterisation recommended by researchers attached to UNESCO [39], by ICOMOS [40] and by the Spanish PNPC [26]. These features were used as external descriptors describing the three landscape types previously obtained. The procedure was performed using a mean comparison test that allowed us to characterise a qualitative variable through quantitative variables. Therefore, Fisher's F-test (k > 2) was used to determine the statistical significance of the variables (heritage attributes) in the landscape types (landscape clusters obtained). The more the mean of a variable in a group is significantly different from the mean of that variable in the whole group, the stronger the link between the characterising quantitative variable and the qualitative category [41].

The landscape group corresponding to high intensity of change was analysed by means of a third PCA considering its detailed structure variation over time (complete temporal trajectory: 1990, 2000, 2006 and 2012). The date of the declaration of protected landscapes and / or that of their specific management plans were taken into account as external categorical variables of the calculated spatial-temporal patterns.

Table 3. Main cultural features used for the study.

Heritage Categories	Attribute Classes	Description
Main human uses and activities	Farming systems	Large areas of wine, olive or rice crops
	Mosaics of crops	Heterogeneous agricultural areas with mixed crops in smaller areas than the previous category
	Agroforestry systems	Combined land use management system in which trees or shrubs are grown around or among crops or pastureland
	Agriculture-livestock	Any of the above categories combined with livestock
	Hunting	Self-explanatory
	Mining	Self-explanatory
	Specific manufacture	Local economy oriented to the production of a specific product
Main historical features	Infrastructure of irrigation and/or water transportation	Self-explanatory
	Defence infrastructures	Remains of historical walls, fortresses, castles, bunkers or trenches, among others
	Prominent geographical references	Salient landscape features with a symbolic meaning or aesthetic value, such as mountains, hills and other geomorphological landmarks
	Prominent building and/or monument	Self-explanatory
	Rural character	Self-explanatory
	Artistic manifestations	Significant artistic representations of the landscape (in paintings or other media) Presence of artistic expressions in the landscape (such as cave paintings or dolmens)

3. Results

3.1. Landscape Dynamics inside and outside Protection

The analysis performed on the entire landscape data matrix, composed by the selected protected landscapes and their surrounding unprotected areas (PCA1; see Figure 2), allowed us to identify the trajectories of landscape change throughout the study period (Figure 3). The PCA plane reveals the main variation tendencies of landscape composition and structure, according to the loadings of the landscapes' variables. PCA axis 1 (explained variance: 29.72%) expresses the dynamics of the landscape in the studied time spam. The indicators of the landscape change (variables with the highest factor loadings) have allowed us to identify the more significant variation in its structure and composition (Table 4a). Thus, the surrounding territorial matrix has undergone a transformation process towards greater spatial heterogeneity linked to land use richness and fragmentation (LMs characterising the positive end of axis 1: SHDI, DIVISION, SHEI, MSIEI, SIEI, PRD, ED and SPLIT; see Appendix A and Table 4a), in which the dominant land uses are those related to the development of woodland systems, mainly composed of forests and *dehesas*. On this PCA axis 1, the dynamics of protected landscapes have followed in part (*n* = 9) the same spatial heterogeneity trend as that of the landscape matrices in which they are immersed, and in part (*n* = 8) the opposite tendency (negative end of axis 1), towards spatial homogenization, land use spatial contiguity and connectivity (landscape structure indicators: LPI, CONTAG, ENN, CAI, COHESION; see Appendix A and Table 4a) and agricultural systems as prevalent land uses (Table 4a; Figure 3).

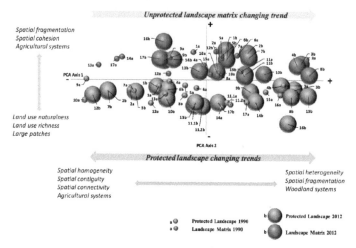

Figure 3. Landscape dynamics inside and outside protection. Principal Component Analyses (PCAs) plane. Indicator variables (higher factor loadings) of the main variation tendencies in the territory are shown at the end of the axes. Codes of studied landscapes are indicated in Table 2.

Table 4. Factor loadings of landscape variables on the main axes of the multi-approach PCAs performed (Figure 2). (a) PCA 1. Inside and outside the boundaries of protected landscapes; (b) PCA 2. Inside protected landscapes; (c) PCA 3. Highly changing landscapes (Landscape cluster 3). Variables with higher factor loadings at the end of the axes are in bold. See Appendix A for landscape metric description.

Landscape Variables	PCA 1 inside and outside Protected Landscapes		PCA 2 inside Protected Landscapes		PCA 3 Landscapes of High Intensity of Change
Land Metrics	Axis 1	Axis 2	Axis 1	Axis 2	Axis 1
CONTAG	−0.137	0.042	**−0.875**	0.136	**0.822**
CAI	−0.122	0.071	**−0.746**	−0.197	**0.695**
COHESION	−0.120	0.128	**−0.723**	0.136	0.427
ENN	−0.109	0.114	**−0.543**	0.078	0.572
LPI	−0.071	−0.101	**−0.806**	**−0.447**	**0.889**
CONTIG	−0.013	−0.026	−0.368	0.049	-
TE	−0.006	0.164	0.272	**0.910**	-
NP	0.002	0.161	0.298	**0.826**	−0.167
PRD	0.045	−0.171	0.239	**−0.676**	0.572
SPLIT	0.052	0.064	0.528	0.612	**−0.824**
DIVISION	0.071	0.100	**0.808**	0.448	-
SHDI	0.085	0.082	**0.852**	0.298	**−0.735**
SIEI	0.119	0.019	**0.902**	0.055	-
ED	0.121	−0.030	**0.772**	0.337	-
MSIEI	0.126	−0.013	**0.854**	−0.061	-
SHEI	0.128	−0.023	**0.856**	−0.153	-
Land Uses					
Agricultural systems	−0.058	0.117	0.048	0.096	-
Shrublands	0.000	−0.050	−0.204	0.158	-
Urban systems	0.009	0.064	0.039	0.190	-
Rocky areas	0.026	−0.052	0.077	−0.267	-
Wetlands/Water bodies	0.027	−0.036	0.331	−0.238	-
Forest systems	0.029	−0.052	−0.158	0.125	-
Dehesas	0.032	−0.034	0.196	−0.168	-

It is remarkable that PCA axis 2 (explained variance: 21%) shows a noticeable difference between the protected landscape and the surrounding landscape matrix, in each case (Figure 3). The indicators of this disparity express a greater naturalness in protected landscapes than in unprotected lands, with large patches of different natural systems (variables with higher factor loadings at the negative end of axis 2: PRD, LPI; forest systems, shrublands and rocky areas; Table 4a; Figure 3). The positive end of axis 2 is related to unprotected agricultural landscapes characterised by areas with a high number of patches and spatial cohesion, indicating a greater degree of aggregation among fragments in agricultural landscapes than in woodland systems (detected through landscape indicators by PCA axis 1; landscape structure indicators: NP, TE, COHESION; Appendix A and Table 4a; Figure 3).

3.2. Spatial-Temporal Variation inside Protected Landscapes

The PCA carried out on the sub-matrix collecting the temporal data of the protected landscapes allowed us to identify that the two main axes of the analysis explained the same variation in their composition and structure, highlighting changes over time in their heterogeneous-homogeneous spatial patterns and land-use fragmentation processes, as indicated by the variables with the greatest loadings at the ends of each axis (Figure 4; Table 4b). Thus, at the negative end of axis 1 (explained variance: 32.95%), landscape homogeneity is expressed by means of LPI and CAI, and metrics such as ENN, CONTAG, COHESION and CONTIG are mainly linked to low fragmentation, contiguity of land uses and landscape connectivity. Spatial contiguity of land uses and landscape connectivity are indicated by COHESION and CONTIG. Forests and shrublands land uses with high loadings at this end of PCA axis 1, are representative of the naturalness of the landscape. In contrast, the positive end of axis 1 is characterised by indicators of spatial heterogeneity (SIEI, SHEI, MSIEI, SHDI) and landscape fragmentation (DIVISION, ED, SPLIT). PCA 2 (explained variance: 17.33%) discriminates among the most natural protected landscapes from those mainly characterised by spatial heterogeneity and land-use fragmentation. In this axis 2 (negative end), PRD and LPI metrics are the indicators of landscapes with large patches of natural land uses (rocky areas and wetlands/water bodies) and SPLIT, DIVISION, TE and NP indices describe processes of landscape heterogeneity and fragmentation (positive end).

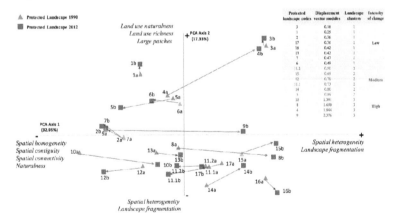

Figure 4. Spatial-temporal variation inside protected landscapes. PCA plane. Indicator variables (higher factor loadings) of the main landscape tendencies are shown at the end of the axes. The length of the arrows indicates the value of the displacement vector modules of protected landscapes from 1990 to 2012. The colour of the arrows characterises the type of landscape according to its intensity of change (black: low intensity of change; purple: intensity medium change rate; red: high change intensity). See associated legend (landscape cluster 1: low intensity of change; landscape cluster 2: medium intensity of change; landscape cluster 3: high intensity of change). Codes of studied landscapes are indicated in Table 2.

The calculation on the PCA plane of the displacement vectors between the coordinates of protected landscapes from the initial to the final times studied (1990 and 2012, respectively) enabled us to determine the direction (spatial homogeneity versus spatial heterogeneity) and magnitude (modules of displacement vectors) of landscape changes. The vector modules indicate high variability in the intensity of landscape changes. We identified three types of landscapes with different intensity of change by classifying the displacement vectors into three clusters according to the magnitude of their modules: (i) Landscape cluster 1. Composed of 8 of the 17 protected landscapes studied and characterised by a low intensity of change; value of the vector modules $= 0 \geq \vec{DV_i} < 0.5$. The direction and sense of the change trajectories of this group of vectors are variable; (ii) Landscape cluster 2. Composed of 5 protected landscapes with a medium intensity of change; value of the vector modules $= 0.5 \leq \vec{DV_i} < 1$. Temporal landscape trajectories do not follow a particular pattern; (iii) Landscape cluster 3. Composed of 4 protected landscapes of high intensity of change; value of the vector modules $= 1 \leq \vec{DV_i}$. The landscape change trend is directed towards spatial heterogeneity and fragmentation (Figure 4).

These landscape clusters were statistically characterised by the heritage features described in Table 3 (human activities and historical remains), which prevailed in the territories studied at the time of their declaration as protected landscapes. Significant statistical attributes considered for protection were those related to cultural heritage, rurality and traditional land uses and practices, as well as landscape features, indicators of aesthetic quality (Table 5). Protected landscapes with the lowest dynamics of change in the period studied (Landscape cluster 1; low intensity of change) were characterised by attributes related to agrarian systems and an economy oriented to local production. Landscapes characterised by their aesthetic or symbolic values showed a medium rate of change (Landscape cluster 2; medium intensity of change). Rural landscapes with traditional forms of land use, agricultural practices and water management have experienced a high intensity of change since their declaration as protected landscapes (Landscape cluster 3).

Table 5. Characterisation of the landscape types by their main heritage features. Statistically significant values of Fisher F-test are indicated in bold and *p*-values in parentheses.

Heritage Categories	Class	Landscapes of Low Intensity of Change	Landscapes of Medium Intensity of Change	Landscapes of High Intensity of Change
Main human uses and activities	Farming systems	1.486 (0.069)	0.987 (0.162)	−0.852 (0.803)
	Mosaics of crops	−1.175 (0.880)	0.541 (0.294)	**2.583** (0.005)
	Agroforestry systems	**1.871** (0.031)	0.837 (0.201)	−1.486 (0.932)
	Agriculture-livestock systems	**2.364** (0.009)	0.182 (0.428)	−1.238 (0.892)
	Hunting	0.821 (0.206)	1.383 (0.083)	−0.448 (0.673)
	Mining	**1.998** (0.023)	0.235 (0.407)	−0.487 (0.673)
	Specific manufacture	**1.567** (0.001)	−0.614 (0.730)	−0.136 (0.554)
Main historical features	Infrastructure of irrigation and/or water transportation	0.299 (0.382)	−1.326 (0.908)	**2.941** (0.002)
	Defence infrastructures	0.223 (0.412)	0.690 (0.245)	1.238 (0.108)
	Prominent geographical references	−0.899 (0.816)	**2.280** (0.011)	0.299 (0.382)
	Prominent building and/or monument	**1.857** (0.032)	−1.647 (0.95)	1.383 (0.083)
	Rural character	1.005 (0.150)	−0.837 (0.799)	**1.886** (0.050)
	Artistic manifestations	1.486 (0.069)	−0.614 (0.730)	1.238 (0.108)

The PCA performed on the data matrix containing descriptive variables of the 4 protected landscapes identified as highly dynamic (Landscape cluster 3), is shown in Figure 5. The LMs selected for this analysis were some of the ones that were detected as the most representative indicators of the changes in the landscape structure in the previous PCAs (NP, COHESION, ENN, PRD, CAI, SHDI, CONTAG, SPLIT and LPI). Since, at this scale, the variance explained by the PCA axis 1 was very high (51.27%), only the variation expressed by this first axis has been considered. LMs characterising the ends of the axis (greater factor loadings; Table 4c) express the spatial-temporal variation of these changing landscapes from homogeneity and connectivity (as indicated by LPI and CONTAG metrics) to heterogeneity and fragmentation (processes represented by the land metrics SHDI and SPLIT). Likewise, the analysis of the complete temporal trajectory registered has allowed us to identify that the transformation of these territories towards fragmentation and spatial heterogeneity has been significantly greater since their declaration as Protected Landscapes (see Figure 5, Declaration date). In Figure 5 the length of the arrows indicates the magnitude of the landscape change.

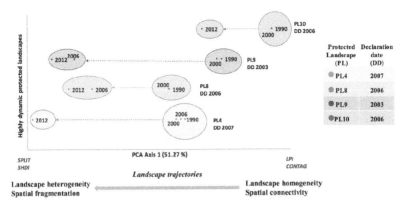

Figure 5. Detailed temporal trajectories of protected landscapes with high intensity of change (Landscape cluster 3; see Figure 4). Indicator variables (higher factor loadings) of the main landscape tendencies are shown at the end of the axes. The date of declaration as Protected Landscape is indicated. The length of the arrows indicates the intensity of landscape change. Codes of studied landscapes are indicated in Table 2.

4. Discussion

The essential rationale for this paper is to know the main reasons behind the decision to preserve landscapes and formally designate them under the protected landscape category. The design inside and outside the protection limits has been successful in achieving the established objectives and has allowed us to evaluate the management effectiveness of the protected landscapes studied. This inside-out approach to estimating protected area effectiveness is a frequently adopted strategy that provides useful insights and knowledge about processes related to land protection [12,32,33,42–46]. The multi-approach analysis performed considering the inside and outside design and the landscape metrics calculated [38,47] has provided a useful methodological tool and an effective set of spatial indicators to quantitatively describe the spatial-temporal patterns of change of the protected and unprotected landscapes studied. Furthermore, this approach has allowed us to verify how the dynamics of the landscape inside and outside the protection boundaries has had an evident effect on landscape configuration.

Comparing variations in landscape composition and structure inside versus outside protection shows a great difference between protected and unprotected land (detected by PCA axis 2; Figure 3). The landscape indicators identified evidence the naturalness of protected landscapes, mainly composed of large patches of different forest systems, shrublands and rocky areas, which characterise and differentiate them from the matrix of mainly agricultural unprotected landscapes in which they are immersed. The great disparity observed between protected landscapes and their surrounding unprotected land matrices indicates the trend towards the protection of certain peculiar features highlighted in human-dominated landscapes, mainly linked to landscape naturalness and uniqueness. Protection of naturalness has traditionally been a central objective of conservation efforts [48,49]. However, in cultural landscapes with a long history of human-maintained systems, it is controversial to consider naturalness as a point of reference to design conservation and management plans [32]. Especially for the Protected Landscape category, whose main reason is to prevent the loss of landscapes, that represent a valuable natural and cultural heritage [50].

The detected indicators of the landscape spatial-temporal change also express a remarkable transformation process, especially pronounced in unprotected landscapes (Figure 3; PCA axis 1). This transformation reflects higher land-use fragmentation and spatial heterogeneity, associated with the disruption of landscape connectivity and to the development of woodland systems as a consequence of the abandonment of agricultural land uses. Modified landscapes are often presented as patches of native vegetation within a matrix of different characteristics [51]. Spatial fragmentation is a complex and progressive process that leads to the division of continuous habitat into smaller and more isolated patches that are separated by dissimilar land matrices [52,53]. Worldwide, habitat loss and fragmentation are some of the main drivers of change in landscape structure and in the intensity of ecological interactions and are linked to environmental exploitation by humans [54]. However, in cultural landscapes habitat fragmentation means the transformation and disconnection of multifunctional agricultural landscapes, which are abandoned or intensified, affecting their functionality and ecological integrity [55]. The loss of traditional land uses and agricultural economic base has resulted in the observed fragmentation of agricultural lands and the consequent change of character and loss of visual quality [56].

The establishment of protected areas has been the primary management measure to preserve landscapes from this abandonment process [57]. Nevertheless, a significant number of the Protected Landscapes studied have followed a trend of transforming their spatial patterns through fragmentation, breaking contiguous landscapes (Figure 3; PCA axis 1). Thus, the spatial-temporal analysis of the studied protected landscapes show the same dynamic described above of landscape transformation towards spatial fragmentation and, as a by-product, spatial heterogeneity (Figure 4). Fragmentation resulting from habitat loss inevitably leads to greater heterogeneity, directly through a change of state or fragment conversion, or indirectly through edge effects [58]. Along with this process, alterations in spatial connectivity occur, since as fragmentation increases, connectivity values become critical [59]. This has happened with particular intensity in most rural landscapes. Indeed, the results obtained through vector analysis highlight that while some of the landscapes studied remain relatively unchanged or little altered, traditional rural landscapes experienced a great intensity of change and that their marked fragmentation process coincides in each case with the date of their designation as a Protected Landscape (Landscape cluster 3; Figure 4; Figure 5, respectively).

The literature on the role of protected areas in the conservation of cultural landscapes and the effectiveness of their management is abundant. These aspects, however, are especially critical in human-dominated landscapes since their protection has traditionally also involved their abandonment and the loss of their heritage values [10,60]. Similar results have been found in Central Spain, where relict hedgerow networks follow similar trajectories of abandonment inside and outside protected areas, revealing a lack of effectiveness of conservation measures [33,46]. In the same region, Sarmiento-Mateos et al [31] found incoherencies in the regulatory schemes of a protected area network, that inhibit rural inhabitants to continue with their traditional activities causing negative consequences for the cultural landscape whose protection is intended.

In summary, there has been a tendency to protect unique landscapes of a more natural character, which contradicts the initial objective of the Protected Landscape category. Furthermore, this category was conceived as part of what IUCN calls a "protected area system" [61]. Protected Landscapes are a very specific category and are not supposed to be declared as isolated entities. Instead, they should be part of a broader conservation plan that includes other categories. In fact, the Natural Heritage and Biodiversity Law (42/2007), that defined Protected landscapes, was modified partially in 2015 by another law [62]. Although it did not change their definition, it introduced the concept of Green Infrastructure and its necessity of being considered jointly with existing protected areas. This follows an initiative for establishing protected landscape networks similar to other European actions [63,64].

And so, Protected Landscapes in Spain have proven to be particularly inefficient to protect rural landscapes. Although the important role assigned to protected areas has driven their establishment, the focus of their conservation and management plans, which favours the constant change of protected sites, makes them vulnerable to accusations of not achieving some of the conservation objectives. Reliance on regulatory schemes and their effective application in landscape management is difficult and too often fails [65], which triggers that some protected areas to be just "paper parks" [66,67]—protected only in name.

5. Conclusions

The methodological approach developed has provided us with an effective set of indicators to quantitatively describe and compare the spatial-temporal patterns of change in the protected and unprotected landscapes studied. The marked disparity observed inside and outside protection highlighted the trend towards protecting the naturalness and uniqueness of the landscape. The quantitative indicators identified also detected a general transformation of the landscape towards spatial heterogeneity and land-use fragmentation, mainly related to the abandonment of agricultural land uses and the expansion of woodlands. This transformation process has been especially pronounced in unprotected landscapes, although it has also occurred in a large part of protected landscapes.

The analyses performed considering the trajectories of the spatial-temporal changes of Spanish Protected Landscapes allowed us to recognise types of landscapes, quantify their intensity of change and identify the different heritage characteristics considered as base values for their protection. Furthermore, through this procedure, we were able to detect the protected landscapes that had maintained their characteristics over time and those that had undergone an accelerated change. In this regard, it is noteworthy that after the declaration of protection, traditional rural landscapes experienced a great intensity of change towards non-rurality characteristics, as opposed to the primary objective of the category of Protected Landscape. Thus, in line with our starting hypothesis, the designation of the Protected Landscape category has not been an effective tool to protect rural landscapes.

The applied methodological design has proven to be useful in providing empirical information on the main reasons taken into account to protect the cultural landscape under the category of Protected Landscape, as well as on the effectiveness of its management guidelines. From the results obtained, we state the need to develop protection schemes in the Spanish legislation focused on traditional knowledge and on the essential requirements of rural populations to effectively protect the valuable heritage of cultural landscapes, mainly shaped by traditional rural activities.

Author Contributions: Conceptualization, N.M., C.A.-S., M.R.d.l.O.C. and D.E.; methodology, N.M., C.A.-S., M.F.S. and C.H.-J.; software, N.M. and C.A.-S.; formal analysis, C.A.-S.; investigation, N.M., C.A.-S., M.F.S. and C.H.-J.; writing—original draft preparation, N.M., C.A.-S., M.F.S. and C.H.-J.; writing—review and editing, N.M., C.A.-S., M.F.S., C.H.-J., M.R.d.l.O.C. and D.E.; funding acquisition, M.R.d.l.O.C. and D.E. All authors have read and agreed to the published version of the manuscript.

Funding: This research was funded by the project LABPA-CM: Contemporary Criteria, Methods and Techniques for Landscape Knowledge and Conservation (H2019/HUM-5692), funded by the European Social Fund and the Madrid regional government.

Acknowledgments: The authors would like to thank Guillermo Sotelo Santos and Paula Nogueira Losada for its valuable help in data gathering.

Conflicts of Interest: The authors declare no conflict of interest. The funders had no role in the design of the study; in the collection, analyses, or interpretation of data; in the writing of the manuscript; or in the decision to publish the results.

Appendix A

Table A1. Landscape metrics and patch metrics used to calculate landscape spatial patterns [35]. A brief description of the metrics used is provided, as well as their calculation methods and ranges of variation.

Landscape Metrics	Formula	Range and Description
Shannon's Evenness Index	$SHEI = \frac{-\sum_{i=1}^{m}(P_i \ln P_i)}{\ln m}$ P_i = proportion of the landscape occupied by patch type (i) m = number of patch types (i) present in the landscape, excluding the landscape border if present	SHEI > 0, without limit It is expressed so that an even distribution of area among patch types results in maximum evenness. As such, evenness is the complement of dominance.
Shannon's Diversity Index	$SHDI = -\sum_{i=1}^{m}(P_i \ln P_i)$ P_i = proportion of the landscape occupied by patch type (i)	SHDI > 0, without limit SHDI = 0 when the landscape contains only 1 patch (i.e., no diversity). SHDI increases as the number of different patch types (i.e., patch richness, PR) increases and/or the proportional distribution of area among patch types becomes more equitable.

Table A1. *Cont.*

Landscape Metrics	Formula	Range and Description
Patch richness density	$PRD = \frac{m}{A}(10000)(100)$ m = number of patch types present in the landscape a = Total landscape area	PRD > 0, without limit. Number of different patch types present within the landscape boundary.
Splitting index	$SPLIT = \frac{A^2}{\sum_{j=1}^{n} a_{ij}^2}$ aij = area (m2) of patch ij. A = total landscape area (m2)	1 ≤ SPLIT ≤ number of cells in the landscape squared. Increases as the landscape is increasingly subdivided into smaller patches and achieves its maximum value when the landscape is maximally subdivided; that is, when every cell is a separate patch.
Euclidean nearest neighbour distance	$ENN = h_{ij}$ h_{ij} = distance (m) from patch ij to nearest neighbouring patch of the same type (class), based on patch edge-to-edge distance, computed from cell center to cell center	ENN > 0, without limit. Distance (m) to the nearest neighbouring patch of the same type, based on the shortest edge-to-edge distance. It has been used extensively to quantify patch isolation
Largest Patch Index	$LPI = \frac{Max(a)}{A} \times (100)$ aij = area (m2) of patch ij. A = total landscape area (m2)	0 < LPI < 100 Percentage of the total landscape comprising the largest patch
Contagion Index	$CONTAG = $ $\left[1+\frac{\sum_{i=1}^{m}\sum_{k=1}^{m}[(P_i)]\frac{g_{ik}}{\sum_{k=1}^{m}g_{ik}}\left[\ln(P_i)\frac{g_{ik}}{\sum_{k=1}^{m}g_{ik}}\right]}{2\ln m}\right](100)$ P_i = proportion of the landscape occupied by patch type i.gik number of adjacencies (joins) between pixels of patch types (classes) i and k based on the double-count method. m = number of patch types (classes) present in the landscape.	0 < CONTAG ≤ 100 It approaches 0 when the patch types are maximally disaggregated and interspersed. CONTAG = 100 when all patch types are maximally aggregated; i.e., when the landscape consists of a single patch.
Patch Cohesion Index	$COHESION = \left[1-\frac{\sum_{j=1}^{n} p_{ij}}{\sum_{j=1}^{n} p_{ij}\sqrt{a_{ij}}}\right]\left[1-\frac{1}{\sqrt{A}}\right]^{-1}(100)$ p_{ij} = perimeter of patch ij in terms of number of cell surfaces aij = area of patch ij in terms of number of cells A = total number of cells in the landscape	0 ≤ COHESION < 100 It measures the physical connectedness of the corresponding patch type
Core Area Index	$CAI = \frac{a_{ij}^c}{a_{ij}}$ a_{ij}^c = core area (m2) of patch ij based on specified edge depths (m). aij = area (m2) of patch ij.	0 ≤ CAI < 100 CAI approaches 100 when the patch, because of size, shape, and edge width, contains mostly core area. A patch with no core area has the highest edge effect and consequently, the ecological processes of the patch may not function properly
Contiguity Index	$CONTIG = \frac{\left[\frac{\sum_{r=1}^{z} c_{ijr}}{a_{ij}}\right]-1}{v-1}$ c_{ijr} = contiguity value for pixel r in patch ij. v = sum of the values in a 3-by-3 cell template (13 in this case). aij = area of patch ij in terms of number of cells	0 ≤ CONTIG ≤ 1 This index assesses the spatial connectedness, or contiguity. Equals 0 for a one-pixel patch and increases to a limit of 1 as patch contiguity, or connectedness, increases. Thus, large contiguous patches result in larger contiguity index values
Edge density	$ED = \frac{E}{A}(10000)$ E = total length (m) of edge in landscape. A = total landscape area (m2).	ED ≥ 0, without limit. ED = 0 when there is no edge in the landscape; that is, when the entire landscape and landscape border, if present, consists of a single patch.
Modified Simpson's Evenness Index	$MSIEI = \frac{-\ln\sum_{i=1}^{m} P_i^2}{\ln m}$ P_i = proportion of the landscape occupied by patch type (class) i. m = number of patch types (classes) present in the landscape.	0 ≤ MSIEI ≤ 1 MSIDI = 0 when the landscape contains only 1 patch (i.e., no diversity) and approaches 0 as the distribution of areas among the different patch types becomes increasingly uneven (i.e., dominated by one type). MSIDI = 1 when the distribution of areas among patch types is perfectly even.
Simpson's Evenness Index	$SIEI = \frac{1-\sum_{i=1}^{m} P_i^2}{1-\left(\frac{1}{m}\right)}$ P_i = proportion of the landscape occupied by patch type (class) i. m = number of patch types (classes) present in the landscape, excluding the landscape border if present.	0 ≤ SIEI ≤ 1 SIDI = 0 when the landscape contains only 1 patch (i.e., no diversity) and approaches 0 as the distribution of areas among the different patch types becomes increasingly uneven (i.e., dominated by one type). SIDI = 1 when the distribution of areas among patch types is perfectly even.
Total edge	$TE = E$ E = total length (m) of edge in landscape.	TE ≥ 0, without limit. TE = 0 when there is no edge in the landscape; that is, when the entire landscape consists of a single patch.
Number of patches	$NP = n_i$ n_i = number of patches in the landscape of patch type i.	NP ≥ 1, without limit. NP = 1 when the landscape contains only 1 patch of the corresponding patch type. It is a simple measure of the extent of subdivision or fragmentation of the patch type.
Landscape Division Index	$DIVISION = \left[1-\sum_{i=1}^{m}\sum_{j=1}^{m}\left(\frac{a_{ij}}{A}\right)^2\right]$ aij = area (m2) of patch ij. A = total landscape area (m2).	0 ≤ DIVISION < 1 DIVISION = 0 when the landscape consists of a single patch. DIVISION achieves its maximum value when the landscape is maximally subdivided. It is similar to a diversity index

References

1. IUCN. *Guidelines for Protected Area Management Categories*; IUCN: Gland, Switzerland; Cambridge, UK, 1994.
2. UNESCO. Report of the World Heritage Committee Sixteenth Session (Santa Fe, United States of America, 7–14 December 1992). In *Convention Concerning the Protection of the World Cultural and Natural Heritage*; UNESCO: Santa Fe, NM, USA, 1992.
3. Council of Europe. *European Landscape Convention*; Council of Europe Publishing Division: Strasbourg, France, 2000.
4. Mitchell, N.; Rössler, M.; Tricaud, P. *World Heritage Cultural Landscapes: A Handbook for Conservation and Management*; UNESCO: Paris, France, 2009.
5. De Montis, A. Impacts of the European Landscape Convention on national planning systems: A comparative investigation of six case studies. *Landsc. Urban Plan.* **2014**, *124*, 53–65. [CrossRef]
6. García-Martín, M.; Bieling, C.; Hart, A.; Plieninger, T. Integrated landscape initiatives in Europe: Multi-sector collaboration in multi-functional landscapes. *Land Use Policy* **2016**, *58*, 43–53. [CrossRef]
7. Janssen, J. Sustainable development and protected landscapes: The case of The Netherlands. *Int. J. Sustain. Dev. World Ecol.* **2009**, *16*, 37–47. [CrossRef]
8. Nogueira Terra, T.; Ferreira dos Santos, R.; Cortijo Costa, D. Land use changes in protected areas and their future: The legal effectiveness of landscape protection. *Land Use Policy* **2014**, *38*, 378–387. [CrossRef]
9. De Montis, A. Measuring the performance of planning: The conformance of Italian landscape planning practices with the European Landscape Convention. *Eur. Plan. Stud.* **2016**, *24*, 1727–1745. [CrossRef]
10. Agnoletti, M. Rural landscape, nature conservation and culture: Some notes on research trends and management approaches from a (southern) European perspective. *Landsc. Urban Plan.* **2014**, *126*, 66–73. [CrossRef]
11. Marull, J.; Tello, E.; Fullana, N.; Murray, I.; Jover, G.; Font, C.; Coll, F.; Domene, E.; Leoni, V.; Decolli, T. Long-term bio-cultural heritage: Exploring the intermediate disturbance hypothesis in agro-ecological landscapes (Mallorca, c. 1850–2012). *Biodivers. Conserv.* **2015**, *24*, 3217–3251. [CrossRef]
12. Vlami, V.; Kokkoris, I.P.; Zogaris, S.; Cartalis, C.; Kehayias, G.; Dimopoulos, P. Cultural landscapes and attributes of "culturalness" in protected areas: An exploratory assessment in Greece. *Sci. Total. Environ.* **2017**, *595*, 229–243. [CrossRef]
13. Campedelli, T.; Calvi, G.; Rossi, P.; Trisorio, A.; Florenzano, G.T. The role of biodiversity data in High Nature Value Farmland areas identification process: A case study in Mediterranean agrosystems. *J. Nat. Conserv.* **2018**, *46*, 66–78. [CrossRef]
14. Janssen, J.; Knippenberg, L.W.J. From landscape preservation to landscape governance: European experiences with sustainable development of protected landscapes. In *Studies on Environmental and Applied Geomorphology*; Piacentini, T., Miccadei, E., Eds.; IntechOpen: London, UK, 2012; pp. 241–266. [CrossRef]
15. Beresford, M.; Phillips, A. Protected landscapes: A conservation model for the 21st century. *Georg. Wright Forum* **2000**, *17*, 15–26.
16. Lacitignola, D.; Petrosillo, I.; Cataldi, M.; Zurlini, G. Modelling socio-ecological tourism-based systems for sustainability. *Ecol. Model.* **2007**, *206*, 191–204. [CrossRef]
17. De Aranzabal, I.; Schmitz, M.F.; Aguilera, P.; Pineda, F.D. Modelling of landscape changes derived from the dynamics of socio-ecological systems: A case of study in a semiarid Mediterranean landscape. *Ecol. Indic.* **2008**, *8*, 672–685. [CrossRef]
18. Jacques, D. The rise of cultural landscapes. *Int. J. Herit. Stud.* **1995**, *1*, 91–101. [CrossRef]
19. Mitchell, N.; Buggey, S. Protected Landscapes and Cultural Landscapes: Taking Advantage of Diverse Approaches. *Georg. Wright Forum* **2000**, *17*, 35–46.
20. Phillips, A. Cultural Landscapes: IUCN's Changing Vision of Protected Areas. In *Cultural Landscapes: The Challenges of Conservation*; Ceccarelli, P., Rössler, M., Eds.; UNESCO World Heritage Centre: Paris, France, 2003; pp. 40–49.
21. Phillips, A. Landscape as a meeting ground: Category V Protected Landscapes. In *The Protected Landscape Approach. Linking Nature, Culture and Community*; Brown, J., Mitchell, N., Beres, M., Eds.; IUCN: Gland, Switzerland; Cambridge, UK, 2005; pp. 19–36.

22. Foster, J. *Protected Landscapes: Summary Proceedings of an International Symposium, Lake District, United Kingdom, 5–10 October 1987*; IUCN: Cambridge, UK, 1988; Available online: https://www.iucn.org/es/node/19466 (accessed on 15 August 2020).

23. Mulero Mendigorri, A. Significado y tratamiento del paisaje en las políticas de protección de espacios naturales en España. *Boletín Asoc. Geógrafos Españoles* **2013**, *62*, 129–145. [CrossRef]

24. Cañizares Ruiz, M.C. Paisajes culturales, Ordenación del Territorio y reflexiones desde la Geografía en España. *Polígonos Rev. Geogr.* **2014**, *26*, 147–180. [CrossRef]

25. De la Cabrera, R.O.; Marine, N.; Escudero, D. Spatialities of cultural landscapes: Towards a unified vision of Spanish practices within the European Landscape Convention. *Eur. Plan. Stud.* **2019**. [CrossRef]

26. Consejo de Patrimonio Histórico. Plan Nacional de Paisaje Cultural [National Plan for Cultural Landscape]. Madrid: Gobierno de España, 2012. English Version. 2012. Available online: http://www.culturaydeporte.gob.es/planes-nacionales/dam/jcr:a08b4444-4929-4033-ac38-68e8f3c2080e/05-paisajecultural-eng.pdf (accessed on 15 August 2020).

27. Ley 42/2007, de 13 de Diciembre. Available online: https://www.boe.es/buscar/act.php?id=BOE-A-2007-21490 (accessed on 15 August 2020).

28. Herrero-Jáuregui, C.; Arnaiz-Schmitz, C.; Herrera, L.; Smart, S.M.; Montes, C.; Pineda, F.D.; Schmitz, M.F. Aligning landscape structure with ecosystem services along an urban–rural gradient. Trade-offs and transitions towards cultural services. *Landsc. Ecol.* **2019**, *34*, 1525–1545. [CrossRef]

29. Castro, A.J.; Martín-López, B.; López, E.; Plieninger, T.; Alcaraz-Segura, D.; Vaughn, C.C.; Cabello, J. Do protected areas networks ensure the supply of ecosystem services? Spatial patterns of two nature reserve systems in semi-arid Spain. *Appl. Geogr.* **2015**, *60*, 1–9. [CrossRef]

30. Arnaiz-Schmitz, C.; Schmitz, M.F.; Herrero-Jáuregui, C.; Gutiérrez-Angonese, J.; Pineda, F.D.; Montes, C. Identifying socio-ecological networks in rural-urban gradients: Diagnosis of a changing cultural landscape. *Sci. Total. Environ.* **2018**, *612*, 625–635. [CrossRef]

31. Sarmiento-Mateos, P.; Arnaiz-Schmitz, C.; Herrero-Jáuregui, C.; D Pineda, F.; Schmitz, M.F. Designing Protected Areas for Social–Ecological Sustainability: Effectiveness of Management Guidelines for Preserving Cultural Landscapes. *Sustainability* **2019**, *11*, 2871. [CrossRef]

32. Schmitz, M.F.; Matos, D.G.G.; De Aranzabal, I.; Ruiz-Labourdette, D.; Pineda, F.D. Effects of a protected area on land-use dynamics and socioeconomic development of local populations. *Biol. Conserv.* **2012**, *149*, 122–135. [CrossRef]

33. Schmitz, M.F.; Herrero-Jáuregui, C.; Arnaiz-Schmitz, C.; Sánchez, I.A.; Rescia, A.J.; Pineda, F.D. Evaluating the role of a protected area on hedgerow conservation: The case of a Spanish cultural landscape. *Land Degrad. Dev.* **2017**, *28*, 833–842. [CrossRef]

34. DECRETO 71/2010, de 13 de abril, del Gobierno de Aragón, de declaración del Paisaje Protegido de las Fozes de Fago y Biniés. Available online: http://www.boa.aragon.es/cgi-bin/EBOA/BRSCGI?CMD=VEROBJ&MLKOB=518586860707 (accessed on 15 August 2020).

35. Peng, J.; Wang, Y.; Zhang, Y.; Wu, J.; Li, W.; Li, Y. Evaluating the effectiveness of landscape metrics in quantifying spatial patterns. *Ecol. Indic.* **2010**, *10*, 217–223. [CrossRef]

36. Uuemaa, E.; Mander, Ü.; Marja, R. Trends in the use of landscape spatial metrics as landscape indicators: A review. *Ecol. Indic.* **2013**, *28*, 100–106. [CrossRef]

37. Su, S.; Xiao, R.; Jiang, Z.; Zhang, Y. Characterizing landscape pattern and ecosystem service value changes for urbanization impacts at an eco-regional scale. *Appl. Geogr.* **2012**, *34*, 295–305. [CrossRef]

38. McGarigal, K.; Cushman, S.A.; Ene., E. FRAGSTATS v4: Spatial Pattern Analysis Program for Categorical and Continuous Maps. Computer Software Program Produced by the Authors at the University of Massachusetts, Amherst. 2012. Available online: http://www.umass.edu/landeco/research/fragstats/fragstats.html (accessed on 15 August 2020).

39. Fowler, P.J. *World Heritage Cultural Landscapes 1992–2002. A Review*; UNESCO World Heritage Centre: Paris, France, 2003.

40. ICOMOS. *The World Heritage List: Filling the Gaps—An Action Plan for the Future*; ICOMOS: Paris, France, 2004.

41. Levart, L.; Morineau, A.; Piron, M. *Statistique Exploratoire Multidimensionnelle*; Dunot: Paris, France, 2000.

42. Alo, C.A.; Pontius, R.G., Jr. Identifying systematic land-cover transitions using remote sensing and GIS: The fate of forests inside and outside protected areas of Southwestern Ghana. *Environ. Plan. B Plan. Des.* **2008**, *35*, 280–295. [CrossRef]

43. Western, D.; Russell, S.; Cuthill, I. The status of wildlife in protected areas compared to non-protected areas of Kenya. *PLoS ONE* **2009**, *4*, e610. [CrossRef]
44. Leisher, C.; Touval, J.; Hess, S.M.; Boucher, T.M.; Reymondin, L. Land and forest degradation inside protected areas in Latin America. *Diversity* **2013**, *5*, 779–795. [CrossRef]
45. Gray, C.L.; Hill, S.L.; Newbold, T.; Hudson, L.N.; Börger, L.; Contu, S.; Hoskins, A.J.; Ferrier, S.; Purvis, A.; Scharlemann, J.P. Local biodiversity is higher inside than outside terrestrial protected areas worldwide. *Nat. Commun.* **2016**, *7*, 1–7. [CrossRef]
46. Arnaiz-Schmitz, C.; Herrero-Jáuregui, C.; Schmitz, M.F. Losing a heritage hedgerow landscape. Biocultural diversity conservation in a changing social-ecological Mediterranean system. *Sci. Total. Environ.* **2018**, *637*, 374–384. [CrossRef] [PubMed]
47. Turner, M.G.; Gardner, R.H. *Landscape Ecology in Theory and Practice: Pattern and Process*; Springer: New York, NY, USA, 2015.
48. Cole, D.N.; Yung, L.; Zavaleta, E.S.; Aplet, G.H.; Stuart Chapin, F.; Graber, D.M.; Higgs, E.S.; Hobbs, R.J.; Landres, P.B.; Millar, C.I.; et al. Naturalness and beyond: Protected area stewardship in an era of global environmental change. In *Georg. Wright Forum*; 2008; 25, pp. 36–56. Available online: https://www.fs.usda.gov/treesearch/pubs/31765 (accessed on 15 August 2020).
49. Gibbons, P.; Briggs, S.V.; Ayers, D.A.; Doyle, S.; Seddon, J.; McElhinny, C.; Jones, N.; Sims, R.; Doody, J.S. Rapidly quantifying reference conditions in modified landscapes. *Biol. Conserv.* **2008**, *141*, 2483–2493. [CrossRef]
50. Phillips, A. *Management Guidelines for IUCN Category V Protected Areas: Protected Landscapes/Seascapes*; IUCN: Gland, Switzerland; Cambridge, UK, 2002.
51. Almeida, M.; Azeda, C.; Guiomar, N.; Pinto-Correia, T. The effects of grazing management in montado fragmentation and heterogeneity. *Agrofor. Syst.* **2016**, *90*, 69–85. [CrossRef]
52. Saura, S. Effects of remote sensor spatial resolution and data aggregation on selected fragmentation indices. *Landsc. Ecol.* **2004**, *19*, 197–209. [CrossRef]
53. Martinson, H.M.; Fagan, W.F. Trophic disruption: A meta-analysis of how habitat fragmentation affects resource consumption in terrestrial arthropod systems. *Ecol. Lett.* **2014**, *17*, 1178–1189. [CrossRef]
54. Sala1, O.E.; Chapin, F.S., III; Armesto, J.J.; Berlow, E.; Bloomfield, J.; Dirzo, R.; Huber-Sanwald, E.; Huenneke, L.F.; Jackson, R.B.; Kinzig, A.; et al. Global biodiversity scenarios for the year 2100. *Science* **2000**, *287*, 1770–1774. [CrossRef]
55. Reza, M.I.H. Measuring forest fragmentation in the protected area system of a rapidly developing Southeast Asian tropical region. *Sci. Postprint* **2014**, *1*, e00030.
56. Brabec, E.; Smith, C. Agricultural land fragmentation: The spatial effects of three land protection strategies in the eastern United States. *Landsc. Urban Plan.* **2002**, *58*, 255–268. [CrossRef]
57. Jacobson, A.P.; Riggio, J.; Tait, A.M.; Baillie, J.E. Global areas of low human impact ('Low Impact Areas') and fragmentation of the natural world. *Sci. Rep.* **2019**, *9*, 1–13. [CrossRef]
58. Franklin, A.B.; Noon, B.R.; George, T.L. What is habitat fragmentation? *Stud. Avian Biol.* **2002**, *25*, 20–29.
59. Malanson, G.P.; Cramer, B.E. Landscape heterogeneity, connectivity, and critical landscapes for conservation. *Divers. Distrib.* **1999**, *5*, 27–39. [CrossRef]
60. Agnoletti, M.; Tredici, M.; Santoro, A. Biocultural diversity and landscape patterns in three historical rural areas of Morocco, Cuba and Italy. *Biodivers. Conserv.* **2015**, *24*, 3387–3404. [CrossRef]
61. *Guidelines for Applying Protected Area Management Categories*; Dudley, N. (Ed.) IUCN: Gland, Switzerland, 2008; p. 10.
62. Ley 33/2015, de 21 de Septiembre. Available online: https://www.boe.es/buscar/act.php?id=BOE-A-2015-10142 (accessed on 15 August 2020).
63. Prezioso, M.; D'Orazio, A.; Coronato, M.; Pigliucci, M.; Sargolini, M.; Idone, M.T.; Pierantoni, I.; Omizzolo, A.; Cetara, L.; Streifeneder, T.; et al. LinkPAs—Linking Networks of Protected Areas to Territorial Development. Executive Summary, ESPON, 2018. Version 27/06/2018. Available online: https://www.espon.eu/sites/default/files/attachments/Linkpas%20%20ExecutiveSummary.pdf (accessed on 15 August 2020).
64. Naumann, S.; Davis, M.; Kaphengst, T.; Pieterse, M.; Rayment, M. Design, Implementation and Cost Elements of Green Infrastructure Projects. Final Report to the European Commission, DG Environment, Ecologic Institute and GHK Consulting, 2011. 2011. Available online: https://ec.europa.eu/environment/enveco/biodiversity/pdf/GI_DICE_FinalReport.pdf (accessed on 15 August 2020).

65.   Watson, J.E.; Dudley, N.; Segan, D.B.; Hockings, M. The performance and potential of protected areas. *Nature* **2014**, *515*, 67–73. [CrossRef]

66.   Phillips, A. Turning ideas on their head: The new paradigm for protected areas. *Georg. Wright Forum* **2003**, *20*, 8–32.

67.   Di Minin, E.; Toivonen, T. Global protected area expansion: Creating more than paper parks. *BioScience* **2015**, *65*, 637–638. [CrossRef]

 *sustainability*

Article

The Integration of Campsites in Cultural Landscapes: Architectural Actions on the Catalan Coast, Spain

Xavier Martín *, Anna Martínez and Isabela de Rentería

IAR Group, School of Architecture La Salle, Ramon Llull University, 08022 Barcelona, Spain;
a.martinez@salle.url.edu (A.M.); isabela.derenteria@salle.url.edu (I.d.R.)
* Correspondence: xavier.martin@salle.url.edu

Received: 29 June 2020; Accepted: 6 August 2020; Published: 12 August 2020

Abstract: Over the last 60 years, the development of tourism in Spain has produced an unprecedented occupation of the territory. Urban growth, hotels and infrastructures have transformed much of the natural environment. This phenomenon has irreversibly altered conditions of regions with great landscape value, putting their cultural heritage at risk. Yet, the campsite is a model of tourist settlement based on shared living in the open natural space. It promotes minimal and temporary interventions in the territory, by means of transportable accommodations with precise occupations of place, leaving a slight footprint. Therefore, architecture contributes to affording the order and services that these individual artifacts cannot provide by themselves. In terms of slight land occupation and natural qualities preservation, the campsite has proven to be one of the most responsible tourist models. It is an opportunity for the future: a resource for landscape integration and local dynamics reactivation. We present a set of architectural actions for the integration of campsites in cultural landscapes along the Catalan coast. These are recommendations catalogued by means of a diagnosis tool that proposes strategies at different levels, from enclosure to lodgings. Focusing on end-users, this research fosters cultural identity preservation and responsible communal living in nature.

Keywords: campsites; landscape identity; architectural strategies; itinerant tourism; cultural heritage; Mediterranean tradition

1. Introduction

Over the last 60 years, the development of tourism in Spain has produced an unprecedented occupation of large areas of its geographic territory [1]. Urban growth, hotels, infrastructures or extensive urbanization have transformed much of the natural environment. And in some cases, this phenomenon has irreversibly altered the conditions of domains with great landscape value, putting their cultural heritage at risk [2].

This process has resulted in a great paradox (Figure 1). Landscapes evoke memories of our experiences, they are the result of the tradition and history of their inhabitants, and they make a region desirable [3]. Being the fundamental desire of tourism [4], the effects of this economic activity have incomprehensibly consumed landscape values for its own benefit: much architecture of the tourist boom was designed for contemplating, but very few to be seen [5].

In response to this situation, this article presents research that recognizes nature as a cultural function [6]. Beyond reducing it to a strictly geographical entity or to an economic good with which to speculate, nature is recognized as the addition of actions that has endured over time and has shaped values and meaning in the cultural landscape [7]. In this sense, the study of the role of architecture in the transformation of the environment and the construction of the touristic spaces should not only be a recording of what has already happened, but also a commitment to provide a more responsible and sustainable architecture [8]. New design solutions should be more adapted to the needs of the moment and the place, facilitating spaces of relationship and encouraging the discovery of landscapes' inherent values to foster their preservation for future generations [9].

Figure 1. Influence of formal tourism in contrast to lightweight occupation by campsites. Sea view from the coast of Torredembarra (Catalonia, Spain). (Source: authors' own.)

Landscape, architecture and tourism are the three axes that articulate this research, focused on the drafting of a catalogue of architectural recommendations for the landscape integration of campsites in the territory of Catalonia (Spain). This region is historically one of the most representative of the country, since it is the gateway from Europe, and has gathered a greater concentration of campsites. Holidays intensify leisure in nature, social relationships among people and the aesthetics perception of our surroundings [10]. Therefore, they allow us to experiment with new ways of living outdoors and in community. The campsite is a model of tourist settlement based on these qualities. Using transportable lodgings, it promotes minimal intervention with light and precise occupations at the place. In addition, its flexibility and ability to integrate into the site conditions favor its recognition as a potentially sustainable tourist settlement, in close relationship with nature and with a lightweight footprint [11].

This research proposes a collection of architectural actions or strategies to facilitate the integration of these settlements in the environment. It also sets out a project methodology based on recognizing the importance of the singularities of the place in order to act on it with a coherent proposal related to its geography and culture. It is an applied, analytical and purposeful study that offers guidelines and tools to promote the landscape integration of campsites and the preservation of the natural qualities, local traditions and cultures of the territory—through both its geographical and historical aspects [12].

The following sections develop the methodological research process and focus on the results obtained and collected in the "Good Practices Manual. Architectural Actions in Campsites." [13]. The article reflects on the role of this manual in the future of the sector and its possible effects on the regeneration and enhancement of the cultural landscape. Finally, as a conclusion, it explores the capacity of the campsite to propose new ways of living based on the most primal itinerancy, but still valid and necessary in the dynamic and transformative societies of the 21st century.

2. Tourism Setting of Catalonia

The tourist development of Spain was caused by the confluence of a series of conditions that in the late 1950s triggered a process of great intensity that quickly transformed the economy, life, landscape and customs of the inhabitants [14]. On the one hand, the good climate of the country allowed to share experiences in the open air: running away from cities to foment the contact with nature and to discover singular landscapes with a recognized cultural value. In short, to do during holidays everything that was not possible in everyday life [15]. On the other hand, the low cost of living in Spain offered a very economical tourist option, as a result of more than twenty years of international isolation. In addition, since the late 1950s, the car had given freedom of movement to travelers. According to the International Road Federation, the number of vehicles in Europe increased by 72% over the previous decade and made the whole territory accessible to tourism [16].

From the early years, one of the Spanish regions with the highest incidence of tourism was the coast of Catalonia. In addition to being a land that meets the above conditions, its privileged location in contact with France allowed it to act as a gateway for international tourism in the country [17]. This phenomenon accelerated the cultural exchange with travelers from all over the world—mainly from Europe and the United States—and introduced new trends, experiences and lifestyles linked to the freedom of leisure and holidays [18]. Suddenly, towns and environments of eminent rural character saw the need to adapt their features and activities to be able to assimilate the large number of travelers which continued to increase year after year [19]. In addition to hotels, spas or tourist developments, the demands of travelers to leave the cities and flee the asphalt led to the consolidation of the campsite as a unique temporary tourist model, in close contact with nature [20].

In 1960, according to the Trade Union of Spanish Tourist Camping Sites, there were 115 campsites scattered around the country, of which 70 were located in Catalonia (Figure 2A–C). And only three years later, the Guide Iberocamping already mentioned the existence of 218 campsites in Spain, 139 of them in Catalonia [21]. These tourist settlements in Spain doubled in three years and the Catalan territory already housed more than 60% of them. Two more facts from the evolution of this model stand out: in 1985, 72% of campsites in Spain were located in coastal areas (527 units out of a total of 733) [22]; and currently, the total number of Catalan campsites has increased to 354 which represents 33% of the overnight stays in Catalonia, compared to hotels, tourist apartments and rural tourism [23,24].

Figure 2. Evolution of tourism incidence of the Catalan coast in relation to the whole region. (**A**) Location of the Catalan coast in Spain; (**B**) Campsite locations along the Catalan coast, 1950 (Source: authors' own based on [20]); (**C**) Campsite locations along the Catalan coast, 2019 (Source: authors' own based on [25].)

The first Catalan campsites have already celebrated more than 60 years since their opening, and the main Catalan campsite association recently celebrated their 40-year history [26], with events aimed at promoting and raising awareness of the sustainable values of this model and its integration into nature.

These entities are part of both the Catalan Campsite Federation [27] and the Mediterranean Campsite Confederation [28], in a sample of the great associative capacity of the sector and of the acquired commitment in favor of a touristic model of family origin, strongly rooted in the dynamics and local traditions of the land. In this sense, with up to 203 member campsites, the Catalan Campsite Federation is leading the recognition of this tourism sector values based on its close relation to the natural environment. For this purpose, in December 2019 it commissioned the publication of the Catalan Campsite White Paper to dig in depth in the activity analysis and future challenges, highlighting that 87% of member campsites are located in rural (14%), natural (30%) or periurban (43%) contexts, and only a 13% are located in urban sectors [25].

On the Administration side, since 2018 the Department of Territory and Sustainability of the Generalitat de Catalunya has been developing the Urban Master Plan for Camping Activities (PDUAC), with a proposal to establish the regulation and parameters that must articulate the implantation of camping activities in the Catalan territory [29]. Furthermore, with the aim of regulating the future of the Catalan coast through urban planning and of preserving those most sensitive sectors of its geography—such as potential natural parks—, in 1992 the Generalitat de Catalunya published the Plan of Areas of Natural Interest (PEIN) [30], and in 2014 the Urban Master Plan of the Coastal System (PDUSC) [31]. According to the Department of Territorial Policy and Public Works, in 2004, before the drafting of the PDUSC, from the total amount of 627 km of coastline on the 500 m deep strip, 46.5% (312.6 km) were urban land, 39.6% (266.5 km) were protected as undevelopable land, 8.2% (54.9 km) were undevelopable land without specific protection and 5.7% (38.3 km) were delimited and undelimited developable land [32].

The interpretation of these documents indicates that approximately 50% of the Catalan coast has already been transformed, with a worrying 26.4% of degraded coast [33] (Figure 3). But it also reflects that some of the protected natural areas, or those which have kept their natural conditions, are currently occupied by campsites or have been at some point [34]. Therefore, the territory of Catalonia, and especially its coast, is presented as a geographical area of great importance for the study of this model of tourist settlement and allows to deepen knowledge both on the evolution of its typologies and on the repercussions on the landscape.

Figure 3. Distribution of protected territories by the Plan of Areas of Natural Interest (PEIN) along the Catalan coast. (Source: authors' own.)

3. Conceptual Background. Hypotheses and Objectives

We are currently at a turning point: we have discovered nature when we almost covered it with cement [35], but it is not too late to foster in the campsite its capacity for active response. After more than 60 years of accelerated tourism development, its effects on the landscape have become clear, and its causes can be studied or contrasted with similar actions in other territories [36]. In general terms, the campsite has been consolidated as a paradigmatic tourism model due to its lightweight occupation of the environment [37]. In this way, its qualities may foment it as a preserver of the landscape and local traditions, values of great significance in the cultural heritage of the territories [38].

However, the campsite's current evolution is starting to diminish this great opportunity. Lately, the presence of stable lodgings is increasing, with lodgings such as bungalows, mobile homes or glamping accommodations. In addition, the architectural design of the common facilities is becoming homogeneous due to both building regulations and standardized solutions. These two aspects, in an incipient phase, are altering the campsite relation to the place, converting these settlements into enclosures of artificialized and globalized landscapes. Yet, camping is a tourist model with a consolidated historical evolution over more than a century. Therefore, taking into account what has happened in recent years, considering the campsite as a phenomenon to be investigated from other areas besides tourism studies is necessary.

In this sense, the General Assembly of the United Nations dedicated the year 2017 to Sustainable Tourism for Development and the Institute for Responsible Tourism (RTI) [39] framed this model in the Sustainable Development Goals (SDGs) by developing a specific tourism sustainability certification for campsites—Biosphere Camping—[40]. The following year, the European Union (EU) declared 2018 the European Year of Cultural Heritage, recognizing the importance of its four categories in the sustainable progress of societies—tangible, intangible, natural, digital—[41]. These consecutive events have raised an awareness of the population and the economic sectors and administrations, establishing a new starting point to recognizing the values of our environment and to considering our history as a dynamic process in constant evolution [42].

Therefore, camping can be understood both as a potential sustainable tourism model and also as a preserver of cultural landscapes identities. The campsite, as a camp, arises from the countryside, from the open spaces linked to the rural land [43]. As a social activity it derived from a sport: hiking is based on a deep respect for nature and inspired by the curiosity to discover new places [44]. Thus, its two fundamental axes—naturalness and temporality—favor a type of light and itinerant environment's occupation, of little impact and short duration. Unlike other tourist models, the campsite's users are agents involved in the construction of the settlement, and the accommodations are transportable elementary units. Each inhabitant adapts the plot with light and removable artifacts [45]: fabrics, ropes, sunscreens, folding furniture, and so on. Furthermore, the user implements it under the guidance of an upper order set by architectural design [46] that provides a structure linked to the qualities of the place and fosters landscape integration as a whole.

To achieve this specific adaptability, the campsite provides a series of services that could hardly be supplied by transportable lodgings: toilets, restaurants, supermarkets, recreation areas, and so on. This scheme minimizes constructions and facilities in common sectors and generates a form of habitat in which movement dilutes the boundaries between public and private [47]. It also promotes responsible use of shared outdoor spaces and social relations between users. Therefore, the feelings of belonging and appropriation of the place are generated both from the individual and from the community, and so the transformations of the environment become much lighter, temporary and reversible than those produced in other types of tourist settlement [48].

In relation to cultural heritage, each specific place has its own activities, characteristics, qualities and singularities rooted in the identity of its landscape. All this acquired values directly affect tourist aesthetic judgment in a positive way, as these are main factors for fostering inhabitants' sense of belonging to a place [10]. Beyond the physical characteristics, in a landscape design project it is also convenient to understand its social, historical and cultural dynamics [49]. These aspects are directly related to the uses supported in their evolution and allow to recognize their presence, giving continuity to the heritage value rooted in the memory of the inhabitants [50].

With the certainty that the landscape is the fundamental value of the cultural heritage of each territory [51], any sustainable intervention or occupation should follow the laws and meanings of the site to foster its preservation, as it is clearly stated in the Catalan Urban Regulation [52]. In this sense, the campsite has the capacity to promote outdoor living and leisure, fostering a responsible way of occupying land in natural contexts. Furthermore, through itinerancy and continuous movement around the territory or within the settlement's boundaries, this tourism model allows to recognize the public space as support for social interactions and shared activities in close relation to nature.

In terms of landscape integration and its identity preservation, the campsite has proven to be one of the most appropriate tourism models due to its natural qualities and its low carbon accommodations [53]: for its rational nature management beyond its limits [54], for the fact of promoting a shared way of living outdoors and for its little constructive footprint [22]. Thus, it could be considered as an opportunity for the future, a resource for the historical recognition and conservation of the cultural heritage of each place, as well as for facilitating the sustainable activation and regeneration of local dynamics [55]. Under the initial idea that a campsite design is a landscape design project, the research is based on four objectives that this settlement should achieve in order to strengthen itself as a paradigm of sustainable tourism and a way of preserving the cultural heritage:

- Integration and Identity: one of the fundamental pillars of the research is the recognition of the intrinsic qualities of the place, to enhance them in the design and thus give continuity to the identity of the landscape that surrounds it. The campsite has to be a univocal part of its environment, assimilating pre-existing conditions and reinforcing the values acquired over time (Figure 4A).
- Preservation and Responsibility: the commitment of the sector is to promote landscape integration at all levels. Not only dealing with its possible visual impact, but also with its ability to generate energy, consume less, reuse waste or preserve the characteristics of the ecosystems, while preserving their environmental functions over time (Figure 4B).
- Temporality and Reversibility: one of the differentiating features of the campsite is its ability to adapt to the changing dynamics of the context and users. It is a model where settlement and activity follow cycles of operation and rotation that encourage reuse. Thus, the vocation of non-permanence and its capacity for regeneration as an active natural space must be able to give continuity to existing values and strengthen them as a complement to the new supported activities (Figure 4C).
- Individual and Community: the campsite encourages the use of the natural environment as a shared public space. Pathways, outdoor activities or common pavilions are aspects that reinforce the identity and sense of belonging to a group. But the individual, as an inhabitant, must also be able to solve his or her most basic needs and have a refuge from maintain daily rites (Figure 4D).

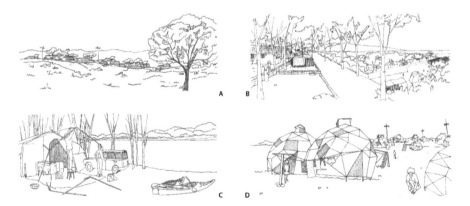

Figure 4. Ideal of a campsite based on the research's main objectives (Source: authors' own.) (**A**) Campsite integration in landscape; (**B**) Sustainable land development based on lightweight occupations; (**C**) Reversible users' adaptations to inhabit nature in leisure; (**D**) Individuals living in community and sharing open spaces.

Thus, this research recovers the historical and architectural values of the campsite as a form of temporary and respectful type of nature dwelling. As an alternative small-scale tourism, it interrelates with the community and contributes to the long-term social well-being, fostering users' interaction and cultural landscapes' regeneration [56]. In order to enhance its qualities and its unique approach, this research identifies the architectural criteria that define the campsite as a potential sustainable settlement, open to the stimuli of its natural environment [57]—with special emphasis on the four announced objectives—. Although tourism has been the most intensive landscape transformation agent in recent decades in Catalonia, especially along the coast, it can also become the engine of change to rectify this situation [58]. Tourists are interested in sun and sea, but also in the natural landscape and the architectural heritage, or the warm human atmosphere of each land, its housing and its inhabitants. Tourism also seeks genuineness, that of the peculiar character of villages and landscapes [59].

4. Methodology

4.1. Mediterranean Strategies Research Line

The objectives of the Mediterranean Strategies research line are the promotion, conservation, regeneration and dissemination of the cultural and architectural heritage of tourist settlements, mainly on Catalonia's Mediterranean coast [60]. Through a technical and critical assessment of the architectural design project, we intend to recover the characteristic mechanisms of Mediterranean construction: the principles of settlement on a place, the use of own materials, traditional construction solutions and climate protection tools [61]. Most of these aspects are considered sustainable passive tools defined by vernacular architecture over time [62].

The present research is based on the inductive methodology followed in previous similar studies [63–65]: obtaining new ways of acting from the study of specific cases and concluding with the definition of a series of strategies applicable in the studied area (Table 1). Through this method, we analyze several reference solutions that are distributed in environments with some similar characteristics, in a process of approaching from the general to the specific. Thus, the analysis focuses on the capacity of the campsite to settle within the context, recognizing the different levels or scales of action: from its landscape integration to the user's occupation of the plot by means of small individual lodgings.

Table 1. Research process based on Mediterranean Strategies inductive methodology.

Methodology	Analytical Stage	Proposal Stage
Analysis of the territory	1) Cultural landscape values. Campsite location.	5) Evolution of campsites [1]. Landscape transformation.
Case study	2) Selection of 5 pioneer examples. Architectural project analysis.	6) Extended sample up to 84 campsites. Characteristics of the place listing.
Comparative approach	3) Five levels of analysis. Conceptual mapping.	7) Topics and concepts definition. Architectural Actions catalogue.
Reflections and conclusions	4) What does landscape integration of campsites mean?	8) What can campsites offer to cultural landscapes?

[1] Graphic catalogue on how both campsites and landscape have influenced each other, from 1954 to 2019.

In a more analytical first stage, once the territory's values and characteristics have been identified, we carry out a comparative analysis between five referent case studies to recognize the characteristics that define them. Subsequently, following from the differences and similarities between the analyzed solutions, there is a deduction of a series of architectural guidelines which are applicable to other locations and similar settlements. At this point, a second stage widens the case studies sample to validate and complete the identified solutions. This research finally concludes with a catalogue of specific architectural actions to foster campsites planning development in relation to place conditions.

Throughout this approach, we keep in mind that a campsite is an element of the landscape and that the landscape is a common good [46]. Therefore, the methodology is aimed at establishing criteria that allow to reinforce the qualities of the natural environment and its public use for future generations, through correct implementation of tourist activities in campsites.

The whole research is seen from the architectural design point of view and is based on graphic records as tools to represent the developed ideas. The specific use of photographs, plans, models or diagrams visually reinforces all the aspects highlighted in the research. This encourages the use of a universal architectural language, with clear informative and communicative qualities that allows to transfer the debate to the public domain. In fact, the social impact on knowledge transfer is a key aspect of the group's research methodology. Administrations, local entities, developers and institutions, public and private, are involved throughout the study with the aim of responding to the territorial needs, while nurturing resources for all agents meant to directly act on it. Therefore, it is not just a matter of establishing collaborations with the institutions for gathering information and data, but of opening a dialogue to recognize the sensitivities of the tourism sector and to establish links with all the stakeholders and end-users involved.

4.2. Analysis of the Territory

We understand the territory as a conjunction between a geography which acts as a support and a history that adds symbolic value on it [66]. Therefore, any intervention on the territory is far from being on a blank sheet where we can erase its memory, but on the contrary: we must recognize its identity and preserve the elements or systems that give it meaning as a place. We must consider that we build on something already built, physically and symbolically [67].

For this reason, the first step in the research is to identify the values of the area throughout several readings of the territory: in-situ field work, digitized air flights, up-to-date data and archives documentation [68]. The results of this data collection are recorded in a thematic map: the Index Map of the Catalan Coast [69]. This document is a graphic catalogue which highlights the geographical conditions of the territory—topography, hydrography—, the infrastructures—highways, train, ports—, the urban frameworks—centers, expansions, urbanizations-, the activities—camping sites, fields, protected zones-, or the administrative scopes of the territory—landscape units, municipalities, counties- (Figure 5). The aim of this study is to place the campsites in relation to the geographical context and to record the transformations undergone over time.

Figure 5. Index Map of the Catalan Coast with landscape interpretation, campsite locations and most relevant camping regions identified. (Source: authors' own.)

4.3. Case Study Campsites. Selection and Analysis

Once the campsites have been identified in relation to their territorial context, we proceed with the assessment of the case studies to analyze. The selection criteria are based on various aspects, such as the quality of the architectural proposals, the type of environment in which they are located, their evolution over time or the access to the necessary information. In this research we have selected five representative cases within the geographical field of study, out of 159 identified cases along the Catalan coast—. These five are considered reference examples as they were pioneers of this typology in Catalonia and were designed by architects with a recognized career. From south to north of the coast, the case studies are:

- COSTA BLANCA campsite (Cambrils, Tarragona, 1962–1989): it was designed by the architects Robert Terradas Via and Jordi Adroer on a plot surrounded by fields on the outskirts of Cambrils. The main communication road between Cambrils and Salou crossed the center of the settlement, which allowed to experiment with the boundaries between public and private. The common buildings were located in the public area nearby the entrance, and the water tank tower rose like a landmark visible from a distance. Once its activity ended, the area was urbanized with apartments (Figure 6A).
- SALOU campsite (Salou, Tarragona, 1956–1987): it was initially designed by the architect José Maria Monravà, but over time other architects also took part in its development, such as Antoni Bonet Castellana, Josep Puig Torné and Jaume Argilaga. It was located below a pine forest, halfway between Salou and Cap Salou, so it had some sectors with sloped topography and sea views, protected by the pines' shade. The urban growth of Salou exerted great pressure on the settlement, which was finally converted into the Municipal Park of Salou (Figure 6B).
- EL TORO BRAVO campsite (Viladecans, Barcelona, 1962–2005): it was designed by the architect Francesc Mitjans under a dense pine forest in the marshy area of the Llobregat Delta. It was one of the four campsites that Mitjans designed in this area of the coast, always with the premise of preserving the forest as a unifying element of the settlement. It had a capacity for about 6000 campers, with singular buildings and a flexible layout to be implemented in that unstable natural environment. The expansion of El Prat airport forced its closure and was declared a protected natural area (Figure 6C).
- CALA GOGÓ campsite (Calonge, Girona, 1961–present): it was designed by the architects Antoni Bonet Castellana and Josep Puig Torné on a land with a topographic difference of about 90 m from sea towards inland. The coastal road crossed the settlement at the bottom, where the main entrance and one of the first nightclubs in the area were located. For its implementation they used the development of existing agricultural terraces. The buildings were built on stone platforms and with a common system based on the traditional Catalan vault. It is still active today with some adaptations motivated by new trends in the sector (Figure 6D).
- LAGUNA campsite (Castelló d'Empúries, Girona, 1968–present): it was designed by the architect Josep Maria Pla Torras in an isolated estate between marshes and located at the end of a 5 km road, which runs between agricultural fields towards the beach. This natural context contrasts with the upcoming tourist development of Empuriabrava, right on the opposite bank of La Muga river. The public buildings follow a common system formed by vaults. The plots are located around a lagoon that structures the settlement. Nowadays, it is still active within the Integral Protection Zone of the Aiguamolls de l'Empordà Natural Park and it maintains its natural character in relation to the landscape (Figure 6E).

The available documentation on the design projects has included mainly the original plans and the architects' writings. Unfortunately, the small size of this type of settlement means that in many cases the information is scarce and inaccurate. The documents' sources were municipal historical archives, collections of the Architects Association of Catalonia and the Cartographic and Geological Institute of Catalonia, as well as private archives of the architects that designed them.

At the same time, we have consulted books, magazines, audiovisuals and tourist promotion campaigns. Most of them are documents published during the first years of the campsites and they show an original reading of the project, without the subsequent interventions they have undergone over the years. In addition, the fieldwork has been a source of great value, since we have been able to visit in-situ the locations of the analyzed cases, even though some have already disappeared. Finally, historical photographs have been one of the most relevant complements to be able to identify projects in their original context and to understand the evolution of the effects of tourism on the landscape. Photographs, and especially postcards [70], have helped us to know how landscape has been transformed by the activities that have taken place there: from walking or sailing, to cultivating, fishing, moving or living [71].

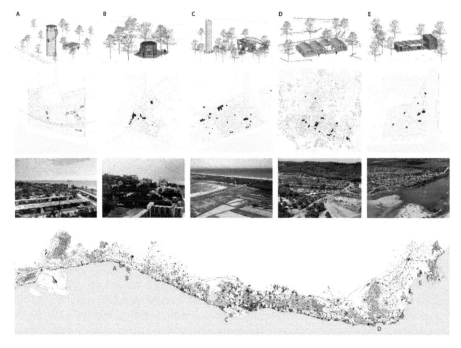

Figure 6. Selection of five case studies designed by relevant architects during 1950s and 1960s along the Catalan coast. (Source: authors' own.) (**A**) Costa Blanca campsite; (**B**) Salou campsite; (**C**) El Toro Bravo campsite; (**D**) Cala Gogó campsite; (**E**) Laguna campsite.

The analysis of the case studies, from the design to the construction, has been essentially based on the graphic documents of the architectural proposals, therefore, it is built on qualitative data. The plans have been redrawn, interpreting the project from the scale of the compound layout to the building's construction details. This process of reconstruction has involved the identification of the pre-existences of the site, on which the architects based their decisions. This exercise of restitution has been done by contrasting documents of varied formats, completing what is missing by assumptions that solve the existing inconsistencies between what was designed and what was finally built.

4.4. Comparative Analysis. Approaches

The analysis settled five sections according to the different levels of approach: the territory (Enclosure), the context (Layout), the settlement (Clusters), the buildings (Pavilions) and the accommodation (Artifacts). Focusing on research objectives, a set of categories was established for the analysis of each of these approaches. Thus, the information obtained is summarized graphically in some diagrams or conceptual schemes to be studied (Figure 7). The methodology's purpose was to provide a reading to every design project in each of the categories, but also to establish a comparative analysis between the five case studies. For that purpose, they were represented schematically following the same graphic criteria: colors, line types, orientation and scale.

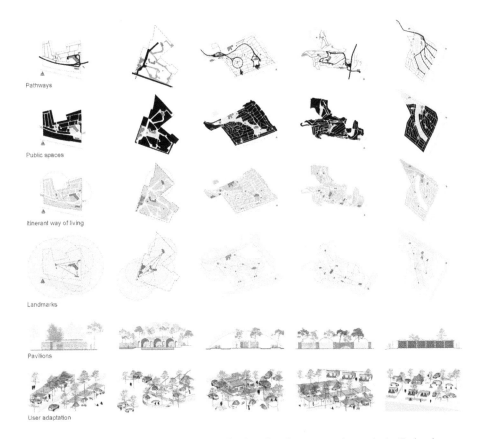

Figure 7. Sample of six different categories developed in the comparative analysis. Each column belongs to a case study keeping the same order as in Figure 6. Each row belongs to a different category, drawn with a specific scheme to emphasize its concepts. From top to bottom: Pathways, Public spaces, Itinerant way of living, Landmarks, Pavilions, User adaptation. (Source: authors' own.)

This comparative graphic process provides a cross view on the general design project conception: on their common decisions, but also on their singularities. It allows us to identify a set of architectural criteria—actions or strategies—that may be suitable for other case studies, or for new design proposals. For instance, if some case studies were set by a clear road hierarchy and these have fostered a slight mobility—bicycle, pedestrian—we can set that a road system organized in relation to context is appropriate for landscape integration of campsites. Furthermore, we can expand on this conclusion by analyzing a wider sample of campsites to identify which layers may define that hierarchy, which permeable materials avoid soil alteration and limit paths, or which vegetation provides adequate shadows and protection.

During this comparative analysis we have followed a similar process in all of the identified topics, taking into account both campsite current evolution and other similar examples to validate the reliability of each architectural action. These architectural project guidelines do not have a regulatory vocation, rather the opposite: they are offered as a tool for diagnosis and decision-making to respond to the specific conditions of each site, knowing that the more they can be applied, the more capacity of landscape integration the campsite will have.

5. Results

5.1. A Study Applied to the Site

In landscape interventions, there is no certainty that generic or universal norms can meet the same needs in places with different dynamics [72]. Each specific place has its own characteristics, its qualities and singularities which have given it a specific and recognized historical memory. Therefore, while designing, the role of the architect or the landscape architect is considered a continuous negotiation between nature and artifice, between social activities and the places that host them, between the individual and the space [73]. For this reason, the campsite design project must be inserted into the attributes of the place considering that in many cases only a slight intervention might be needed: a minimal architectural action that reinforces the meanings of the place [74].

The results of this research are catalogued in six thematic blocks. The first one contextualizes and sets out the content of this research: its convenience, its essence, the importance of the place in the design project and the aspects that have guided the methodology. The other five thematic blocks show the collection of architectural actions, grouped into the five levels of approach following the comparative analysis—from the territory scale to the user one—. Each one of these sections is divided into different topics that provide a further detailed classification of concepts. The Architectural Actions are catalogued by means of the following sequence (Table 2):

Table 2. Classification of Architectural Actions with Sections, Topics and Concepts.

Section	Topic	Concept	Actions (n.)
A_ENCLOSURE It expands on the relationship between the settlement and the landscape.	Context	Visuals	8
		Mobility	6
		Climate	8
		Permeability	8
	Access	Arrival	7
		Equipped area	7
		Car park	11
	Limits	Natural elements	9
		Artificial elements	8
		Edges	14
B_LAYOUT It develops the internal structure of the campsite.	Road	Main routes	12
		Secondary routes	10
		Network	11
		Treatment	12
	Free spaces	Squares	9
		Axes	8
		Indefinite areas	7
		Facilities	7
	Landmarks	Vertical	7
		Horizontal	5
		Punctual	7
		Functional	6

Table 2. *Cont.*

Section	Topic	Concept	Actions (n.)
C_CLUSTERS It outlines the potential display of elements and uses at the settlement.	Upper order	Natural	7
		Buildings	9
		Elements	13
	Common Buildings	Reception	7
		Restaurant	8
		Commercial	7
		Warehouses	7
		Toilets	13
		Sports	11
	Lodging	Dynamic situations	8
		Settled situations	6
		Distributions	12
D_PAVILIONS It explores the common buildings' architectural and constructive typology.	Exterior	Open buildings	9
		Passive systems	8
		Platforms	11
	Shape	Geometry	7
		Modulation	7
		Flexibility	7
		Horizontality	7
	Construction	Tradition	12
		Unique schemes	8
		Industrialization	9
E_ARTIFACTS It deals with the individual adaptation of users in the accommodation.	Plots	Minimum conditions	13
		Variety	6
		Delimitation	10
	Itinerant	Situation	8
		Rotation	8
		Adaptation	13
	Permanent	Situation	6
		Rotation	6
		Adaptation	14

The research and dissemination of the Architectural Actions are designed to facilitate their understanding by a wide range of stakeholders: architects, entrepreneurs, administration, campers, maintenance managers, and so on. For this reason, each concept is developed following multiple simultaneous speeches that can lead to parallel readings, focusing on one or several aspects. Altogether, as a result of the research, a total of 464 architectural actions are classified into 54 concepts and 15 topics. In addition, the graphic record's methodological importance is shown in the 117 published photographs and in the 147 plans, drawings or icons included.

5.2. A Cross-Reading with the Characteristics of Place

The construction of landscape has been the consequence of man's intervention on nature, and landscape's identity has become the fundamental feature for the recognition of its cultural heritage [75]. For this reason, an integrative and valid action for a specific place can be absolutely contrary to the needs of a different environment, where processes, conditions and significantly different meanings have probably intervened over time [76]. Therefore, this research highlights the importance of "reading the site" as a process for linking design requirements with the characteristics of the context. It is a procedure which consists of identifying the pre-existing elements of the site from the analyzed case studies, as well as its conditions and the values that give it meaning. The recognition of these aspects makes it possible to draw an x-ray of the environment, its shortcomings and, above all, its strengths. The following list shows pre-existing characteristics to consider in this previous reading of the site, selected for their direct influence on design decisions (Table 3):

Table 3. Pre-Existing Features and Identifiable Elements throughout the "Reading the Site" Process.

Characteristics of Place	"Reading the Site"	Elements
Architectural heritage	The built elements often belong to local tradition, reinforce landscape identity and can be reused.	Walls, paths, fountains, farmhouses, buildings, etc.
Cultural heritage	Intangible qualities enhance cultural dynamism and discovery of toponyms, local products and customs.	Legends, memory, experiences, characters, gastronomy, etc.
Activities	Existing or previous uses of the site relate the settlement to its context and encourage the rooting.	Agricultural, livestock, leisure, commercial, residential, educational, etc.
Layouts	Linear elements of the environment provide an upper territorial order which can be followed.	Borders, fences, irrigation, pathways, infrastructures, green corridors, etc.
Hydrography	The water introduces a reference to the place, allows distant views and is a source of biodiversity.	Streams, canals, rivers, lagoons, lakes, sea, etc.
Topography	Traditional terrain transformations have given rise to systems of great value and low visual impact.	Dunes, slopes, walls, margins, terraces, platforms, etc.
Flora	The existing vegetation is a sample of the activities the site has endured over time.	Undergrowth, bushes, tress, monumental trees, unique species, cycles, etc.
Fauna	The animal life reflects natural cycles of the habitat and it directly relates to landscape dynamics.	Insects, fishes, reptiles, mammals, birds, migrations, farms, etc.
Climate and Energy	Site conditions and technological innovations facilitate low consumption and waste treatment.	Orientation, sunshine, winds, rainwater, waste, drainage, etc.

5.3. The Matrix. A Diagnosis Tool.

The identification of these pre-existing elements on the site, and their deep analysis in relation to the context, provides valuable information on what actions need to be implemented in order to remain part of the cultural landscape of the place [77]. With the aim of highlighting architects' decisions on the case studies' design projects, the comparative analysis also expands on their specific site conditions. By means of a cause-effect sequence these design solutions are related to the issue they overcome (Table 4). Furthermore, the comparison between these different aspects clarifies the most adequate classification for all different topics identified throughout the five levels of approach.

All the architectural responses to each of these pre-existing features are raised as potential actions to be implemented in forthcoming developments. The clue of this proposal stage of the research is validating the capability of these specific architectural solutions to announce them as design recommendations in other locations with similar conditions. Thus, once a first collection of actions is set from the referent case studies, a second round is developed through other examples from the same geographic frame. In this sense, from the five initial case studies, the sample is expanded up to 84 campsites located in different contexts along the Catalan coast. At this point, both pioneer and current campsites are selected if their evolution can be traceable enough to determine the adequacy of their architectural solutions.

Table 4. Synthesis of the Comparative Analysis. Architectural Actions Identified in the Case Studies.

Characteristics of the Place	Architectural Actions Identified in the Case Study Campsites [1]				
	Costa Blanca	Salou	El Toro Bravo	Cala Gogó	Laguna
Architectural heritage	A road crosses the settlement and sets a shared public space	Traditional stone walls for terraces and buildings	A house converted into a common building	Buildings based on a traditional vault system	Buildings shape based on a local vault system
Cultural heritage	Traditional agricultural sheds define territory scale	Nearby tourism facilities foster leisure offer	Place of isolation provides a naturist camping zone	Typical local farmhouse become a landmark	Light footprint construction to be integrated in the Natural Park
Activities	Water tank tower sets a landmark for road tourists	Seafront and tourism set main access towards beach	Fishing and water ski as commercial birdcall	Existing agricultural terraces facilitate plot setting	Nature is preserved from nearby formal tourist sprawl
Layouts	Agricultural fields and a road set access and limits	Agricultural fields and fences define plot pattern	Fields and irrigation canals define plot pattern	Main road goes through and around plot terraces	Marshes, fields and trails define plot pattern
Hydrography	Existing reservoir used for rainwater collection	A stream defines side edge and beach connection	A lagoon and the beach define edges	Two streams and the beach define edges	Inner lagoon fosters nature quality and free spaces
Topography	Upper order on flat terrain set by wattle pergolas	Low stone walls create platforms for setting plots	Flat terrain facilitates bike and pedestrian mobility	Stone walls create terraces and exterior buildings	River and lagoon embankments define edges
Flora	Cane margins foster main road protection	Combination of existing and planted trees	Pine forest ease itinerant plots in shadow	Addition of wattle pergolas for lack of trees	Plantation of local tamarinds ease shading
Fauna	No permanent lodging to foster natural regeneration	Natural habitat protection from urban sprawl context	Marshes drainage to prevent mosquito	Varied marine fauna as a commercial birdcall	Natural protected land for bird migrations
Climate and energy	Centralization of buildings and facilities	Lavatories and pool nearby water tank	Irregular plot layout preserves pine trees	South-oriented permanent plots over sea	Solar power and reuse of rainwater

[1] Highlighted words are related to the 15 topics of the architectural actions catalogue.

Once the topics classification is defined and initial architectural actions are validated and completed through other selected cases, a diagnosis tool can be established to simplify the "reading the site" process prior to design: the Matrix. This chart directly relates each of the pre-existing elements in the context to the topics of the Architectural Actions (Figure 8). Therefore, following a visual source, we can quickly identify which are the most relevant recommendations to consider in any particular place, according to the characteristics detected and taking into account the classification by different topics and levels of approach.

In this sense, the Matrix offers a view focused on results and in response to the qualities of each landscape. Thus, it is a typological chart which follows design guidelines from the analyzed case studies, and which can evolve globally. Columns show each of the identified concepts, in their five scales in which project actions will be developed. Rows collect the typologies of pre-existing detectable elements on the site. The relation among concepts is classified by levels of incidence, discerning the relevance that an element may have for developing a specific architectural action.

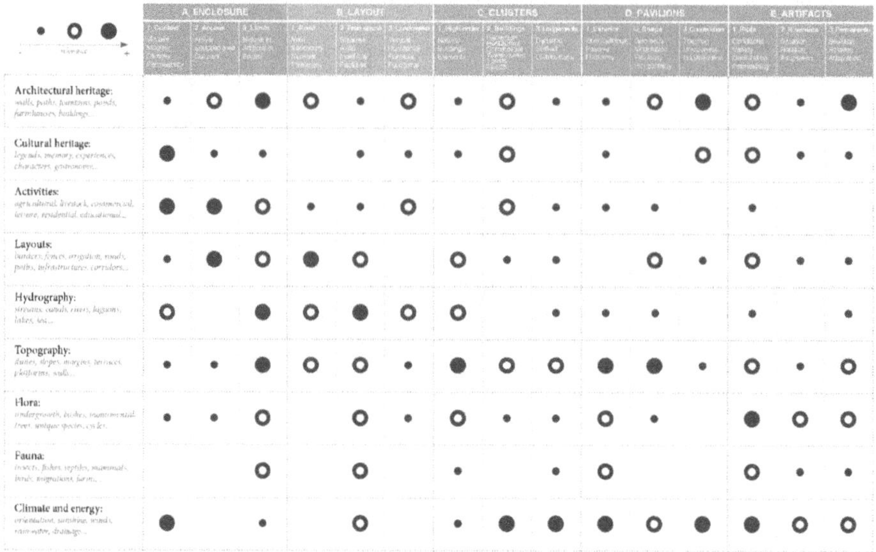

Figure 8. Representation of the Matrix. Each column belongs to a specific topic within the five sections. Each row belongs to a pre-existing feature of a place. This design tool shows the relevance of the site elements to focus on the most adequate architectural actions in each concept. (Source: authors' own.)

5.4. Architectural Actions

Once the pre-existing elements of the site have been identified and their impact on the project has been assessed through the Matrix, the architectural guidelines can be introduced both in the design process and the decision-making to face any action that may be taken in that place. Following the methodology, this catalogue of recommendations is organized into the five thematic sections defined by the levels of approach. In addition, each section addresses three specific topics, and each of these topics develops the proposed architectural actions highlighting the main one as a reference.

These strategic actions guide the great challenges of the design process to achieve a campsite model that guarantees the qualities defined in the research objectives: its landscape integration, its preservation of nature, the promotion of reversibility and the identification of the individual in the community. The following sub-sections set out the considerations for each of these topics and detail the concepts developed. In addition, each one of these reference actions is exemplified by means of a specific campsite solution which is illustrated by a picture and conceptualized in a diagram.

5.4.1. Enclosure

The campsite is an outdoor unit that forms a settlement by means of a layout of boundaries that define its perception as a private area. Despite being an enclosure, the implementation criteria must generate proposals to reduce its tightness and minimize the image of a hermetic sector. Incorporating the qualities of the context can serve to facilitate the reaction to the dynamic conditions of the environment and enhance integration into the natural landscape, avoiding the consolidation of occupations in sensitive areas of public interest.

The topics, concepts and reference actions developed in the Enclosure approach are:

- Context: visuals, mobility, climate, permeability. The campsite must recognize the qualities of the place, be part of it and assimilate its own dynamics. Integration in the landscape involves giving value to its characteristics—geographical, visual, historical, functional, etc., introducing and enhancing them in the design. For instance, in littoral campsites, beach dunes free from users' occupation enhance autochthonous flora preservation and flood prevention (Figure 9A).

- Access: arrival, equipped area, car park. The arrival to or departure from the campsite is a transition process that requires an area of relation between the inside and the outside. Access is a breaking point of the boundaries that define the site, so it is an important space of interaction between users and the general public, as it may contain shared activities and act as a reference of the settlement. For instance, a vertical water tank in close proximity to a welcoming reception pavilion provides a unique accessing experience and fosters place belonging (Figure 9B).
- Limits: natural elements, artificial elements, edges. The campsite must be a permeable enclosure, which is directly related to the elements of the site. The boundaries must be transformed into edges, as areas with thickness that meet all the requirements and incorporate multiple superimposed filters, both natural and artificial [78]. For instance, in rural contexts, a sequence of low embankment, vegetation, water and a transparent fence provide a slight but secure separation which improves the quality of the views (Figure 9C).

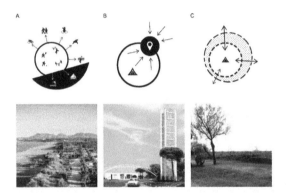

Figure 9. Enclosure: sample of architectural actions represented by icons and pictures. (Source: authors' icons and authors' or public domain pictures.) (**A**) Context in Las Dunas campsite; (**B**) Access in El Toro Bravo; (**C**) Limits of Laguna campsite.

5.4.2. Layout

On an overall scale, the campsite is organized with a structure that links the characteristics of the environment with the needs of the inhabitants. It also has the capacity to do so in a way that is flexible enough to dynamically adapt to the context's changing conditions and the daily needs. The layout, far from imposing, is subtly introduced into the context and guides the actions of users towards the domestic, recreational and social. The campsite is organized from the arrangement of the whole: a new order or structure ensures that the temporary prevails over the permanent.

The topics, concepts and reference actions developed in the Layout approach are:

- Road: main routes, secondary routes, network, treatment. The basic structure of the campsite must recognize a hierarchy of pathways and displacements of different types and formats. Vehicles and transportable lodgings must be able to coexist with pedestrians, bicycles or scooters; thus, the establishment of a road gradation according to the intensity of use adds order to the campsite and organizes the different areas according to their character and materiality. For instance, irregular narrow paths with permeable pavements and abundant vegetation soften traffic around plots and foster the use of public spaces (Figure 10A).
- Free spaces: squares, axes, indefinite areas, facilities. In the campsite, all unoccupied space must be part of the shared natural environment. The open spaces organize recreational activities within the settlement, so their distribution must recognize the pre-existences and introduce the character

of the uses that relate to leisure in nature. For instance, in topographic developments, a sequence of buildings related to different terraces create sight views that reinforce coexistence of multiple activities in a public axis (Figure 10B).

- Landmarks: vertical, horizontal, punctual, functional. The structure of the campsite must be clear enough to facilitate the user's orientation and be a birdcall for visitors. Landmarks are the elements that define a known environment. These also guide displacements and dynamics of the settlement, with clear visuals and suggestive architectural solutions that are introduced into the site dynamics. For instance, outdoor facilities—showers, lavatories, sports areas—become daily-use places which reinforce social experiences and highlight specific areas (Figure 10C).

Figure 10. Layout: sample of architectural actions represented by icons and pictures. (Source: authors' icons and authors' or public domain pictures). (**A**) Road in Molí Serradell campsite; (**B**) Free spaces in Cala Gogó campsite; (**C**) Landmarks in Salou campsite.

5.4.3. Clusters

Campsites are based on social relations, characteristic of an inhabited place. Besides, campsites have a temporary vocation, with dynamic or time-limited activities and transportable lodgings that generate itinerant occupations. This temporary nature makes it necessary to provide the settlement with qualities that facilitate the users' orientation and dwelling, reinforcing the shared use of public space and the sense of belonging. Therefore, clusters are a resource to provide the place with common spaces for interaction which reinforce its identity.

The topics, concepts and reference actions developed in the Clusters approach are:

- Upper order: natural, buildings, elements. The campsite must shape its own identity and relate it to that of the environment, following its qualities and singularities. The upper order generates a common thread that facilitates the recognition of the campsite as an autonomous entity, with a hierarchical organization at different levels, and with a series of unique elements—natural or artificial—that give it its character. For instance, the distribution of a singular pattern based on everyday elements such as low walls, benches or fountains defines communal outdoor spaces and fosters users' sense of belonging (Figure 11A).
- Common buildings: reception, restaurant, commercial, warehouses, toilets, sports. The equipment must complement the domestic functions of the lodgings and contribute to provide character and leisure value to open spaces. The organization of buildings in the campsite must be balanced and consider the structure defined by paths and open spaces, in order to promote shared activities, social relations and a sense of belonging to the place. For instance, in isolated campsites located in rural landscapes, centralization of main buildings around the access frees the rest of the enclosure for both accommodations and leisure activities in close proximity to nature (Figure 11B).

- Lodging: dynamic situations, settled situations, distributions. The layout of the plots must allow groupings of lodgings to foment shared dynamics. Different types of accommodation generate situations and ways of living of very varied characteristics, but they can be adapted following organizations around a shared space, a twisted alignment or using low topographical differences. For instance, in topographic terrains, irregular stone-made terraces provide a series of misaligned plots which improve privacy, sight views and landscape integration due to their fragmentation and materiality (Figure 11C).

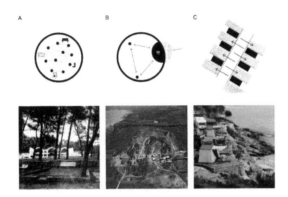

Figure 11. Clusters: sample of architectural actions represented by icons and pictures. (Source: authors' icons and authors' or public domain pictures.) (**A**) Upper order in Salou campsite; (**B**) Common buildings in Delfín Verde campsite; (**C**) Lodging in Torre de la Mora campsite.

5.4.4. Pavilions

In the campsite, the common buildings are constructions open to natural environment in order to reinforce the original values of camping. These conditions determine the singular typology of the pavilion. This is a set of qualities that, beyond proposing specific solutions, resolve the implantation of the settlement as a system, considering the pre-existences and singularities of each place. They are effective, simple and unique objects that infiltrate in leisure and shared contexts following construction systems of the local tradition. They propose innovative and suggestive formal designs that resolve the contact with the ground and the dynamic adaptation to climate conditions. Due to their unique and effective composition, they have a great capacity for attraction, without the need to be apparent, but accompanying the user at all times.

The topics, concepts and reference actions developed in the approach of the Pavilions are:

- Exterior: open buildings, passive systems, platforms. In the campsite, buildings must follow the typology of the pavilion, as elements that are related to the outside and add value to it. The pavilions are buildings open to nature and adaptable to the topographic conditions of the place, with terraces, courtyards, porches and large openings. For instance, open facades at ground level extend pavilion activities to the open spaces, reinforcing their public character (Figure 12A).
- Shape: geometry, modulation, flexibility, horizontality. The geometric control in the pavilions is key to being able to propose unique and innovative solutions with efficient use of resources. Buildings are conceived as containers of activity, open spaces and are easy to enlarge so that they can accommodate multiple uses over time. Furthermore, these require clear and recognizable designs, with basic geometries and easily identifiable in their context. For instance, the combination of buildings with different rectangular volumes creates a slight urban character with an inner shared courtyard for social interactions surrounded by nature (Figure 12B).
- Construction: tradition, unique schemes, industrialization. In the campsite, buildings must reinforce the identity of the place, with the use of local materials and construction systems.

Traditional solutions respond to the climate and comfort needs following architectural typologies that have evolved over time; therefore, their current implementation facilitates the design of efficient, innovative and site-based proposals. For instance, the Catalan Vault is a traditional brick-made dome that enhances landscape integration due to its modular conditions and its capacity of aggregation and subtraction of elements (Figure 12C).

Figure 12. Pavilions: sample of architectural actions represented by icons and pictures. (Source: authors' icons and authors' or public domain pictures.) (**A**) Exterior in Salou campsite; (**B**) Shape of La Ballena Alegre campsite; (**C**) Construction in El Toro Bravo campsite.

5.4.5. Artifacts

Life during holidays, in a carefree context, allows us to explore a habitat without restrictions and while linked to the outdoors and nature. In terms of ways of living, dynamic and leisure contexts are more permissive than stable and routine ones. Therefore, they can be developed in accommodations with less demanding conditions, where comfort and basic needs can be supplied by the whole settlement. Besides, home at the camping is not a unique space but it is distributed throughout the land and is diluted between paths, squares and buildings. Each action of the dweller can take place in a different location, through a continuous pathway, but always with a fixed reference in the most personal unit of use: the plot. This represents a small portion of the landscape, a fragment of nature that the user temporarily transforms into a dwelling place, giving rise to spontaneous and ephemeral situations. The inexistence of physical boundaries enhances the shared use of open space, relationships and itinerant occupations. Therefore, an unoccupied plot has the condition of free space, and remains as such once the user leaves it.

The topics, concepts and reference actions developed in the approach of the Artifacts are:

- Plots: minimum conditions, variety, delimitation. In the campsite, the plots should meet the right conditions to enjoy a leisure habitat in nature, based on horizontality, shade and connectivity. In this way, the individual adaptations of the users will be of little entity and able to be reversible, according to the temporality of their stays. Yet, there are other factors that also influence and configure a wide range of choices for users: orientation, privacy, dimensions, supplies, etc. For instance, in littoral campsites, front line plots offer sea views but a lack of privacy which can be solved by means of topography and autochthonous vegetation (Figure 13A).
- Itinerant: situation, rotation, adaptation. Occupations with transportable lodgings should be facilitated and enhanced due to their temporary and environmentally nature friendly character. Itinerant accommodation belongs to the users, who install it, adapt it and move with it. Thus, at the end of the stay it disappears and frees up areas of the settlement that can be easily renewed. For instance, users' adaptation by means of tents, caravans or fabrics in relation to the natural

elements on-site—trees, bushes—provides different levels of intimacy and social interaction, from public space to private beds (Figure 13B).

- Permanent: situation, rotation, adaptation. Stable lodgings remain on-site and must be designed according to pavilion typology criteria. The qualities of openness to the outside, contact with the ground, geometric control and local construction solutions allow the design of permanent accommodation to be linked to their environment, enabling individual adaptation by means of the use of light elements. For instance, modular lodgings such as mobile homes are stand-alone elements that follow Passivhaus strategies for lowering energy consumption [79], foster waste re-use and enhance user comfort adaptation in relation to climate conditions (Figure 13C).

Figure 13. Artifacts: sample of architectural actions represented by icons and pictures. (Source: authors' icons and authors' or public domain pictures.) (**A**) Plots in Torre de la Mora campsite; (**B**) Itinerant artifacts in Cala Gogó El Prat campsite; (**C**) Permanent artifacts in Laguna campsite.

6. Discussion

6.1. The Role of Campsites in the Preservation of Cultural Landscapes

As a result of the research, Architectural Actions are not only set as a catalogue of recommendations, but as a methodological framework for developing a campsite design project integrated into the landscape. These recommendations are not proposed as mandatory regulations but as a guide to facilitate the decision-making. In addition, the Matrix eases the initial diagnosis and consolidates the importance of understanding the site as a place with specific qualities that must be identified and enhanced within the design process.

Thus, this study proposes a way of doing, a guided path to achieve the recognition and preservation of the cultural landscapes' values through campsites. As recommendations, the more they can be introduced into the dynamics of the settlement, the more likely this will be integrated into its context. But the determining point of the process is to correctly identify which actions are appropriate for each place by means of the Matrix. On the other hand, the approach to the field of study from different scales of work—from the territory to the user—and the incorporation of different readings and graphic resources with an informative character—actions, images, icons, descriptions, etc.—facilitates the understanding of its contents.

This document is aimed at anyone who is sensitive to or involved in promoting the enjoyment of nature while preserving its essential values—what makes each place unique—in a process that must increase the quality of the campsites and their landscape integration. Therefore, beyond this study, it is pertinent to convey to citizens, administrations and businesspeople the relevance of cultural landscapes in recognition of the history and the result of our actions on nature. It is important to make the debate public, to incorporate different sensitivities and to realize that the responsibility is shared, as it is to do with the enjoyment of the natural environments that surround us.

In that sense, the Catalan Campsite Federation and related associations have recently launched several initiatives to explore and publicize the current situation of the sector: the Good Practices Manual is one of them, as well as the drafting of the Catalan Campsite White Paper, the Girocamping PRO congress, the participation in international fairs or the holding of technical seminars, round tables and open debates. All these activities seek to publicize the role of the campsite as a nature-friendly tourist model, as well as generate synergies to promote new points of view on the sector, and diagnose the current situation to improve its capacity to act as a preserver of the landscape and culture of the places.

The more than 60 years of campsite development in Catalonia—since the country's tourist boom—provide an evolution that must be taken into account, placing more emphasis on the paradigmatic cases of the early years. In its beginnings, this was a field of exploration of new architectural typologies, without restrictions or regulations. In addition, architects followed a very strong desire to combine modernity imported by international tourism and the tradition imposed by the economic precariousness of the time. This duality has conditioned the campsite's relationship with the environment during all these years.

The symbolic value of the places is the factor that generates more expectations for the travelers, in their desire to discover new territories [80]. Therefore, it is a paradox that the action of tourism on the landscape can irreversibly alter its qualities when these are its raison d'être. The first identified campsites in the country had their origin in the evolution of previous agricultural uses. They were very small scale interventions, often precarious, but they facilitated the basic conditions for inhabiting nature without altering its qualities. This fact, which originally occurred due to a lack of resources, is now becoming a responsible and indispensable way to ensure sustainable tourism and the preservation of cultural heritage (Figure 14).

Figure 14. Conceptualization of a campsite surrounded by nature. Once occupation finishes, artifacts may disappear, and the site can keep its landscape identity. (Source: authors' own.)

The landscape has been built physically and symbolically over time using local techniques from tradition and productive activities, mainly agricultural [81]. Therefore, tourism actions are not the first to transform nature. In fact, there is no certainty that nature is still intact [82]: the different activities are deposited on the site in a specific way and add new meanings that, over time, shape their identity and end up being adopted as genuine. Thus, in order to preserve the uniqueness of each environment and its balance between artifice and nature it is necessary to establish a link between the campsite and the place, to give it continuity. From the scope of this research, these considerations become fundamental to ensure the sustainable development of these settlements and their contexts.

The campsite has qualities that set it apart from other tourist settlements. It can accommodate shared dynamics in an open system linked to outdoors space, with the active participation of users. In addition, it has a temporary vocation and is flexible enough to disappear without leaving a trace, as has happened in many campsites transformed into municipal parks or located in a respectful way in natural parks and other protected areas [83]. For this reason, as the initial hypothesis of the research has advanced—and as it can be deduced from its results, campsites have the capacity to stand as active agents for the regeneration of cultural landscapes, where temporality and adaptability are key aspects.

Due to their size, campsites may have a positive impact on the sustainable resource management capacity, for example, the generation of renewable energies, the control of water cycles, the reuse of waste, the conservation of large unbuilt areas or the preservation of vegetation and autochthonous fauna. They can also be actively introduced in other sectors such as the productive—agriculture, livestock—or education—school camps, nature discovery, considering new activities for the lower tourist impact seasons. On the other hand, these settlements make it possible to rethink the links between rural and urban contexts, recovering the continuity of the green corridors that connect inland forests with the sea. In this sense, campsites can also diminish the homogeneity of the built coastal front and strengthen the importance of a continuous green littoral.

Therefore, it is a question of recognizing nature as an element in continuous transformation and of adapting the campsite to the climatic conditions while encouraging its dynamic activities. The set of these causes of unsteadiness is likely to alter the existing balance between social activities and their environment, which must constantly introduce new solutions to the demands of the moment. Thus, we assume that the landscape should not be understood as a simple stage to be preserved, but above all as an element to work with, a genuine instrument of a new and emerging design discipline [84]. This means a way of proceeding that recognizes the place and enhances it, but also one that introduces the users' dynamics into the environment's ones to link the two them. This process recognizes the passage of time as a definition of its own history and turns the place into one of cultural value.

Beyond freezing the natural environment in its pretended essence, the new lines of debate must be based on this contrast between artifice and nature [85]. Therefore, the natural environment must also be understood as an architectural element that can favor the proposal of new spatial relationships [86]. In this sense, architects make landscape: just by building, a landscape is being made, since it is adapted as the physical support, and new traces are added to the memory of the place. One of the quintessential Mediterranean towns, Cadaqués, would be an inhospitable place without the constructions that shaped and characterized it [87]. Architecture—art, culture—has formed the landscapes we know through the modeling of nature (Figure 15). In fact, the landscape leaves a mark and prints it inside us, in the same way that it also receives our footprints [88]. Even in those territories without constructions, architecture has also acted over time in a symbolic way, by recognizing its values, giving meaning to places and inciting curiosity in visitors.

Figure 15. Conceptualization of a series of plots located on platforms made by traditional stone walls. These pre-existing elements used to be a topographic adaptation for agricultural purposes. Their conservation relates touristic activities to site history and cultural experiences for landscape discovery. (Source: authors' own.)

6.2. The Individual and the Community. Ways of Living in Balance with Nature

In campsites, both the layout on the site and the users' temporary occupations go hand-in-hand throughout their evolution. When the camper withdraws their accommodation and leaves the plot, the natural environment remains intact or slightly altered, with a high degree of reversibility. In this way we can follow the dynamic cycle of the settlement based on the close contact between nature and tourism—and previously conceived from architecture.

This wide range of possibilities and cross-relationships opens the door to considering the current situation of campsites and what their role should be in the future of cultural landscapes. If the focus remains on the leisure and respectful use of nature, these environments may continue to act as preservers of the identity and meanings of each place. In fact, the most pertinent question is not so much whether landscapes should change or not, but how they can adapt to the passage of time and remain recognizable [89]. Trees, topography, beaches or the climate, as well as gastronomy, music, traditions or postcards, are all elements that have forged a shared ideal of what holiday life means.

The future of the campsite depends on its ability to incorporate these elements of the environment and to generate areas of shared relationships. It is not too late to explore new ways of inhabiting nature and reinventing our experiences [90]. We still are on time to recognize the cultural value of a place and nurture it with some resources to highlight its qualities. In fact, campsites encourage temporary occupations that can reinforce the inherent values of the landscape, as well as dissolve the physical barriers imposed by alien actions, and extend part of community life to the urban structures that support them. In terms of landscape integration and cultural identity preservation, sustainable tourism is evolving through campsites in their nomadic way of inhabiting nature, based on the temporality and informality of primal leisure occupations [91].

Finally, holiday life, free and carefree, may enhance the interaction between users allowing to create such close links between them. However, the generation of this feeling of belonging to a group is definitely influenced by the way in which outdoors space is shared. These settlements must ensure support for the temporary habitat of large groups of people away from their usual homes. For this reason, the environments in which they are located must facilitate community development and increase the multiple relationships that are established with the site.

In order to preserve the landscape's cultural heritage, users must feel part of it and collaborate in caring for it and recognizing the values that make it desirable. Therefore, campsites must be able to guide the experience of the inhabitants, while preserving some of the rituals that link them to the everyday life from which they are momentarily separated. These developments not only must provide a private shelter or community structure, but also enhance individual initiatives and the possibilities for establishing interactions with each other (Figure 16). Individuals, in any community, must be able to contextualize their situation in the group and in the place, in order to relate their personal experiences to the collective memory that links them. Thus, they can strengthen the desire to recognize themselves within the identity of the environment that welcomes them.

Figure 16. Itinerant dwelling under a forest. Some low stone walls create a pattern under the homogeneous leaves ceiling. By means of the artifact and some fabrics, each inhabitant adapts the plot to their own needs. The open space free of users' occupation can support plenty of activities, such as playing games, walking, having lunch, teleworking, etc. The campsite shape is flexible and dynamic, directly related to people's interactions. (Source: authors' own.)

7. Conclusions

The results of this research allow us to advance on several future lines. First, the catalogue of architectural actions provides criteria for determining the campsite level of integration to the site. This taxonomy becomes a tool for the analysis and measurement of the architectural and landscape qualities of the settlement. For instance, it allows us to determine the state of the current campsite stock and to establish guidelines for the stakeholders and the administration to improve its quality.

Secondly, the revision of this tourist model confirms its validity as a paradigm, and its potential as a sustainable asset from the architecture and landscape point of view. Most of these camps were established with very few means and have lasted to the present day with slight changes. Over time, characteristics such as the economy of resources and the use of local techniques and materials have become proper criteria for current sustainable design. In addition, due to its own configuration, the campsite maintains an important area of highly preserved land, with the potential to be a collector and distributor of energetic resources: solar, wind, geothermal, biomass, etc. The preservation of undeveloped large areas among highly built sectors positions them as forthcoming green corridors that favor the protection of biodiversity, the passage of fauna, the regulation of rainwater or green mobility, among other possibilities.

Finally, the research also highlights the capacity of these settlements to preserve the identity of the place and its culture, organizing communities based on itinerancy and temporality with a high degree of respect for the environment. Furthermore, the architectural criteria of the actions afforded are also applicable to other types of temporary camps, such as cultural events, refugee camps or pilgrimages. This research offers a fundamental tool in establishing the settlement by prioritizing the recognition of the context when defining its form. From here on, the scale's gradation in the design of the camp ensures the appropriation of the place by the user: everyday rituals and interior routes relate to the neighborhood scale, while the urban structure refers to the community spaces and to the buildings acting as landmarks. The application of these guidelines from the architectural point of view ensures individual and community well-being, protects privacy and fosters social relationships in a natural environment. Thus, the Architectural Actions' contribution should be to favor the construction of scenarios for the generation of community life in balance with nature, empowering both users and entities in the sustainable development of cultural landscapes.

Author Contributions: Conceptualization, X.M. and A.M.; methodology, X.M. and A.M.; formal analysis, X.M. and A.M.; visualization, X.M. and A.M.; writing—original draft preparation, X.M.; writing—review and editing, X.M., A.M. and I.d.R. All authors have read and agreed to the published version of the manuscript.

Funding: This research project was financed by Federació Catalana de Càmpins and EURECAT—Technology Centre of Catalonia. The first stage of this research was funded by Secretaria d'Universitats i Recerca del Departament d'Empresa i Coneixement de la Generalitat de Catalunya, grant numbers 2018-URL-Proj-023 and 2017 SGR 1327.

Acknowledgments: The authors would like to express their gratitude to the stakeholders that participated in this study and contributed with their valuable insights.

Conflicts of Interest: The authors declare no conflict of interest. The funders had no role in the design of the study; in the collection, analyses, or interpretation of data; in the writing of the manuscript, or in the decision to publish the results.

References and Notes

1. Nogueira, B. Ensayos para un Atlas de la Costa del Sol. In *Turismo líquido*; Pié, R., Rosa, C., Eds.; IHTT (UPC): Barcelona, Spain, 2013; pp. 298–337.

2. Vázquez Montalbán, M. Turismo, medio de incomunicación de masas. *CAU Construcción Arquit. Urban.* **1970**, *0*, 10–13.

3. Poullaouec-Gonidec, P. Evocaciones. In *Landscape + 100 Palabras Para Habitarlo*; Colafranceschi, D., Ed.; Gustavo Gili: Barcelona, Spain, 2007; p. 26.

4. Breiby, M.A. Exploring Aesthetic Dimensions in a Nature-based Tourism Context. *J. Vacat. Mark.* **2014**, *20*, 163–173. [CrossRef]

5. Rosa, C. El Turismo como futuro: La ciudad del ocio. In *Turismo líquido*; Pié, R., Rosa, C., Eds.; IHTT (UPC): Barcelona, Spain, 2013; pp. 38–49.

6. Spengler, O. *Le Déclin de l'Occident*; Gallimard: París, France, 1964; p. 167.

7.   Hernández, J.M. Sobre el Paisaje Cultural. In Proceedings of the 4th European Congress on Urban and Architectural Research (EURAU'08), Madrid, Spain, 16–19 January 2008.
8.   Ahern, J. Urban landscape sustainability and resilience: The promise and challenges of integrating ecology with urban planning and design. *Landsc. Ecol.* **2013**, *28*, 1203–1212. [CrossRef]
9.   Pié, R.; Rosa, C. Un turismo sin arquitectura o una arquitectura sin argumento. In *Turismo líquido*; Pié, R., Rosa, C., Eds.; IHTT (UPC): Barcelona, Spain, 2013; p. 7.
10.   Kirillova, K.; Fu, X.; Lehto, X.; Cai, L. What makes a destination beautiful? Dimensions of tourist aesthetic judgment. *Tour. Manag.* **2014**, *42*, 282–293. [CrossRef]
11.   Milohnic, I.; Cvelic-Bonifacic, J. Sustainable Camping Management: A Comparative Analysis between Campsites and Hotels in Croatia. In Proceedings of the 3rd Tourism in Southern and Eastern Europe (ToSEE), Opatija, Croatia, 14–16 May 2015.
12.   Elden, S. Land, terrain, territory. *Prog. Hum. Geogr.* **2010**, *34*, 799–817. [CrossRef]
13.   Martín, X.; Martínez, A. Manual de Bones Pràctiques. In *Accions Arquitectòniques En Els Càmpings*; La Salle-URL: Barcelona, Spain, 2019.
14.   AMCAM. *L'arribada del Turisme a Cambrils*; Cambrils Municipality Archive: Cambrils/Tarragona, Spain, 2012.
15.   Barba, R. Les peces mínimes del turisme. In *Debat Urbanístic Sobre la Costa Brava*; Coromines, E., Sola, P., Eds.; COAC: Girona, Spain, 2003; pp. 71–79.
16.   Jiménez, E. El turismo y los modos de transporte. In *Turismo líquido*; Pié, R., Rosa, C., Eds.; IHTT (UPC): Barcelona, Spain, 2013; pp. 98–103.
17.   Garay, L.A.; Cànoves, G. El desarrollo turístico en Cataluña en los dos últimos siglos: Una perspectiva transversal. *Doc. D'anàlisi Geogràf.* **2009**, *53*, 29–46.
18.   NO-DO. Camping. La caravana Wally Byam en Castelldefels. In *Noticiarios Y Documentales Cinematográficos*; Documentary; Radio Televisión Española: Madrid, Spain, 10 October 1960; Available online: https://www.rtve.es/filmoteca/no-do/not-927/1469759/ (accessed on 8 August 2020).
19.   Díaz, P. Transformación y urbanización del frente costero español. In *Turismo Líquido*; Pié, R., Rosa, C., Eds.; IHTT (UPC): Barcelona, Spain, 2013; pp. 190–203.
20.   Caparrós, J.A. *El ABC del Camping*; Ediciones Gaisa: Valencia, Spain, 1960.
21.   Feo, F. Los campings en España. *Cuad. Tur.* **2003**, *11*, 83–96.
22.   Miranda, M.J. El camping, la forma más reciente de turismo. *Cuad. Geogr.* **1985**, *37*, 157–174.
23.   IDESCAT Catalan Institute of Statistics. Available online: https://www.idescat.cat/ (accessed on 15 December 2019).
24.   INE Spanish National Institute of Statistics. Available online: https://www.ine.es/ (accessed on 5 September 2017).
25.   Calabuig, J. *Llibre Blanc del Sector del Càmping a Catalunya. Anàlisi de L'activitat, Tendències i Reptes de Futur*; EURECAT: Reus, Tarragona, Spain, 2019; Available online: https://www.campingsdecatalunya.org/wp-content/uploads/llibre-blanc-campings-catalunya-doc-sintesi.pdf (accessed on 2 March 2020).
26.   The Catalan Campsite Associations Are: Girona, Costa Daurada and Terres de l'Ebre, Lleida, Muntanya and Pirineu Català, and Barcelona. Available online: http://www.catalunya.com/federacio-catalana-de-campings-i-ciutats-de-vacances-20-16-103 (accessed on 29 March 2020).
27.   Catalan Campsite Federation. Available online: www.campingsdecatalunya.org (accessed on 12 January 2020).
28.   Mediterranean Campsite Confederation. Available online: https://campingsdelmediterraneo.org/ (accessed on 10 October 2019).
29.   Urban Master Plan for Camping Activities (PDUAC), Catalan Government. Available online: http://territori.gencat.cat/ca/06_territori_i_urbanisme/planejament_urbanistic/pla_director_urbanistic_pdu/en_curs/Catalunya/pdu_campings/index.html (accessed on 27 January 2020).
30.   The Catalan Government; Pla d'Espais d'Interès Natural (PEIN), regulatory Decret 328/1992. Available online: http://sig.gencat.cat/visors/enaturals.html (accessed on 29 July 2020).
31.   The Catalan Government; Pla Director Urbanístic del Sistema Costaner (PDUSC), regulatory DOGC 6722. Available online: https://portaldogc.gencat.cat/utilsEADOP/PDF/6722/1374519.pdf (accessed on 29 July 2020).
32.   Tarroja, A. *Anuari Territorial de Catalunya 2004*; Institut d'Estudis Catalans: Barcelona, Spain, 2004.
33.   VVAA. *Catalunya, Destrucción a Toda Costa. Informe Sobre La Situación Económica Y Ambiental Del Litoral*; Greenpeace: Barcelona, Spain, 2012.

34. Martín, X. Arquitectura del turismo informal. El camping Como Modelo de Ocupación Temporal En El Litoral Mediterráneo de Catalunya. Ph.D. Thesis, La Salle—Ramon Llull University, Barcelona, Spain, 17 September 2018.

35. Rubert de Ventós, X. El Mediterráneo como mito local. *DC Pap.* **2003**, *9*, 17–27.

36. Ollé, M.; Mataix, S. Interview to architect J. Prada Poole. In *Camping, Caravaning, Arquitecturing*; Documentary; July 2011; Available online: http://www.miquelolle.com/camping.html (accessed on 5 June 2020).

37. Hunter, C.; Shaw, J. The ecological footprint as a key indicator of sustainable tourism. *Tour. Manag.* **2007**, *28*, 46–57. [CrossRef]

38. Font, X.; Garay, L.; Jones, S. Sustainability motivations and practices in small tourism enterprises in European protected areas. *J. Clean. Prod.* **2016**, *137*, 1439–1448. [CrossRef]

39. Responsible Tourism Institute (RTI). Available online: https://www.responsibletourisminstitute.com/en (accessed on 25 February 2020).

40. Biosphere Camping Certification. Available online: https://www.biospheretourism.com/en/biosphere-camping-certification/98 (accessed on 17 July 2020).

41. European Year of Cultural Heritage. Available online: https://europa.eu/cultural-heritage/european-year-cultural-heritage_en.html (accessed on 14 March 2020).

42. Saarinen, J. Critical Sustainability: Setting the limits to Growth and Responsibility in Tourism. *Sustainability* **2014**, *6*, 1–17. [CrossRef]

43. Hailey, C. *Camps: A guide to 21st-Century Space*; The MIT Press: Cambridge, MA, USA, 2009.

44. Careri, F. *Walkscapes. El Andar Como Práctica Estética*; Gustavo Gili: Barcelona, Spain, 2014.

45. Monteys, X.; Fuertes, P. *Casa Collage. Un Ensayo Sobre la Arquitectura de la Casa*; Gustavo Gili: Barcelona, Spain, 2001.

46. Gustafson-Melka, K. *Paysage de Camping*; IAURIF: Paris, France, 1981.

47. Careri, F. *Pasear, Detenerse*; Gustavo Gili: Barcelona, Spain, 2016.

48. Mehrotra, R.; Vera, F. *Ephemeral Urbanism: Cities in Constant Flux*; Ediciones ARQ Escuela de Arquitectura Pontificia Universidad Católica de Chile: Santiago, Chile, 2016.

49. Martínez, A. Restablecer el carácter del lugar. In *Estrategias Mediterráneas, IAM Group*; Claret: Barcelona, Spain, 2015; pp. 92–95.

50. Saarinen, J. Traditions of Sustainability in Tourism Studies. *Ann. Tour. Res.* **2006**, *33*, 1121–1140. [CrossRef]

51. VVAA. *Catàlegs de Paisatge de Catalunya*; Observatori del Paisatge: Barcelona, Spain, 2019.

52. The Generalitat de Catalunya. *Catalan Urban Regulation 2002*. Article 3—Concept of Sustainable Urban Development. Chapter II—General Principles of Urban Action. Available online: https://www.parlament.cat/document/cataleg/47926.pdf (accessed on 27 July 2020).

53. Moretto, D.; Branca, T.; Colla, V. Energy efficiency and reduction of CO_2 emissions from campsites management in a protected area. *J. Environ. Manag.* **2018**, *222*, 368–377. [CrossRef] [PubMed]

54. Pogodaeva, S.; Yu, E. Ecotourism as a Factor of Ecological Oriented Civilization (A Case Study Based on French Tourist Discourse). *Adv. Eng. Res.* **2020**, *191*, 224–229. [CrossRef]

55. Breiby, M.A.; Duedahl, E.; Oian, H.; Ericson, B. Exploring Sustainable Experiences in Tourism. *Scand. J. Hosp. Tour.* **2020**. [CrossRef]

56. Weaver, D. Comprehensive and Minimalist Dimensions of Ecotourism. *Ann. Tour. Res.* **2005**, *32*, 439–455. [CrossRef]

57. Hogue, M. A Short History of the Campsite. *Places J.* **2011**. [CrossRef]

58. Bellmunt, J. Litoral. In *Landscape*; Colafranceschi, D., Ed.; Gustavo Gili: Barcelona, Spain, 2007; p. 120.

59. VVAA. El turismo en la costa I. *Cuad. Arquit.* **1966**, *64*, 3.

60. IAM Group. *Estrategias Mediterráneas*; Claret: Barcelona, Spain, 2015.

61. Martínez, A.; Vives, L.; IAM. Stratégies Méditerranéenes pour la régéneration architectonique du bord de la mer. *Classeur. Mare Nostrum* **2017**, *2*.

62. Rudofsky, B. *Architecture without Architects. A Short Introduction to Non-Pedigreed Architecture*, 3rd ed.; University of New Mexico Press: Albuquerque, Mexico, 1987.

63. For instance, research projects "Strategies for sustainable regeneration of touristic settlements in the Mediterranean coast" (BIA2011-28297-C02-02), "Research studies for the urban and architectonic development of Palamós harbour" (2015-01-524-2270013), or "Research study for the landscape and urban integration of the mobility infrastructures in the coastal territory of Maresme" (PTOP-2015-562)

64. For instance, Masters theses "La Ciutat residencial de Tarragona" (A. Peguero, 2016), "Un vuelo por la Costa Brava" (J. Gordon, 2013) or "Camping La Ballena Alegre" (A. Luna, 2012)

65. For instance, PhD theses "Arquitectura del turismo informal" (X. Martín, 2018), "Arquitectura, vehicle de modernitat" (M. Bosch, 2017), "L'art de viure. Cases per a artistes a Cadaqués" (M. Arnal, 2016), "El Pati a la ciutat mediterrània" (A. Hosta, 2016), "Detalles en la arquitectura de J.A. Coderch" (I. de Rentería, 2013), "Depth Configurations: Proximity, permeability and territorial boundaries in urban projects" (K. Scheerlinck, 2013) or "La construcción del territorio de Ibiza" (S. Cortellaro, 2010)

66. Delaney, D. *Territory: A Short Introduction*; Blackwell Publishing: Oxford, UK, 2005.

67. Barrios, R. El Concurso Internacional de Elviria. In *Turismo Líquido*; Pié, R., Rosa, C., Eds.; IHTT (UPC): Barcelona, Spain, 2013; pp. 286–297.

68. Gordon, J.; Martin, X.; Peguero, A. El trabajo de campo. Instrumento esencial en el proyecto de investigación. In *Estrategias Mediterráneas*; IAM Group; Claret: Barcelona, Spain, 2015; pp. 100–101.

69. Ferrer, A. El dibujo del territorio como forma de conocimiento. In *Estrategias Mediterráneas*; IAM Group; Claret: Barcelona, Spain, 2015; pp. 80–83.

70. Martín, X.; Martínez, A. Las Postales de Foto Raymond. Testigos viajeros del camping en los años 60. In Proceedings of the INTER—Photography and Architecture, Pamplona, Spain, 2–4 November 2016.

71. AMCAM. *Un Segle de Transformacions del Paisatge de Costa de Cambrils*; Cambrils Municipality Archive: Cambrils, Spain, 2013.

72. Levinthal, D.; Warglien, M. Landscape Design: Designing for Local Action in Complex Worlds. *Organ. Sci.* **1999**, *10*, 3. [CrossRef]

73. Baglivo, C.; Galofaro, L. New ecology. In *Landscape*; Colafranceschi, D., Ed.; Gustavo Gili: Barcelona, Spain, 2007; p. 140.

74. Español, J. Lugar. In *Landscape*; Colafranceschi, D., Ed.; Gustavo Gili: Barcelona, Spain, 2007; p. 122.

75. Barba, R. El projecte del lloc: Entre l'anàlisi de l'entorn i el disseny de l'espai exterior. *Quad. D Arquitectura I Urban.* **1982**, *153*, 60–67.

76. Pallasmaa, J. *Habitar*; Gustavo Gili: Barcelona, Spain, 2016.

77. Betsky, A. Inestabilidad. In *Landscape*; Colafranceschi, D., Ed.; Gustavo Gili: Barcelona, Spain, 2007; p. 100.

78. Sennett, R. *L'espai Públic. Un Sistema Obert, un Procés Inacabat*; Arcàdia: Barcelona, Spain, 2014.

79. Dequaire, X. Passivhaus as a low-energy building standard: Contribution to a typology. *Energy Effic.* **2012**, *5*, 377–391. [CrossRef]

80. Anton, S. Identitat i turisme. Entre imatge i percepció. *Paradigmes* **2010**, *5*, 156–165.

81. Ruisánchez, M. Acción. In *Landscape*; Colafranceschi, D., Ed.; Gustavo Gili: Barcelona, Spain, 2007; p. 22.

82. Moneo, R. The Murmur of the Site. *CIRCo* **1995**, 24. Available online: http://www.arranz.net/web.arch-mag.com/7/circo/24.html (accessed on 22 April 2019).

83. E.g. El Toro Bravo campsite (converted into a natural protected area), Salou campsite (transformed into a Municipal Park in Salou), or Laguna campsite (integrated into the Natural Park Aiguamolls de l'Empordà)

84. Dern, J. Cultural tourism? A few reflections on the Leeuwarden, Holland, symposium. *Quad. D Arquit. I Urban.* **1992**, *194*, 20–27.

85. Gausa, M. Maresmes. Quaderns d'Arquitectura i Urbanisme. *Infiltracions* **1992**, *195*, 76–87.

86. VVAA. Colonitzacions. In *Quaderns D'Arquitectura I Urbanisme*; COAC: Barcelona, Spain, 1992.

87. Pla, A. *Paesaggio e Architettura*; Public Debate in ETSA La Salle–URL on 17 April 2018; Available online: https://www.salleurl.edu/es/conferencia-y-mesa-redonda-de-arquitectura (accessed on 8 August 2020).

88. Zanini, P. Confín. In *Landscape*; Colafranceschi, D., Ed.; Gustavo Gili: Barcelona, Spain, 2007; p. 40.

89. Girot, C. Identidad. In *Landscape*; Colafranceschi, D., Ed.; Gustavo Gili: Barcelona, Spain, 2007; p. 95.

90. Sansot, P. Autour de la frénêsie paysagère. *L Architecture D Ajourd Hui* **1992**, *279*, 38.

91. Codina, P. *Camping. Tècnica I Orientacions Generals I Coneixements Necessaris*; Rafael Dalmau: Barcelona, Spain, 1964.

Article

The Sustainable Development of Urban Cultural Heritage Gardens Based on Tourists' Perception: A Case Study of Tokyo's Cultural Heritage Gardens

Ge Chen [1], Jiaying Shi [2], Yiping Xia [1] and Katsunori Furuya [2,*]

[1] Institute of Landscape Architecture, College of Agriculture and Biotechnology, Zhejiang University, Hangzhou 310058, China; dovexxc@163.com (G.C.); ypxia@zju.edu.cn (Y.X.)

[2] Department of Environmental Science and Landscape Architecture, Graduate School of Horticulture, Chiba University, Chiba 271-8510, Japan; jsntsjy@gmail.com

*   Correspondence: k.furuya@faculty.chiba-u.jp

Received: 30 June 2020; Accepted: 2 August 2020; Published: 5 August 2020

Abstract: For the cultural heritage gardens in the urban environment, modern high-rise buildings inevitably change their original landscape and form a new landscape experience with visual impact. Whether cultural heritage gardens and modern cities can coexist harmoniously is one of the critical issues to achieve their sustainable development. This research aimed to find an indicator of landscape morphology, which can predict the visitor's cognition for such cultural landscape forms. This study surveyed tourists' preferences in six selected cultural heritage gardens in Tokyo. We used hemispheric panoramas to calculate the view factors of certain elements of the landscape at the observation points. The results showed that Sky View Factor was a positive predictor of tourists' preference, and this predictability did not change significantly with the attributes of tourists. We also found that tourists' attitudes towards the high-rise buildings outside the gardens have become more tolerant and diverse. These findings could be applied to predict visitors' perception preference of cultural heritage landscape in the context of urban renewal, contributing to the sustainable development of cultural heritage landscape and urbanization.

Keywords: cultural heritage garden; tourists' preference; sustainable development; Tokyo; view factor

1. Introduction

Cultural heritage landscape that remains in the urban environment represents the preservation and embodiment of the historic culture and people's wisdom of the city and the country [1]. It is also an attractive point for urban tourism development and an essential resource for urban economic growth [2]. However, urbanization after the Industrial Revolution has caused a rapid increase in the urban population. At the same time, a large number of modern high-rise buildings have emerged, and the social and cultural atmosphere has changed. As a result, the city's structure and landscape have undergone tremendous changes. Worldwide, the cultural heritage landscape in many cities is facing a significant threat of destruction and even gradually disappearing due to urbanization [3–6]. Therefore, nowadays, how to achieve the sustainable development of the urban heritage landscape has become a worldwide focus of attention. The Sustainable Development Goals (SDGs) adopted by the member states of the United Nations (UN) in 2015 clearly state that it is necessary to strengthen the protection and maintenance of the sustainable development of world cultural heritage and natural heritage [7].

Despite the sharp image of asphalt and skyscrapers, Tokyo still preserves nine metropolitan cultural heritage gardens today. They are not only green public open spaces but also the historical mark of the entire city [8]. These gardens were built from the Edo period to the Taisho era (from the 17th century to the beginning of 20th century). As precious and fragile resources, they are registered as cultural assets of Tokyo and Japan by the Japanese government [9]. Under the provisions of Japan's National Cultural Property Protection Law [10], the internal landscapes of these heritage gardens are well preserved. Still, the surrounding areas of the gardens are not included in the areas covered by the protection law. Thus, some of these gardens are surrounded by modern high-rise buildings now. The original designers of the garden would never be expected that tall modern buildings outside the gardens have now replaced their bright design idea as "shakkei" (borrowed scenery). The landscape space seen by tourists visiting the gardens now is very different from the original appearance of the gardens.

While urban renewal changes the spatial form of cultural heritage landscapes, it also continually alters the people's perception of it [11,12], and affects the research on the protection and management of heritage landscapes [13]. For visitors to the Tokyo metropolitan cultural heritage garden, modern high-rise structures that break into their horizons may bring about visual conflict, but at the same time, may also bring about a new viewing experience. The cultural landscape is the result of the constant interaction between human civilization and the natural environment, which is a dynamically changing process [14]. Previous research has tended to criticize the negative impact of urbanization on heritage landscapes. However, with the development of the times, people's aesthetic preferences for cultural heritage have become more and more diverse. Therefore, studies on the perception of tourists should pay more attention to the sustainable development of cultural heritage landscapes [15,16].

This current study aims to find a visibility indicator of landscape morphology which can predict the perception of tourists in the cultural heritage gardens in the urban context. We took the Tokyo Metropolitan Cultural Heritage Gardens as the research objects. On-site questionnaires were carried out to investigate the tourists' perception preferences, and the view factors of different landscape elements at the observation points were calculated. Spearman correlation coefficients and stepwise linear regression analysis were used to examine the relationship between tourists' preference and view factors. Specifically, this study focuses on the following questions:

1. What are the attitudes and preferences of tourists nowadays towards the influence of high-rise buildings outside traditional Japanese gardens in the urban context?
2. Can the view factors (including sky, garden, and building view factors) be considered as a reliable predictor of tourists' preferences?
3. Do attitudes and preferences of this phenomenon change due to demographic attributes of visitors?

To our best knowledge, the study is one of the first investigations to focus on the predictability of landscape morphological indicators for the cultural heritage landscape. Additionally, based on the results of the investigation, we attempt to discuss whether the urbanization and the protection of heritage in the city must conflict and whether they can achieve positive common development [2,17]. It will also provide some reference for the study of the sustainable development of historical heritage landscape in the city from the perspective of historical landscape perception and evaluation.

2. Literature Review

2.1. Heritage Landscape in the Context of Modern Cities

The visual relationship between cultural heritage and its surroundings has been an active topic of sustainable research for many years. In the context of urbanization, the most visible influence of modern high-rise buildings on the traditional landscape is that it has broken into the background of the heritage garden and become a new form of view [18]. The visual impact of the outside high-rise modern buildings on the cultural heritage garden has already attracted the attention of many scholars. Most of the previous studies held a negative attitude. Antrop pointed out in 1997 that the destructive modern impact after World War II changed the structure and function of the traditional landscape [19]. Some original scenes have even been completely erased, while the newly formed modern landscapes lack individuality [20]. Oh proposed in 1998 that urban expansion brings the existing landscape in the city with the accumulated visual pressure [21]. Swensen and Jerpåsen suggested in 2008 that the heritage landscape and its surrounding environment should be taken as a whole and protected by better legal planning [22]. UNESCO also proposed in 2017 that urban development threatened the visual integrity of the historical heritage landscape [23].

The cultural heritage gardens in Tokyo is full of visual and cultural conflicts influenced by the modern urban environment and are typical cultural heritage landscape remain in the city. Japanese scholars Shinji et al. proposed in 1989 the negative impact of modern buildings on the landscape of Tokyo Metropolitan Cultural Heritage Gardens [24]. They claimed that modern high-rise buildings destroyed cultural landscape and caused visual invasion. Many subsequent studies have also argued that those modern high-rise buildings have harmed the gardens [25–28]. More recently, other opinions have appeared which believe that although modern buildings affect the landscape of traditional gardens, the contrast between modern and classic may not necessarily be negative, and may even stimulate people's imagination [29]. Such opinions, however, have not been confirmed through any research.

To date, however, there has been little empirical evidence of whether the attitude preference of tourists is related to any landscape morphological indicators of such conflict landscape. The study by Shinji et al. only measured the heights of the visible buildings outside the gardens, and the distances between them and the observation points in five Tokyo Metropolitan Cultural Heritage Gardens [24]. They then determined the elevation angle of the line of sight to the building, judged its impact on tourists' perceptions. The judgments were based on theories without conducting visitors' surveys, and the correlation between spatial features and visitor preferences were not analyzed. Senoglu et al. studied the impact of the high-rise buildings outside the Hama-rikyu Gardens in 2018 and collected visitors' attitudes and preferences through questionnaires [29]. They verified the prospect–refuge theory in Hama-rikyu Gardens but failed to find a correlation between geometric proportions of the buildings and tourists' preferences. Therefore, inspired by the previous research, this study surveyed the various choices of tourists and tried to find out the landscape morphological indicators that are related and predictive to the preferences of tourists in the cultural heritage gardens.

2.2. Landscape Morphological Indicators

View factor is also called form factor or shape factor. The most commonly used one in urban morphology is Sky View Factor (SVF), which is defined as the percentage of the visible area of sky at a particular location, with a value from 0 to 1, indicating a completely closed space to a fully open space [30,31]. SVF is often used to study the geometric characteristics of urban canyons, urban temperature distribution, and urban thermal comfort [32]. The methods for calculating SVF include the photographic process, GPS signal-based method, simulation method, among others [33]. Among them, the photographic process is the classic, where a hemisphere diagram azimuthal projected from panoramas taken by spherical cameras are used to calculate the SVF [34,35]. This method is more suitable to measure the view factor of various landscape elements at discrete observation points for the garden landscapes. As a visual physical indicator, the concept of SVF is similar to some landscape visibility indicators such as the "Visible Green Index," which was proposed by Japanese scholar Aoki in 1987, based on visual psychology [36]. It has become one of the regular greening evaluation indexes recognized by the Japanese government since 2004 [37].

In 1981, Takei and Oohara suggested that the proportion of sky occupied by buildings had a clear correlation with the visual pressure of high-rise buildings on people [38]. As an index of landscape space, sky proportion is often used as an index of landscape visual quality evaluation [39]. Jiang et al. proposed a relationship between tree cover density and landscape preferences in 2015 [40]. Based on the previous research, Gong et al. further estimated the SVF, tree view factor (TVF), and building view factor (BVF) in high-density urban environments to extract street features in 2018 [34]. Therefore, inspired by the previous research, this study attempted to introduce view factor into the spatial morphology description of cultural heritage garden landscapes in cities, and at the same time, quantify the SVF, BVF, and garden view factor (GVF), defined as the percentage of the visible area of garden landscape at the observation points, of the observation points.

3. Materials and Methods

3.1. Selection of Observation Points

The Tokyo Metropolitan Cultural Heritage Gardens are an important cultural heritage from the Meiji and Edo periods of Japan (17th to 20th century). There are nine gardens in Tokyo (Figure 1, Table 1), where the modern urban environment conflicts with classical natural landscapes. Among them, Kyu-Iwasaki-tei Garden, Kyu-Furukawa Garden, and Tonogayato Garden are private gardens with western-style historical buildings as the main view. They are not the type of traditional Japanese-style gardens that we focused on in this study and thus were excluded. Observation points were selected in Hama-rikyu Garden, Kyu-Shiba-rikyu Garden, Koishikawa Korakuen Garden, Rikugien Garden, Kiyosumi Garden, and Mukojima-Hyakkaen Garden.

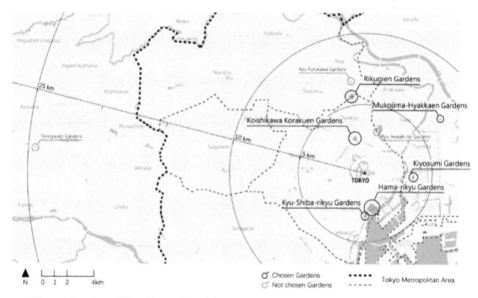

Figure 1. Location of Tokyo Metropolitan Cultural Heritage Garden (drawn based on Google map and mapbox open source map).

Table 1. Basic information of Tokyo Metropolitan Cultural Heritage Garden.

Name of Gardens	Area (m²)	Construction Century	Property
Hama-rikyu Garden	250,215.72	17th century	Stroll type Daimyo garden of Edo period
Kyu-Shiba-rikyu Garden	43,175.36	17th century	Stroll type Daimyo garden of Meiji period
Koishikawa Korakuen Garden	70,847.17	17th century	Stroll type Daimyo garden of Edo period
Rikugien Garden	87,809.41	17–18th century	Stroll type Daimyo garden of Edo period
Kiyosumi Garden	43,656.95	19th century	Stroll type Daimyo garden of Meiji period
Mukojima-Hyakkaen Garden	10,885.88	19th century	Private garden of Edo period
Kyu-Iwasaki-tei Garden	18,235.47	19th century	western-style historical buildings with Private garden
Kyu-Furukawa Garden	30,780.86	20th century	western-style historical buildings with Private garden
Tonogayato Garden	21,123.59	20th century	western-style historical buildings with Private garden

The "heat maps" based on big data were selected to decide the research points to ensure that the selected points were in line with the subjective preferences of most garden visitors. The points with "visible outside high-rise buildings" were selected from the hotspots as the research points, which also ensured that the influence of outside buildings here on the viewing of garden visitors was widespread, rather than random or rare.

The specific observation points were selected from the heat maps generated from the photos with geographic coordinate information [41]. After writing the relevant code through IDEA, we used the Flickr API for keyword tag search (Japanese and English names of the six target gardens), sourced 6513 photos (2004–2018), and set the geographic coordinate data in EXCEL. Then, the data were imported into QGIS to generate the heat maps. The selection of observation points was based on the following principles: (1) The skyline seen in the observation points should be coherent; (2) Landscape elements in the observation points should be varied; (3) If there are several observation points in one garden, the perspective of the buildings outside the gardens seen in the observation points should be different; and (4) The site of the observation point should be safe and undisturbed. Finally, nine observation points were selected (Figure 2).

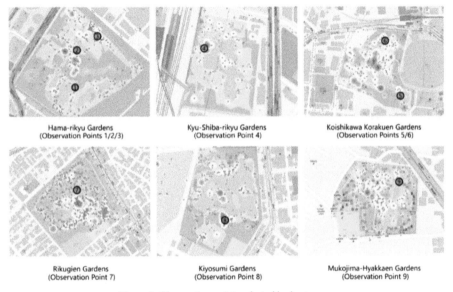

Hama-rikyu Gardens
(Observation Points 1/2/3)

Kyu-Shiba-rikyu Gardens
(Observation Point 4)

Koishikawa Korakuen Gardens
(Observation Points 5/6)

Rikugien Gardens
(Observation Point 7)

Kiyosumi Gardens
(Observation Point 8)

Mukojima-Hyakkaen Gardens
(Observation Point 9)

Figure 2. Observation points selected by heat maps.

3.2. The Questionnaire Survey

Trilingual questionnaires in Chinese, Japanese, and English were printed for tourists to judge the proposition of "whether modern buildings outside gardens have a positive impact on the overall landscape of the garden." The attitude of the respondents was measured on a 6-point Likert scale, with 1–6 points representing: totally disagree, disagree, somewhat disagree, somewhat agree, agree, fully agree.

The questionnaire-based survey was conducted from 25–28 March 2019, with similar weather conditions; that is, clear sky with little clouds. In the target gardens, visitors were randomly invited to have a full-angle observation of the landscape and fill out questionnaires at the selected observation points. Respondents were required to fill out questionnaires separately to ensure independent opinions. A total of 388 tourists (175 females; 231 East Asian (including Japanese) and 157 non-East Asian) participated in the survey, with an average of 43.1 respondents for each observation site (minimum 26 and maximum 47). Respondents ranged in age from 16 to 82, with an average age of 39.8.

3.3. View Factor

While conducting the questionnaire survey, panoramic images were taken at the observation point by a GoPro Fusion spherical camera (Figure 3a). The camera was fixed at the height of 170 cm to ensure that the shooting height was consistent with the viewing angle of the human eye. The image obtained was fed into RayMan software [42,43]. An R package was used to project the panoramic image from a cylindrical projection to an azimuth projection (Figure 4), and then used to generate a hemisphere diagram (Figure 3b) [44]. Figure 3c shows the fisheye diagram segmented according to landscape elements, with extraction of sky (in blue), garden landscape (in green), and buildings (in yellow), based on which the view factor of each landscape elements were calculated.

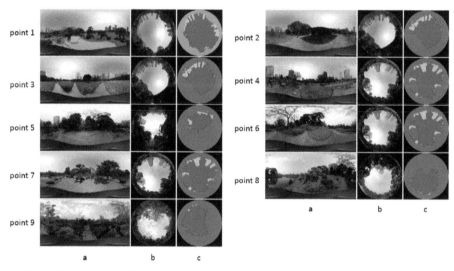

Figure 3. Panoramas (**a**) cylindrical projected panoramas, (**b**) fish-eye hemisphere, and (**c**) segmented fish-eye hemisphere.

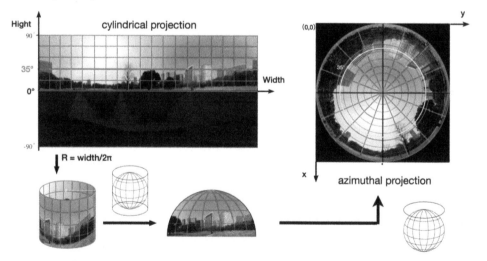

Figure 4. Geometric model from fish-eye hemisphere with panoramic view from cylindrical projection to azimuth projection (redrawn from study of [33]).

3.4. Analysis Method

The dependent variables of this study were visitors' preference score of the influence of modern high-rise buildings that are outside, on the overall landscape of the historic garden. The independent variables of SVF, GVF, and BVF were computed from the hemisphere diagram. The statistical analysis software JASP (0.9.2.0 release) was used for data analysis.

4. Results

4.1. Reliability Analysis and Descriptive Statistics

Unstandardized Cronbach's α was calculated for the collected preference scores to ensure that the data can be used for further analysis (Cronbach's α of all data higher than 0.788) (Table 2). The descriptive statistical results of the respondents' preference score for each observation point and the results from the calculation of the view factors are also shown in Table 2.

Table 2. Reliability coefficients, number of questionnaires, mean scores, and view factors of each observation point preference scores.

	1	2	3	4	5	6	7	8	9	Mean
Cronbach's α	0.894	0.851	0.871	0.888	0.896	0.918	0.850	0.788	0.829	/
Valid	47	46	45	46	45	45	46	41	26	/
Mean	3.894	3.783	3.600	3.435	3.178	3.356	4.130	3.122	3.808	/
Std. Deviation	1.088	1.052	1.304	1.500	1.628	1.334	1.392	1.166	1.357	/
SVF	0.562	0.403	0.494	0.443	0.197	0.160	0.375	0.344	0.362	0.371
GVF	0.357	0.531	0.408	0.455	0.771	0.823	0.620	0.650	0.633	0.583
BVF	0.081	0.066	0.098	0.102	0.032	0.017	0.005	0.006	0.005	0.046

The results show that the average of respondents' preference scores of all observation points of the impact of outside high-rise buildings on the overall landscape is 3.59. The most preferred site was spot-7, while the lowest preference was for spot-8, with preference scores of 4.13 and 3.12, respectively. Of the 388 respondents, 210 (54.12%) agreed with the proposition that the high-rise buildings outside the garden had a positive impact on the overall landscape. Among them, 108 respondents (27.84%) had a strongly positive attitude (5 Points and above). On the other hand, 178 respondents (45.88%) believed that the buildings outside the garden had a negative impact on the overall landscape, of which 89 respondents (22.94%) had a strongly negative attitude (2 points and below) (Figure 5).

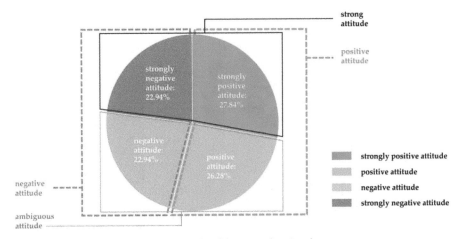

Figure 5. The results of the respondents' preference.

4.2. Correlation between View Factor and Tourist Preference

The Pearson correlation coefficient was employed to test the correlation between respondents' preferences and view factors. Table 3 indicates that the SVF and GVF are significantly correlated with respondents' preferences. SVF is positively correlated with respondents' preferences, while GVF is negatively correlated.

Table 3. Pearson correlation between view factors and respondents' preferences.

		SVF	GVF	BVF
Respondents'	Pearson's r	0.131 **	−0.112 *	0.019
preference	*p*-value	0.010	0.028	0.707

$* p < 0.05, ** p < 0.01.$

Then, stepwise multiple linear regression was employed to analyze the predictability of highly correlated view factors for tourist preferences. The results show that SVF has strongly positive predictability on the respondents' preferences (Table 4). The adjusted R^2 of the regression model is 0.015, and ANOVA analysis of stepwise multiple regression is significant ($p = 0.010$).

Table 4. Predictors of view factors in stepwise multiple linear regression.

	Regression Model						
Model	Unstandardized Coefficients		Standardized	t	Sig.	Correlations	
	B	Std. Error	B			Partial	Part
SVF	1.411	0.544	0.131	2.594	0.010 **	0.131	0.131
	$R = 0.131; R^2 = 0.017;$ Adjusted $R^2 = 0.015; F(385) = 6.731; p = 0.010$						

$** p < 0.01.$

4.3. The Influence of Tourist Attributes on Preferences

Analysis of variance was used to test whether the attributes of respondents affected their preferences. The respondents' preferences were the dependent variables, while their nationality (East Asian countries/other foreign countries), gender (male/female), and age (10 years as a group) were the independent variables. The results show that generation affected tourist preferences ($p < 0.05$, $\eta^2 = 0.055$) (Table 5).

Table 5. Analysis of variance of respondents' nationality and gender on cognitive attributes.

Dependent Variable	Fixed Factor	Sum of Squares	df	Mean Square	F	p	η^2
	Nationality	279.30	287	279.27	3.794	0.052	0.013
Respondents' Preferences	Gender	202.00	287	202.04	2.735	0.099	0.009
	Age	29.53	281	4.219	2.349	0.024 *	0.055
Respondents' Preferences	Nationality	289.75	285	289.75	3.962	0.047 *	0.013
(age as a potential	Gender	255.48	285	255.48	3.493	0.063	0.012
influencing factor)	Nationality * Gender	76.83	285	76.83	1.051	0.306	0.004

$* p < 0.05.$

A multivariate analysis of variance was then employed with the nationality (East Asian countries/other foreign countries) and gender (male/female) of the respondents as independent variables, and age as a potential influencing factor. It was found that under this condition, nationality had an effect on the dependent variable ($p < 0.05$, $\eta^2 = 0.013$). However, the effect size of both age and nationality is less than 0.06, judged to be a small effect according to Cohen's d effect size [45], indicating that tourists' preference for modern buildings outside the gardens will not be significantly affected by their attributes.

5. Discussion

5.1. Tourists' Attitude Towards Architecture Outside the Garden

Among the 388 questionnaires, 210 (54.12%) held a positive attitude; that is, the buildings outside the garden had a positive impact on the landscape of the garden. A total of 178 (45.88%) of the questionnaires held a negative attitude. This shows that the respondents' attitude towards the influence of modern high-rise buildings outside the cultural heritage garden is not necessarily negative. Even respondents who think that the high-rise buildings outside the garden have a positive impact on the garden landscape are slightly more than those who disagree. The attitude of nearly half of the respondents is not strong (three to four points account for 49.22% of the total questionnaire), which is a more neutral attitude.

In terms of each garden, the gardens with higher scores of respondents' preference are: Rikugien Garden, Hama-rikyu Garden, Mukojima-Hyakkaen Garden and Kyu-Shiba-rikyu Garden (mean of preference scores >3.4). It can be found that the SVFs of these gardens are all higher (SVF > 0.36). Among them, Rikugien Garden and Hama-rikyu Garden have the largest garden area and wider garden landscape, so the SVF is also higher. Hama-rikyu Garden and Kyu-Shiba-Rikyu Garden are located in the same coastal area with a broad view of the city, so the SVF of Kyu-Shiba-Rikyu Garden is also high.

In the process of conducting the questionnaire, through discussions with the tourists, we obtained many different opinions from various traditional concepts. For example, two tourists from Spain believed that both the urban background and the traditional landscape of observation point 4-Kyu-Shiba-rikyu Garden were rare for them, and the traditional Japanese garden and modern architecture together were new landscape forms, rather than destroyers of the traditional landscape. Additionally, the building complex outside Hama-rikyu Garden and Kyu-Shiba-rikyu Garden is a very famous skyline view of Tokyo [46]. Many respondents said that under such a modern visual background, the uniqueness of traditional landscapes could be more prominent and form an impressive visual impact. Another typical example is observation point 9 at Mukojima-Hyakkaen Garden. The architecture outside the garden includes the Tokyo Sky Tree, which is a famous landmark of Tokyo [47]. According to statistics from a Japan Tourism Website, after the Tokyo Sky Tree was built in 2012, the shooting spot in Mukojima-Hyakkaen Garden, where the Tokyo Sky Tree can be seen, has become the most popular spot among tourists. Therefore, we believe that the impact of outside high-rise buildings on the overall landscape of the traditional garden can be discussed with a more diverse and tolerant perspective. In other words, urbanization does not necessarily have a destructive negative impact on the heritage landscape in the city. The modern cultural atmosphere created by urbanization and the historical features embodied in the heritage landscape may inspire each other under specific circumstances, creating compelling visual and cultural conflicts.

In addition, in Hama-rikyu Garden and Koishikawa Korakuen Garden, we chose two to three research points in the same garden, and different respondents' preference scores were obtained. For Hama-rikyu Garden, the buildings outside the garden of the three research points were the same, and the difference was the garden landscape. Research points 1–3 represented: plant landscape of cherry blossoms and pine of Japanese garden, Japanese traditional garden pavilion landscape, and rape flower field landscape, with respondents' preference scores decreasing in order. This illustrates the influence of the garden landscape itself on tourists' preference for buildings outside the gardens.

For Koishikawa Korakuen Garden, the biggest difference between the two observation points was that the Tokyo Dome, one of Tokyo's landmarks, can be seen at point 5 but not at point 6. However, the results showed that the respondents do not have high preferences for point 5, and the difference between the two points was not obvious. During the questionnaire survey, many respondents said that although it is one of Tokyo's landmarks, the Tokyo Dome is too close to the garden. The huge volume and round shape make people feel visually depressed. Additionally, the landscape of Koishikawa Korakuen Garden was relatively closed compared to other gardens with a low SVF.

Therefore, the buildings outside the gardens cannot be judged solely by building attributes, but also needs to be considered comprehensively in combination with building distance, building height and other factors.

The Sustainable Development Goals (SDGs) also mention the promotion of knowledge and skills development and appreciation of cultural diversity. In the urban context, the existence of heritage landscapes is more precious and meaningful, while its meaning and culture also develop with the changes of the times. The new culture and connotation formed by the heritage landscape in the urban environment have been more widely accepted, which is also supposed to be the goal that sustainable development expects to achieve. There is no doubt that excessive urbanization would cause irreversible damage to the heritage landscape. However, we hope to make the results of inevitable urbanization more positively related to the heritage landscape and to achieve the sustainable development of the heritage landscape and city.

5.2. The Predictability of View Factor on Preferences

The results of the correlation analysis showed that SVF ($r = 0.131$, $p = 0.01$) and GVF ($r = -0.112$, $p < 0.05$) were independent variables that had a strong correlation with the preferences of the respondents. Meanwhile, SVF and tourist preferences were positively related; and GVF was negatively correlated. However, BVF did not show any strong correlation, which also validated the research of Senoglu et al., that the visible building index does not explain tourist preferences [29].

The results of the stepwise multiple linear regression of two substantial correlated factors showed that SVF was a strong positive predictor of respondents' preferences ($F = 6.731$; $p = 0.010$). The ANOVA analysis of stepwise multiple regression was significant ($p = 0.010$), indicating that the regression model had remarkable statistical significance. However, the adjusted R2 interpretation of the regression model was 0.015, which stated that the model did not fit the dependent variable well. We believe that this was partly due to too few observation points and the small sample size of the independent variables, which also made it impossible to determine the optimal value interval of SVF. It also reflected the deficiencies in the design of this study to some degree. In future research, more observation points should be set up to expand the sample size of each view factor. Although the value of R^2 is not an absolute measure of the goodness of fit, the purpose of this study is not to determine the threshold of the linear relationship. Therefore, our findings indicated that the SVF of the cultural heritage garden is a valid positive predictor of the tourists' preference for modern high-rise buildings outside the garden on the overall landscape.

In addition to discovering the predictability of SVF, there were also some other findings from data analysis. Since the research object of this study was the cultural heritage gardens, we focused on the two conflicting objects of "buildings outside the garden" and "garden landscape". Therefore, we did not simply use the concept TVF, but introduced the concept of GVF, which includes both the TVF and garden landscape view factor. In the correlation analysis, SVF and GVF showed inverse correlations with respondents' preference, which was easy to understand. With the parameter calculation method used in this study, the value of BVF was so small that the increase of SVF would inevitably lead to a decrease of GVF. However, it also reflected the disadvantage of the setting of SVF and GVF; that is, the correlation between these two parameters was too strong. Therefore, we added the tree view factor of each observation point and found that it was also strongly negatively correlated with the preferences of the respondents ($r = -0.121$; $p = 0.017$), but did not show predictability for the dependent variable.

It is counterintuitive that there was a negative correlation between the vegetation view factor and the respondents' preferences. Studies by Asgarzadeh et al. show that green trees have a certain degree of relief for people's sense of oppression. Still, at the same time, the type and quality of green plants have a meaningful impact on people's psychological perception [48]. The start date of our study was at the end of March. From the panoramic photos taken in our research, we can see that many tall deciduous trees in the garden had not yet grown new leaves, which had an inevitable influence on the shielding effect of the buildings. Therefore, we believe that in future study of cultural heritage gardens, seasons are an influencing factor that needs to be considered. The time of our experiment was the "cherry blossom season", which is also one of the most representative seasons of Japan. In other words, we also believe that this experiment should be seasonal, and an annual experiment in each season should be conducted. On the other hand, the results of this experiment also indicate that it may not be enough to describe the landscape of cultural heritage gardens with simple visual proportion. On this basis, future studies should pay more attention to the embodiment of landscape culture in the description.

In addition, it is found that age and nationality (East Asian or not) had a certain degree of influence on the predictability of SVF, which is consistent with the research results of Senoglu et al. [29]. We believe that people of different ages will have different opinions on the renewal and sustainable development of cultural heritage gardens. Additionally, garden visitors who were more familiar with East Asian culture and East Asian traditional garden culture than Western tourists, tended to have different attitudes towards the impact of buildings outside the gardens. Although this impact was very small, it also indicated that the sustainable development of cultural heritage gardens requires multiple considerations.

6. Conclusions

This study took the cultural heritage gardens in Tokyo as the research object and investigated the tourists' perception preference of the impact of urban high-rise buildings on the historical garden landscape in the context of urbanization. The findings show that tourists' attitudes towards the high-rise buildings outside the traditional gardens were increasingly diversified, and the impact of this phenomenon was not necessarily negative. Therefore, in future policy setting for the protection and management of the cultural heritage landscape and its surrounding environment in the city, the relationship between the city and the cultural heritage landscape should be viewed from a dynamic perspective. The managers of the urban heritage landscape should try to associate the results of urbanization with the heritage landscape more positively to achieve the goal of sustainable development of cultural heritage landscape and city together.

Furthermore, this exploratory study introduced the urban morphological description parameter SVF into the evaluation of traditional garden landscapes in an urban context and attempted to quantify tourists' preferences and landscape spatial form in Tokyo Metropolitan Cultural Heritage Gardens. The results of this study showed that Sky View Factor and tourists' preferences were significantly positively correlated, while Garden View Factor was significantly negatively related. Meanwhile, SVF had predictability for tourists' preferences, which means that, to a certain extent, and within a specific range, the wider the landscape of the observation point, the more tourists would tend to think that the outside modern high-rise buildings have a less negative impact on the traditional landscape. However, the building view factor did not show any correlation with the dependent variable. Additionally, the age and nationality of tourists caused differences in tourist preferences, but the impact was minimal.

Our research explores the coexistence and win–win challenges between the protection and management of cultural heritage landscapes and urban development in cities from a novel perspective. From the standpoint of people's continually changing and updated views and artistic perspectives over time, the sustainable development of heritage gardens in the city is discussed. The novelty of this paper is in its extension of the research and evaluation of the perception of urban cultural heritage landscapes to the field of urban morphology and proposition of a new method of quantifying landscape morphology, using spherical cameras to calculate the SVF, GVF, and BVF. This study also takes the lead in analyzing and discussing the correlation between the view factors of the landscape elements and the tourists' perception preference, which provides a new idea for the research of urban heritage landscape protection and sustainable development in the context of urbanization.

Author Contributions: Conceptualization, G.C. and J.S.; methodology, G.C. and J.S.; software, G.C. and J.S.; validation, G.C.; formal analysis, G.C.; investigation, G.C. and J.S.; resources, Y.X. and K.F.; data curation, G.C.; writing—original draft preparation, G.C.; writing—review and editing, G.C. and J.S.; visualization, G.C.; supervision, J.S.; project administration, J.S.; funding acquisition, Y.X. and K.F. All authors have read and agreed to the published version of the manuscript.

Funding: The second author gratefully acknowledges financial support from the China Scholarship Council (201706230230).

Acknowledgments: The authors thank all interviewees for providing valuable feedback to questionnaires, and the experts for their suggestions on this paper.

Conflicts of Interest: The authors declare no conflict of interest.

References

1. Boniface, P.; Fowler, P. *Heritage and Tourism in the Global Village*; Routledge: Abingdon, UK, 2002; ISBN 978-1-134-90843-1.
2. Nuryanti, W. Heritage and postmodern tourism. *Ann. Tour. Res.* **1996**, *23*, 249–260. [CrossRef]
3. Udeaja, C.; Trillo, C.; Awuah, K.G.B.; Makore, B.C.N.; Patel, D.A.; Mansuri, L.E.; Jha, K.N. Urban Heritage Conservation and Rapid Urbanization: Insights from Surat, India. *Sustainability* **2020**, *12*, 2172. [CrossRef]
4. Hosagrahar, J.; Soule, J.; Girard, L.F.; Potts, A. Cultural Heritage, the Un Sustainable Development Goals, and the New Urban Agenda. *BDC Boll. Cent. Calza Bini* **2016**, *16*, 37–54. [CrossRef]
5. Jigyasu, R. The Intangible Dimension of Urban Heritage. In *Reconnecting the City*; John Wiley & Sons, Ltd.: Hoboken, NJ, USA, 2014; pp. 129–159, ISBN 978-1-118-38394-0.
6. Harvey, D.C. Heritage Pasts and Heritage Presents: Temporality, meaning and the scope of heritage studies. *Int. J. Herit. Stud.* **2001**, *7*, 319–338. [CrossRef]
7. *Culture: Urban Future. Global Report on Culture for Sustainable Urban Development*; UNESCO: Paris, France, 2016; ISBN 978-92-3-100170-3.
8. Burgess, J.; Harrison, C.M.; Limb, M. People, Parks and the Urban Green: A Study of Popular Meanings and Values for Open Spaces in the City. *Urban Stud.* **1988**, *25*, 455–473. [CrossRef]
9. Havens, T.R.H. *Parkscapes: Green Spaces in Modern Japan*; University of Hawai'i Press: Honolulu, HI, USA, 2011; ISBN 978-0-8248-3477-7.
10. Cultural Properties | Agency for Cultural Affairs. Available online: https://www.bunka.go.jp/english/policy/cultural_properties/ (accessed on 30 June 2020).
11. Hough, M. *Out of Place: Restoring Identity to the Regional Landscape*; Yale University Press: London, UK, 1990; ISBN 978-0-300-05223-7.
12. Chiesura, A. The role of urban parks for the sustainable city. *Landsc. Urban Plan.* **2004**, *68*, 129–138. [CrossRef]
13. Jones, M.; Daugstad, K. Usages of the "cultural landscape" concept in Norwegian and Nordic landscape administration. *Landsc. Res.* **1997**, *22*, 267–281. [CrossRef]
14. Antrop, M. Why landscapes of the past are important for the future. *Landsc. Urban Plan.* **2005**, *70*, 21–34. [CrossRef]
15. Wu, J. Urban ecology and sustainability: The state-of-the-science and future directions. *Landsc. Urban Plan.* **2014**, *125*, 209–221. [CrossRef]
16. Von Wirth, T.; Grêt-Regamey, A.; Moser, C.; Stauffacher, M. Exploring the influence of perceived urban change on residents' place attachment. *J. Environ. Psychol.* **2016**, *46*, 67–82; [CrossRef]
17. Winter, T. *Post-Conflict Heritage, Postcolonial Tourism: Tourism, Politics and Development at Angkor*; Routledge: Abingdon, UK, 2007; ISBN 978-1-134-08494-4.
18. Jerpåsen, G.B.; Larsen, K.C. Visual impact of wind farms on cultural heritage: A Norwegian case study. *Environ. Impact Assess. Rev.* **2011**, *31*, 206–215. [CrossRef]
19. Antrop, M. The concept of traditional landscapes as a base for landscape evaluation and planning. The example of Flanders Region. *Landsc. Urban Plan.* **1997**, *38*, 105–117. [CrossRef]
20. European Environment Agency Task Force. United Nations Economic Commission for Europe. In *Europe's Environment: The Dobris Assessment*; Stanners, D.A., Bourdeau, P., Eds.; European Communities: Copenhagen, Denmark, 1997; ISBN 978-92-826-5409-5.
21. Oh, K. Visual threshold carrying capacity (VTCC) in urban landscape management: A case study of Seoul, Korea. *Landsc. Urban Plan.* **1998**, *39*, 283–294. [CrossRef]

22. Swensen, G.; Jerpåsen, G.B. Cultural heritage in suburban landscape planning: A case study in Southern Norway. *Landsc. Urban Plan.* **2008**, *87*, 289–300. [CrossRef]
23. Labadi, S. UNESCO, world heritage, and sustainable development: International discourses and local impacts. In *Collision or Collaboration*; Springer: Cham, Switzerland, 2017; pp. 45–60, ISBN 978-3-319-44515-1.
24. Shinji, I.; Shimizu, T.; Takemata, T. The Present Situation of Landscape Destructions on the Cultural Property Gardens in Tokyo. *J. Jpn. Inst. Landsc. Archit.* **1988**, *52*, 43–48. [CrossRef]
25. Arifin, N.H.S.; Masuda, T. Visitors' Judgements on the Scenery of Ritsurin Garden. *J. Jpn. Inst. Landsc. Archit.* **1997**, *61*, 259–262. [CrossRef]
26. Koizumi, M.; Ishikawa, M. A Study of Landscape Structure in Dai-sensui and Yokobori Area of Hamarikyu Garden. *J. Jpn. Inst. Landsc. Archit.* **2007**, *70*, 497–500. [CrossRef]
27. Shinobe, H. A study on the scenery protection around the garden in the city. *J. City Plan. Inst. Jpn.* **2012**, *47*, 625–630. [CrossRef]
28. Lin, L.; Homma, R.; Iki, K. Visual Impact Analysis and Control Method of Building Height for Landscape Preservation of the Traditional Gardens: A Case Study on the Suizenji Jōjuen in Kumamoto City. In *Proceedings of the Smart Growth and Sustainable Development: Selected Papers from the 9th International Association for China Planning Conference, Chongqing, China, 19–21 June 2015*; Pan, Q., Li, W., Eds.; GeoJournal Library; Springer International Publishing: Cham, Switzerland, 2017; pp. 115–125, ISBN 978-3-319-48296-5.
29. Senoglu, B.; Oktay, H.E.; Kinoshita, I. An empirical research study on prospect–refuge theory and the effect of high-rise buildings in a Japanese garden setting. *City Territ. Archit.* **2018**, *5*, 3. [CrossRef]
30. Oke, T.R. *Boundary Layer Climates*; Psychology Press: Hove, UK, 1987; ISBN 978-0-415-04319-9.
31. Oke, T.R. Canyon geometry and the nocturnal urban heat island: Comparison of scale model and field observations. *J. Climatol.* **1981**, *1*, 237–254. [CrossRef]
32. Svensson, M.K. Sky view factor analysis – implications for urban air temperature differences. *Meteorol. Appl.* **2004**, *11*, 201–211. [CrossRef]
33. Li, X.; Ratti, C.; Seiferling, I. Quantifying the shade provision of street trees in urban landscape: A case study in Boston, USA, using Google Street View. *Landsc. Urban Plan.* **2018**, *169*, 81–91. [CrossRef]
34. Gong, F.-Y.; Zeng, Z.-C.; Zhang, F.; Li, X.; Ng, E.; Norford, L.K. Mapping sky, tree, and building view factors of street canyons in a high-density urban environment. *Build. Environ.* **2018**, *134*, 155–167. [CrossRef]
35. Hämmerle, M.; Gál, T.; Unger, J.; Matzarakis, A. Comparison of models calculating the sky view factor used for urban climate investigations. *Theor. Appl. Climatol.* **2011**, *105*, 521–527. [CrossRef]
36. Aoki, Y. Relationship between percieved greenery and width of visual fields. *J. Jpn. Inst. Landsc. Archit.* **1987**, *51*, 1–10. [CrossRef]
37. Xiao, X.; Wei, Y.; Li, M. The Method of Measurement and Applications of Visible Green Index in Japan. *Urban Plan. Int.* **2018**, *33*, 98–103. [CrossRef]
38. Takei, M.; Ohara, M. A Study on Measurement of the Sense of Oppression by A Building: Part-4 Estimation of A Permissible Value of the Sense of Oppression and Conclusion of this Study. *Trans. Archit. Inst. Jpn.* **1981**, *310*, 98–106. [CrossRef]
39. Bulut, Z.; Yilmaz, H. Determination of landscape beauties through visual quality assessment method: A case study for Kemaliye (Erzincan/Turkey). *Environ. Monit. Assess.* **2008**, *141*, 121–129. [CrossRef]
40. Jiang, B.; Larsen, L.; Deal, B.; Sullivan, W.C. A dose–response curve describing the relationship between tree cover density and landscape preference. *Landsc. Urban Plan.* **2015**, *139*, 16–25. [CrossRef]
41. Sun, Y.; Fan, H.; Helbich, M.; Zipf, A. Analyzing Human Activities through Volunteered Geographic Information: Using Flickr to Analyze Spatial and Temporal Pattern of Tourist Accommodation. In *Progress in Location-Based Services*; Lecture Notes in Geoinformation and Cartography; Krisp, J.M., Ed.; Springer: Berlin/Heidelberg, Germany, 2013; pp. 57–69, ISBN 978-3-642-34203-5.
42. Matzarakis, A.; Rutz, F.; Mayer, H. Modelling radiation fluxes in simple and complex environments—application of the RayMan model. *Int. J. Biometeorol.* **2007**, *51*, 323–334. [CrossRef]
43. Matzarakis, A.; Rutz, F.; Mayer, H. Modelling radiation fluxes in simple and complex environments: Basics of the RayMan model. *Int. J. Biometeorol.* **2010**, *54*, 131–139. [CrossRef] [PubMed]
44. Honjo, T.; Lin, T.-P.; Seo, Y. Sky view factor measurement by using a spherical camera. *J. Agric. Meteorol.* **2019**, *75*, 59–66. [CrossRef]
45. Cohen, J. The Effect Size Index: D. In *Statistical Power Analysis for the Behavioral Sciences*; Lawrence Erlbaum Associates: Mahwah, NJ, USA, 1988; pp. 20–26, ISBN 0-8058-0283-5.

46. Hamarikyu Gardens | Chūō Ku | Japan | AFAR. Available online: https://www.afar.com/places/bang-li-gong-en-si-ting-yuan-hamarikyu-gardens-tokyo (accessed on 30 June 2020).

47. 向島百花園|公園へ行こう! Available online: https://www.tokyo-park.or.jp/park/format/index032.html (accessed on 30 June 2020).

48. Asgarzadeh, M.; Koga, T.; Yoshizawa, N.; Munakata, J.; Hirate, K. Investigating Green Urbanism; Building Oppressiveness. *J. Asian Archit. Build. Eng.* **2010**, *9*, 555–562. [CrossRef]

 *sustainability*

Article

Land-Use Changes of Historical Rural Landscape—Heritage, Protection, and Sustainable Ecotourism: Case Study of Slovak Exclave Čív (Piliscsév) in Komárom-Esztergom County (Hungary)

Peter Chrastina [1], Pavel Hronček [2], Bohuslava Gregorová [3] and Michaela Žoncová [3,*]

[1] Department of Historical Sciences and Central European Studies, Faculty of arts, University of Ss. Cyril and Methodius Trnava, Námestie J. Herdu 2, 917 01 Trnava, Slovakia; peter.chrastina@ucm.sk
[2] Department of Geo and Mining Tourism, Institute of Earth Resources, Faculty of Mining, Ecology, Process Control and Geotechnologies, Technical University of Kosice, Němcovej 32, 040 01 Košice, Slovakia; pavel.hroncek@tuke.sk
[3] Department of Geography and Geology, Faculty of Natural Sciences, Matej Bel University in Banská Bystrica, Tajovského 40, 974 01 Banská Bystrica, Slovakia; bohuslava.gregorova@umb.sk
* Correspondence: michaela.zoncova@umb.sk

Received: 30 June 2020; Accepted: 22 July 2020; Published: 28 July 2020

Abstract: The landscape surrounding the village of Čív (Piliscsév in Hungarian) in the north of the Komárom-Esztergom County is part of the cultural heritage of the Slovaks in Hungary. This paper discusses the issue of the Čív landscape changes in the context of its use (historical land use). Between 1701 and 1709, new inhabitants began cultivating the desolated landscape of the Dorog Basin, which is surrounded by the Pilis Mountains. This paper aims to characterize the Slovak exclave Čív land use with an emphasis on the period from the beginning of the 18th century (Slovak colonization of the analyzed territory) to 2019. These findings subsequently lead to the evaluation of the stability of the cultural-historical landscape as an essential condition for the development of ecotourism in the cultural landscape. The study results show that a long-term stable cultural landscape has a similar potential for the development of ecotourism as a natural landscape (wilderness). Research conclusions were aimed at creating three proposals for the cultural landscape management of the study area, conceived by the fundamental pillars of ecotourism, which would lead to its stable and sustainable use in ecotourism.

Keywords: historical landscape; cultural landscape; land use; landscape stability; sustainability; ecotourism

1. Introduction

Land-use changes are a constant cycle around the world, whether due to natural, quasi-natural, or anthropogenic processes. They always depend on the size of the driving force that acts on the (historical) landscape-land-use in an earlier period (earlier developmental stage), i.e., how intensively it can cover older layers and to what extent the oldest historical landscape will be preserved in the given area. A cultural landscape is created in a given time horizon if the driving forces are anthropogenically influenced or if they are of anthropogenic origin. Consequently, we can consider the historical-cultural landscape to be stable or relatively stable, if the earlier driving forces are weaker, less intense, or even almost absent. Stable cultural landscapes created in the oldest possible history represent an essential heritage of human society, not only of regional but also of national or European importance.

The landscape of the Slovak exclave Čív (hun. Piliscsév) in the northern part of the Komárom-Esztergom County, which is a part of the cultural heritage of Slovaks in Hungary, is an interesting example. The study presents the local landscape historical development issue in the context of its use (historical land-use) with specific forms that the Slovak ethnic group brought from the territory of their origin (ancestors). The researched landscape with elements preserved from the Roman Empire period, as an already abandoned landscape at the southwestern foothills of Piliš (hung. Pilis), began to develop from the first quarter of the 18th century. The development progressed in the process of changes initiated by the population of the Slovak ethnic group of Roman Catholic denomination from the Counties located north of the Danube.

The study is research into the issue of historical land-use in order to examine the land-use of Slovak exclave Čív from the beginning of the 18th century to 2019. The development of the local landscape's historical land-use after the arrival of Slovak colonists is shown on five thematic maps (from the years 1782–1785, 1841, 1882, 1941, and 2019), a table with areas of land-use classes (LUC), and a diagram. Research results are supplemented by a framework proposal for the management of the local landscape to preserve or adapt its character and values to ecotourism development in the historical-cultural rural landscape.

Geographers and landscape ecologists primarily address the issue of long-term landscape changes or its specific components; e.g., Fescenko, Nikodemus, and Brūmelis [1] reconstructed changes in spatial patterns of forest area during the last 220 years in an agricultural matrix of Zemgale region (Latvia). Olah, Boltižiar, and Gallay [2] analyzed the Slovak landscape (since the 18th century) and its current trends on the example of 16 model areas. Fialová, Chromý, and Marada [3] studied the development of the Vltava riverbanks' functional use in Prague. The work of Trpáková [4] represents a landscape ecologist point of view on the dynamics of landcover changes in the model areas Sokolovsko, Kladruby nad Labem, and Novodvorsko-Žehušicko in the Czech Republic. Wang et al. [5] prepared an analysis of land-use changes related to invasive species from 1996 to 2006 and predicted land-use and landcover changes to 2018, while they also characterized the significant driving forces in the State of Connecticut (USA).

M. Boltižiar, P. Chrastina, and J. Trojan [6], P. Chrastina [7–9], P. Chrastina and M.Boltižiar [10,11], P. Chrastina et al. [12], P. Chrastina, J. Trojan, and P. Valášek [13], and P. Chrastina, K. Křováková, and V. Brůna [14] applied an integrated (interdisciplinary) approach to the historical land-use of Slovak exclaves in Hungary, Romania, and Serbia.

The cultural landscape is increasingly becoming an object of interest for visitors, especially in the intensively economically developed countries of Europe. It has the potential for developing various types of tourism, including ecotourism.

This type of tourism has emerged since the 1970s. It is a very pure form of tourism in terms of the impact on the environment and the population of the local region. Its primary feature is the sustainability and adequacy of the given conditions of the landscape concerning the stability of existing ecosystems and the use of available natural resources. This form of tourism is currently still perceived rather restrictively in terms of economic activities. Ecotourism is only fulfilled substantially by the understanding of the relationship of man and nature and the perception of man as part of the landscape. Ecotourism emphasizes the local population and the local landscape and provides visitors with the natural or cultural heritage of the chosen destination. Being impact-less to the landscape to which they are heading is essential for ecotourists. Therefore, respect for the local landscape, different lifestyles, nature, or culture are the key concepts.

The concept of ecotourism began to become gradually domesticated in the 1980s. Subsequently, the International Ecotourism Society formulated the first generally accepted basic definition of ecotourism: "Responsible travel to natural areas that conserves the environment and improves the well-being of local people" [15]. This initial definition is the backbone of the study of ecotourism and remains a valid concept in general, also from a broader perspective of experts who consider ecotourism as any form of environmentally friendly tourism. With a more precise definition of the term, ecotourism is a form of tourism that contributes to natural resources protection. The abovementioned definition is a base for experts specializing in ecotourism around the world. Table 1 provides selected basic definitions of ecotourism from experts and institutions from different regions of the world.

Table 1. Overview of selected definitions in chronological order, based on the initial definition of ecotourism as responsible travel to natural areas that conserves the environment and improves the well-being of local people.

Source	Definition
Ziffer (1989) [16]	A form of tourism inspired primarily by the natural history of an area, including its indigenous cultures. The ecotourist visits relatively undeveloped areas in the spirit of appreciation, participation, and sensitivity. The ecotourist practices a non-consumptive use of wildlife and natural resources and contributes to the visited areas through labor or financial means aimed at directly benefiting the conservation of the site and the economic well-being of the local residents.
Brandon (1996) [17]	Environmentally responsible travel and visitation to relatively undisturbed natural areas, in order to enjoy and appreciate nature that promotes conservation, has low negative visitor impact, and provides for beneficially active socio-economic involvement of local populations.
Honey (1999) [18]	Travel to fragile, pristine, and usually protected areas that strive to be low impact and (usually) small scale. It helps educate the traveler; provides funds for conservation; directly benefits the economic development and political empowerment of local communities; and fosters respect for different cultures and for human rights.
Weaver (1999) [19]	Interest in ecotourism, now widespread among tourism planners and marketers, is rationalized by a number of popular assumptions regarding the sector's potential economic, environmental, and socio-cultural benefits. Ecotourism is a form of tourism that fosters learning experiences and appreciation of the natural environment, or some component thereof, within its associated cultural context.
Weaver (2008) [20]	Ecotourism as a very specific form is part of the broad concept of nature-based tourism, or it can be said that ecotourism describes a nature-based operation in the field of tourism. "The most obvious characteristic of Ecotourism is that it is nature based".
Kurek et al. (2008) [21]	Ecotourism is travel resp. a visit of areas of natural value and conservation for the purpose of exploring them, while respecting the principles of protection of the visited areas and their ecosystems and the integrity of the local community.
Rahman (2010) [22]	It focuses primarily on experiencing and learning about nature, its landscape, flora, fauna and their habitats, as well as cultural artifacts from the locality. A symbiotic and complex relationship between the environment and tourist activities is possible when this philosophy can be translated into appropriate policy, careful planning and tactful practicum.
The International Ecotourism Society (2015) [23]	Ecotourism is now defined as "responsible travel to natural areas that conserves the environment, sustains the well-being of the local people, and involves interpretation and education". Education is meant to be inclusive of both staff and guests.
Global Ecotourism Network (2019) [24]	Responsible travel to natural areas that conserves the environment, socially and economically sustains the well-being of the local people, and creates knowledge and understanding through interpretation and education of all involved.
Cambridge Dictionary (2019) [25]	The business of organizing holidays to places of natural beauty in a way that helps local people and does not damage the environment.

The issue of ecotourism in general, e.g., [15–17,20,26–30], and the sustainability of ecotourism, e.g., [31–34], is currently being studied by a number of authors.

According to the given definitions, ecotourism as a new progressive global type of tourism stands (respectively stood) on three pillars of sustainability—environmental, social, and economic [35].

Ecotourism understood in this way would be sustainable only in those areas of the world where the wilderness, or the least anthropogenically impacted landscape, is still prevailing today. In areas with dominating cultural landscape, ecotourism perceived in this way would not be able to find its fulfilling application. However, people surrounded by modern (digital) civilization in the globalizing present are looking for an escape into areas with a stabilized cultural landscape with several tangible and intangible cultural (technical) monuments. This trend is starting to apply mainly in Europe, but also in other parts of the world (e.g., in Southeast Asia [36]), where the human civilization has left a reflection in the (natural) landscape since prehistory, but mainly since ancient times.

The idea of shifting the position of sustainable ecotourism to four pillars, i.e., environmental, social, economic, and cultural, is beginning to appear in the works of experts in ecotourism in Europe from the 90s of the 20th century, e.g., in France [37,38], Germany [39,40], Finland [41,42], Croatia [43], Greece [35], Russia [44,45], Romania [29], Scotland [46,47], and Turkey [48,49].

The cultural pillar is an important component of ecotourism in the region of Southeast and East Asia, e.g., [50–52], and Latin America napr. [19,53].

Theoretical and methodological issues of the cultural component (cultural landscape and cultural monuments) in ecotourism were described in the works of Wallace and Russell [42], Donohoe [35], and Campbell [46].

Experts in tourism in Central Europe began to address ecotourism later than the World Scientific Forum. Due to the nature of the landscape in this geographical area, where the cultural landscape significantly dominates, and the natural landscape (wilderness) has been preserved only in the form of relics of small protected areas, ecotourism in the original view of the three pillars would be challenging to develop in practice. In the area of the Western Carpathians, national parks and large protected areas are mostly not wilderness but a cultural landscape (e.g., Duna Ipoly Park in Hungary, Low Tatras National Park, Tatra (s) National Park, Veľká Fatra National Park in Slovakia and Užanskyj nacionaľnyj pryrodnyj park in Ukraine). Therefore, experts have begun to understand sustainable ecotourism in terms of the four pillars, while the cultural pillar is often more important than the environmental (often the only one, because the wilderness is absent). The state and understanding of the issue are also pointed out by the very definitions of ecotourism by experts from Central Europe, or slightly more precisely from the Western Carpathians countries. Hungary: Ecotourism is the environmentally responsible travel and visiting of relatively unspoiled natural areas to please and appreciate their natural as well as current and past cultural values, which preserve them by mitigating the effects of the visit and bringing social and economic benefits to the local population [54]. Ecotourism is a form of tourism, or an alternative form of tourism, which is the opposite of its mass form. It emphasizes the importance of the environment and its cultural component, as well as their protection. Sustainability is an essential feature of ecotourism [55]. Slovakia: Ecotourism is a form of tourism in which tourists, individually or in small groups, visit attractive places that are little-known from a human point of view, whether it is their natural environment or cultural component in a tangible or intangible form [56]. According to B. Gregorová, [57] the co-author of this study, ecotourism is a form of tourism associated with wandering the landscape ("natural" and cultural) and observing, while gaining an authentic experience of learning about nature, as well as local communities and their culture. It develops sustainably; therefore, it minimizes negative impacts on the natural, social, and cultural environment to preserve the natural and cultural diversity and identity of the landscape. It is more economically efficient concerning local communities (rural settlements), despite its non-mass character, as it is provided by local service providers. Czech Republic: Ecotourism is wandering through nature (landscape), and its observation is developing mainly in areas with valuable landscape and nature. It supports natural areas, host communities, and nature conservation authorities. It provides local communities with alternative earning opportunities and raises awareness following the protection of the landscape's natural and cultural values [58]. Ecotourism is a type of tourism that focuses primarily on exploring natural beauty in terms of nature reserves, national parks, protected landscapes, and other natural attractions, intending to ensure that these attractions are not disturbed by tourism. In addition to natural heritage, it also focuses on cultural heritage and its protection. Poland: Ecotourism is an active exploration of valuable tourist attractions of natural and cultural character with intense emotional contact individually or in small groups. Adaptation to local communities, traditions, and culture is essential [59].

The research and the presented study aim to evaluate changes in land-use and stability of the historical (cultural) rural landscape from the beginning of the 18th century until 2019. It was an essential precondition for its preservation and protection in terms of historical heritage necessary for sustainable ecotourism. The researched historical rural landscape represents a cultural heritage as an essential specific element of the Slovak ethnic group living in a landscape inhabited by the dominant indigenous Hungarian population of the then Hungary.

At the same time, our effort is to present Central Europe's cultural landscape as a type of territory suitable for the development of ecotourism, which is currently carried out on relics of the original wilderness protected in the form of large or small protected areas. We also want to point out that the sustainable historical-cultural rural landscape of the Slovak exclave (village and its surroundings) of Čív (in the Dorog Basin) can develop this form of tourism.

We have to note here that the current landscape of Europe, particularly Central Europe, is made up almost exclusively of cultural landscapes, and even the national parks of the Carpathian arch protect mostly the historical-cultural landscape. Directing ecotourism only into the wilderness is problematic in this geographical area, although this type of landscape is a priority for ecotourists. A stabilized historical (rural) cultural landscape could also be a potential replacement for the unpreserved wilderness.

2. Materials and Methods

2.1. Study Area

The territory of Čív with an area of 2490 ha is located approximately 10 km south-southeast of Esztergom, in the Komárom-Esztergom County in the northern part of Hungary (Figure 1). The defined research area of 1771.5 ha lies in the central part of the cadastral area in the core settlement vicinity. The reason for this definition of the research area was the fact that the boundaries of the village were definitively constituted only in the 19th century. Therefore, the study of Čív's land-use and its changes in the earlier periods would not be correct. The choice of the quadrilateral shape of the model area is a proven methodology used by, e.g., Kolejka [60] (pp. 148–151) in the study of landscape changes of the water reservoir Nové Mlýny in the Czech Republic, Gobin, Campling, and Feyen [61] (p. 215, 217) in the study of agricultural aspects of land-use of the city of Ikem in southeastern Nigeria, resp. Kuplich, Freitas, and Soares [62] (p. 2102) in the land-use classification of the city of Campinas in Brazil. The maps of land-use in the Tatras in the Atlas of representative landscape types of Slovakia [63] (pp. 166–167) or transects of the development of Corine Land Cover classes in the Atlas of land cover changes in selected European countries [64]. P. Chrastina et al. [12] also used the same methodology of delimitation of the territory to evaluate the historical land-use of the Tardoš settlement in Hungary.

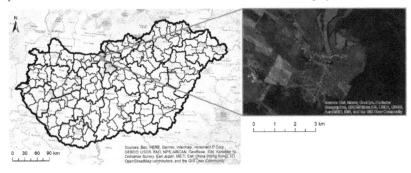

Figure 1. Location of the examined area.

The Dorog Basin (hung. Dorogi-medence), built of sands, loess, and loess clays, occupies most of the area. The Piliš Mountains (hung. Pilis), part of the Dunazug Mountains, belong to the Transdanubian Central Mountains (hung. Dunántúli-középhegység), which is formed by limestone and dolomite rocks [65] (pp. 147–148). The studied area's altitudes range from about 163 m (floodplain of the Čív brook in the northwest of the area) to 418 m (Barina hill, or Barány-hegy). The erosion-denudation georelief of the Dorog Basin has a hilly character with steeper slopes and valleys (Figures 2 and 3).

Figure 2. Location of Čív in the Dorog Basin. View from the north from Tatarské salaše hill (246 m above sea level) Author: P. Chrastina.

Figure 3. The Dorog Basin from Tatarské salaše hill (246 m above sea level). Author: P. Chrastina.

A warm, slightly dry to dry climate with annual total precipitation of up to 600 mm is typical for the studied area. The average annual temperature is around 10 °C. The Piliš Mountains belong to a warm mountain climate with a small temperature inversion and annual total precipitation of 800 to 1000 mm [66] (p. 240, pp. 242–243). Northwestern windflows significantly dominate during the year.

Soil types in the basin part of the area represent brown soils developed on loess and loess clays. The bottom parts of the damp, summer-drying valleys are covered with low-fertile pseudogley. Areas of regosols were created on areas of windblown sands at Homoki and Plešina sites in the south, resp. southwest of the study area. Fluvisols and gleys developed on the floodplain of the Čív brook (hung. Csévi Patak). Rendsinas on permeable limestones and dolomites cover the forested slopes of the Piliš Mts.

The original primeval forests in the basin part of the studied area consisted mainly of thermophilic Pontic-Pannonian oak forests on loess and sand [67] (pp. 10–11), [68] (pp. 93–94). On the slopes of the Piliš Mts. with rendzinas grew grassy oak forest [67] (p. 30). The wet floodplain of the Čív Brook and the adjacent areas were potentially covered by lowland floodplain forests (Ulmenion Association) and wetland alder enclaves [67] (p. 14).

2.2. Methods

The methodological procedure, research methods, and the elaboration of the study were divided into several directly related and concurrently implemented steps or stages, which systematically led to the fulfillment of the research goal and the processing of its results into the original scientific work (Figure 4).

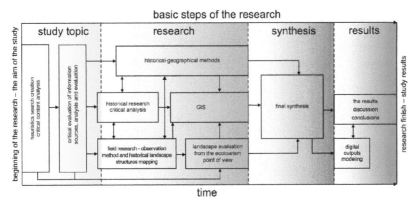

Figure 4. Schematic representation of theoretical and methodological procedures of the research and works. Source: authors' research.

The information database heuristics, the critical analysis of archival and literary sources was the basis, which led to the creation of a search on the researched topic using the bibliometric method [69–71].

Following the results of literary sources critical content analysis and the results of historical and field research, we used a combination of a range of evaluation, comparison, and synthesis methods [72]. As part of the synthesis of the results, we made proposals of the country's management to maintain its stability for the development of ecotourism.

Historical and field research. An essential step in elaborating the topic were the methods of historical research, especially critical analyses of historical sources, and the deductive and comparative-historical method. Following the historical research, we continued in the summer of 2019 with comprehensive field research of the historical landscape [73–80]. The preparation of the author's photo documentation was also a part of the field research.

Map data search, identification, content analysis, georeferencing, and vectorization. Cartographic sources consisted mainly of medium-scale maps (Table 2).

Table 2. Cartographic sources, their characteristics and sources.

Designation of the Map Base/Source	Map Sheet Number	Scale	Year of Production	Source
I. military mapping, *Theil des Pester und Graner Comitat*	*Coll. XIII. Sectio XIX.*	1: 28,800	1782-1785	https://mapire.eu/en/ [81]
II. military mapping, *Königreich Ungarn, District diesseits der Donau, Comitat Gran. Honth. Pesth.; District jenseits - diesseits der Donau, Comitat Komorn. Gran. Pesth.*	Section 48. Colonne XXXI. Section 49. Colonne XXXI.	1: 28,800	1841	https://mapire.eu/en/ [81]
III. military mapping	4961/2 4962/1	1: 25,000	1882	https://mapire.eu/en/ [81]
Topographic map of military mapping	?	1: 75,000	1941	https://mapire.eu/en/ [81]
Orthophoto (satellite image)	-	cca 1: 75,000	(24. 3.) 2019	Google Earth [82]

The I. to III. military mapping maps were taken from the Military Historical Archive in Budapest (Hadtörténelmi Levéltár). The military map from 1941 was obtained digitized (online) from the Mapire web application [81]. According to G. Timár et al. [83,84], resp. G. Timár and S. Biszak [85], the initial georeferencing procedure was realized into the WGS84 coordinate system. Maps were reprocessed into the HD72/EOV coordinate system (EPSG 23700) as part of the digitization itself, in which the area was quantified (calculated) as well. A digital satellite image from Google Earth Pro from 2019 was modified analogously (Figure 5).

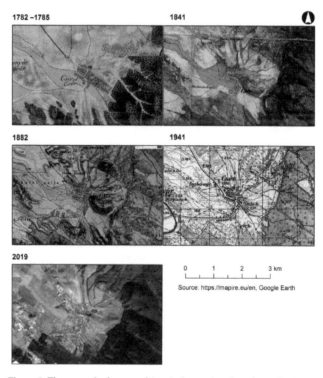

Figure 5. The researched area on historical maps in selected time horizons.

A mask (rectangular section) was created around the built-up area of Čív for the correct comparison of the sections of the examined area, as stated in the previous part of the study. This mask was also applied to digitize the selected areas to perform their comparative analysis (Figure 5).

Comparative Analysis of Cartographic Data

The areas of land-use classes (LUC) with line layers of watercourses and roads were interpreted on each of the georeferenced digital maps using geographic information systems (QGIS3). The final product of the comparative analysis is represented by large-scale maps of historical land-use (1: 25,000), depicting the land-use of the studied area in the relevant period [86–88].

The scales of most cartographic data and satellite imagery were smaller than the resulting classification of land use on thematic maps. Therefore, their content was generalized. The military map from 1941 and the satellite orthophoto image (2019) had the highest degree of content generalization.

The multitemporal analysis consisted of the LUC dynamics study in the QGIS3 environment. Specifically, it evaluated the development of LUC areas in the period given by the presentation of a map or satellite image and the subsequent statistical processing (numerical and graphical analysis).

Interpretation of the historical land use development as an input matrix for evaluating the stability of the historical landscape with use for ecotourism. Framework proposals determination for the management of the cultural landscape of the studied area aimed at its protection (stabilization, preservation) as a historical heritage necessary for the development of sustainable ecotourism.

3. Results

3.1. Prehistorical and Historical Land Use of the Čív Landscape

The initial human interventions in the local landscape occurred in the Eneolithic period (3500–2900 BC) [89]. Therefore, it is probable that the first deforestation of the original oak forests growing on brown soils, which man transformed into a mosaic of fields and pastures with islands of forests, took place in the Dorog basin already in the middle of the fourth and the early third millennium B.C.

In Roman times (around 1 A.D. -375/380, respectively 400 A.D.) the small fortress (kastel) Lacus Felicis lay on the territory of the village [90] (pp. 21–22), which protected the road leading from Brigetio through the territory villages to Aquincum (now Budapest). Remains of a watchtower with a defensive ditch (burgus), which were probably built during the reign of Emperor Valentinian I. (364–375 A.D.), were discovered in the locality Margeta (the northeast part of the studied area). Another monument from this period [89] (p. 30) is a torso of a stone-paved road in the southwest of the village at the Hosszú-rétek II. site, as well as two road milestones.

The economic activities of the Romanized population marked the landscape structure of the studied area, thanks to which the then landscape roughly resembled the current cultural landscape of Čív. The dry localities of the basin were occupied by fields intersected with roads with milestones and military architecture (burgus, potentially also a kastel). Settlements of members of higher social classes with a residential and economic function (so-called villa rustica) were scattered on agricultural land. The vineyards covered slopes with "warm" exposure. The wetland localities in the basin were characterized by a mosaic of extensive meadows and pastures, wetlands with reeds, and enclaves of floodplain forests or wetlands with alders. The deforested edges of the Piliš mountain slopes served potentially as meadows and pastures, but the steep slopes remained forested.

During the Mongol (Tatar) invasion into Hungary (1241–1242 A.D.), their military units allegedly camped at the Tatarské salaše hill [91] (p. 18) After the departure of the Mongol army, which conquered and burned the town of Esztergom in 1242 A.D. (the castle defended itself), the local settlement was decimated, and the areas of agricultural land were overgrown for about a decade. At the same time, the deforested foothills of Piliš were regenerated.

The territory of today's village belonged to the monarch in the second half of the 13th century, but the Esztergom Chapter had already owned the entire property in 1418 [92] (p. 15). The LUC mosaic in the basin part consisted of pastures with islands of non-forest woody vegetation and small fields. Extensive cattle breeding and farming works were performed by subjects from the surrounding villages of Kestúc (hung. Kesztölc) and Alberth [89] (p. 14).

The continuous development of the studied area's cultural landscape was interrupted at the beginning of the Modern Age. Most of the villages in the Dorog Basin were burned down by the Turks from 1526 to 1543. Written sources from 1564 and 1570 prove that the area of today's Čív has become a wasteland, a depopulated area [92] (p. 16), [89] (p. 16).

The economic renewal was made possible by Esztergom's (1685) and Buda's (1686) liberation from the Turks and the conclusion of the Peace of Karlovac (1699). The first settlements of Slovak colonists in the studied area were established in the years 1701 to 1709 [93] (p. 77). The migration of settlers on the Esztergom Chapter's lands caused the later constitution of the village (around 1720). The inhabitants of Veľký Čív (hung. Nagy Csév) were Slovaks and Catholics from the territory of today's western Slovakia; based on the research of the Čív dialect [89] (p. 35), we can assume that they came mainly from the Nitra County and southern areas of the Trenčín County.

Some of the colonists dealt with extensive cattle breeding on pastures. The others devoted themselves to agriculture. The fields on the clay-sandy and sandy-clay brown soils were formed mainly in the dry valleys and erosion-free slopes of the basin hills with a smaller slope. The local plow farms farmed with a two-field system provided up to three times the yield in 1715 [93] (p. 77). The vineyards of south slope of Starí vinohrad hill were established after the deforestation of the slope of the basin uplands in 1712; the vineyards on the southern slope of Vršek (or Na Vršku) hill are probably younger. A church dedicated to the Virgin Mary, Queen of Angels, was built in the middle of the village, at the foot of Vršek hill around the same time (the early 18th century).

Visual reconstruction of prehistorical and historical land use of the Čív landscape in the geographical area of the Dorog basin is clearly presented in the author's scheme of the spatio-temporal axis (Figure 6).

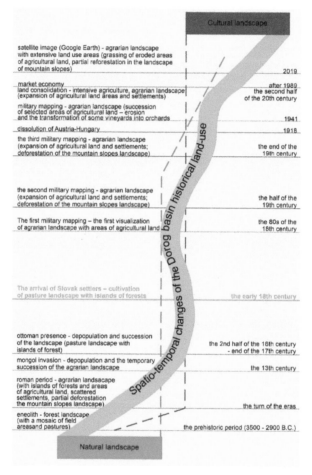

Figure 6. The graphic shows the visual axis of the main historical events in the geographical area of the Dorog Basin, which affected the natural landscape in terms of its spatio-temporal changes aimed at the emergence of the (current) cultural landscape. The distances between the lines of historical events represent their impact on the landscape, instead of time. The bigger the distance between the lines of events, the more intense their impact is on the transformation of the contemporary landscape. Source: authors' research.

In 1732, Veľký Čív had 236 adults, of which were 26 peasants and 12 cotters [89] (p. 17), [92] (p. 18). The Theresian land-register from 1768 provides a general view of the representation of specific LUCs and the intensity of land-use in the studied area in the late 1960s [94] (pp. 263–265). During this period, the name of the village was Čív (hung. Csév). The document in Čív states 59.75 peasant settlements with an area of approximately 7000 ha. A total of 128 heads of families lived in the village in 1767.

They grew vineyards, cabbage, hemp, corn in larger areas, and fruit in the home gardens. The vineyards were located near the village at Vršek, or Na vršku hill, and further from the village at the Starí vinohrad hill. Plowlands and meadows further formed the landscape. A grain mill operated on the Čív Brook, which had very little water.

There was plenty of quality construction and firewood in the surrounding forest. The forest cover of the studied area fell below 40% during this period. The erosion of arable land in the sloping terrain of the basin and at the foot of Piliš Mts. was a disadvantage.

3.2. Visualization and Quantification of Changes in the Historical Land-Use of the Studied Rural Landscape

The origin, or, more precisely, the anthropogenic transformation of the natural landscape into a cultural landscape in space and time must be understood in the interaction of both characteristics. The research needs to choose a historical landscape at a specific time horizon and emphasize its stability, development, and changes compared to the previous period, which are visualized by the authors' scheme (Figure 7), whose time horizons can be imaginarily identified with the selected stages in the years 1782 to 2019 in the area.

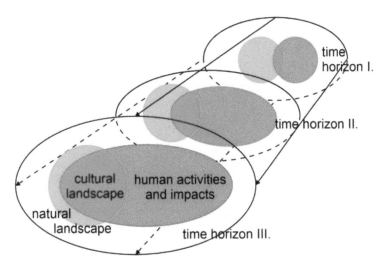

Figure 7. Visualization of man and the natural landscape interaction in time (in different time horizons) leading to the expansion (respectively to the emergence) of the cultural landscape and the defined (researched) territory. Source: authors' research

The visualization of land-use changes in the studied area from 1782–1785 to 2019 was processed using computer digitization. The results are presented in digital "map" outputs from the time horizons in question (Figure 8). The results of land-use dynamics changes and their quantification in these time horizons are presented in Table 3 and Figure 9. The analysis of the obtained results allows a detailed discussion and conclusions leading to the use of historical rural cultural landscapes in ecotourism.

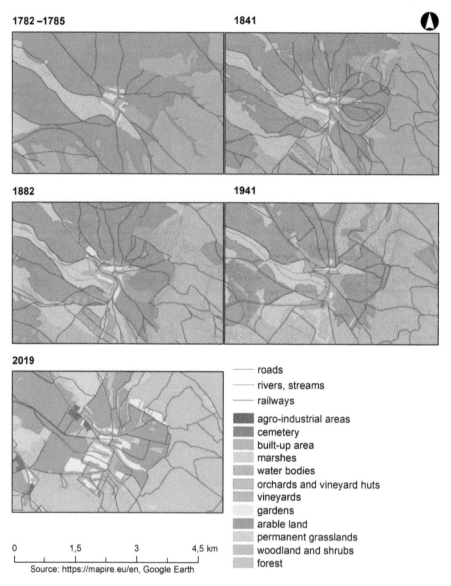

Figure 8. Computer visualization of the historical land use of the studied area in selected time horizons.

Table 3. Development of land use classes (LUC) of the studied area in the years 1782/1785–2019.

Land Use Classes (LUC)	1782–1785		1841		1882		1941		2019	
	ha	%	ha	%	ha	%	ha	%	ha	%
Forest	504.9	28.5	535.5	30.2	481.1	27.2	546.5	30.8	736.0	41.5
Woodland and shrubs	x	x	3.9	0.2	104.5	5.9	224.0	12.6	165.2	9.3
Permanent grasslands	158.1	8.9	364.2	20.6	157.6	8.9	141.3	8.0	150.2	8.5
Arable land	1004.9	56.7	718.9	40.6	764.3	43.1	667.6	37.7	513.8	29.0
Gardens	11.6	0.7	15.2	0.9	17.9	1.0	7.2	0.4	78.7	4.4
Vineyards	69.4	3.9	106.9	6.0	114.4	6.5	x	x	1.9	0.1
Orchards and vineyard huts	x	x	5.4	0.3	10.4	0.6	15.2	0.9	4.3	0.2
Water bodies	x	x	x	x	x	x	x	x	2.4	0.1
Marshes	x	x	x	x	92.9	5.2	97.7	5.5	x	x
Built-up area	22.6	1.3	21.5	1.2	27.8	1.6	70.7	4.0	100.7	5.7
Cemetery	x	x	x	x	0.6	x	1.3	0.1	0.9	0.1
Agro-industrial area	x	x	x	x	x	x	x	x	17.4	1.0
Σ	1771.5	100.0	1771.5	100.0	1771.5	100.0	1771.5	100.0	1771.5	100.0

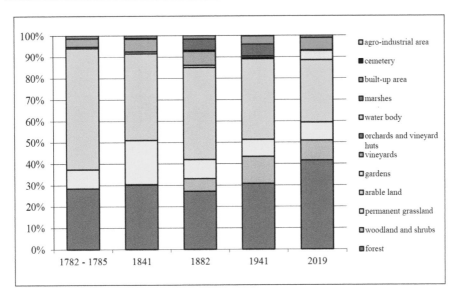

Figure 9. Visualization of the development of land use classes (LUC) of the studied area in the years 1782/1785–2019

3.3. Analysis of Changes in the Historical Land-Use of the Studied Rural Historical Landscape in Terms of Its Stability

The research results point to the importance of natural driving forces [95], which played an essential role in the process of anthropogenic exploitation of the local landscape during the observed periods. The agrarian potential of the studied area is reflected in the relatively stable LUC areas with a high rate of anthropogenic use (arable land, permanent crops), which limited the originality and ecological stability of the local landscape.

The dynamics of the forest area during approximately 237 years (1782–1785 to 2019) reflect the synergistic effect of sustainable management on the slopes of the Piliš Mts. Due to the mountain slopes' inclination, people usually used specific areas extensively and left them forested, in addition to the foothills (with temporal grasslands and arable land). Such development corresponds to the so-called "Forest Transition Theory" [96] (pp. 26–29). Slow and permanent (except in 1882) growth of forest area affected not only previously deforested mountain slopes, but also areas of agricultural land with different intensity of use (grasslands, arable land, and others), and this process was accompanied by the introduction of non-native species (pine, black locust).

The thermophilic oak forest covered about 28.5% (504.9 ha) of the studied area in 1782–1785 (Figure 8). Common pine could have also occurred in some places, on the edges of forest stands, or overgrown areas after extensive logging. The area of forests has changed only minimally (by two percentage points) since that period. Oak forest area has risen to 736 ha (41.5%) in 2019 (Figure 8) since the second half of the 20th century, which represents an increase of almost 70% compared to the first period (Table 3, Figure 9).

There are two aspects of increasing the area of forest in the study area. The first reflects the effect of environmental risk on the intensive use of easily erodible soils on slopes with a greater inclination. Specific localities were gradually afforested, the trend of reforestation has been characteristic since the second half of the 19th century, especially for the mountain and foothill areas of the former Austria-Hungary, respectively the selected successor states in Central Europe [97–100]. Examples from the Slovak exclaves in Hungary-Békešská Čaba (hung. Békéscsaba) [9], Šára (hung. Sári) [6] and [101] suggest that the afforestation of areas in the lowlands has been replaced by systematic planting of non-forest woody vegetation or free spreading black locust, poplar, and other woody plants. Areas of permanent grasslands (PG) and non-forest woody vegetation (NDV) were also created here. The second aspect relates to the change in land-use priorities after 1989.

Grazing of horned cattle on the undergrowth of grassy oak forests supported the transformation of selected localities into pasture forests, e.g., Pánská hora hill [89] (p. 66). Areas of former pasture forests have changed their LUC to forests after the natural succession.

The analyzed LUC mitigates the relatively high values of the coefficient of anthropogenic influence (Cai) and the coefficient of cultural landscape originality (Cclo) of the studied area. The coefficient Kao starts from zero, while the upper limit does not exist. A value of 1 is reached when the area of both areal types is in equilibrium. Values higher than one means that areas with a high intensity of anthropogenic use predominate—V (arable land, settlements, built-up areas, permanent crops). In case of the predominance of less intensive areas—N (forest, permanent grassland, water area, alternatively also woodland and shrubs, wetland), the value of the coefficient approaches 0 [102] (p. 17). The Cclo coefficient expresses the ratio of relatively positive (forest, permanent grasslands) and relatively negative elements of the landscape (arable land). If this ratio exceeds 1, the landscape is stable and vice versa—the landscape becomes unstable when the value approaches 0 [103] (p. 148). The expansion of forest areas since the middle of the 20th century has also had a positive impact on the course of both coefficients, which indicates a gradual improvement in the ecological stability of the landscape (Table 4).

Table 4. Development of coefficient of anthropogenic influence (Cai) and coefficient of cultural landscape originality (Cclo) from 1782-1785 to 2009.

Coeficient	1782–1785	1841	1882	1941	2019
$\text{Cai} = {}^V/_N$	1.7	0.9	1.1	0.8	0.7
$\text{Cclo} = {}^{\text{forest + grasslands}}/_{\text{arable land}}$	0.7	1.3	0.8	1.0	1.7

Woodland and shrubs. The given LUC was represented by opened or connected stands of woody plants (e.g., black locust, pine, willow, poplar, mulberry) and shrubs (hawthorn and similar). People have often cultivated these communities by pruning branches and shoots (pollarding) in the vicinity of vineyards, orchards, or fields. Woodland and shrubs have appeared in the landscape structure of the area since 1841 (3.9 ha). Copses were present near vineyards on the southwestern edge of the area (Figure 8). The area was 104.5 ha (approx. 6%) in the next period (1882). The area of the woodland and shrubs on the forest edges, potholes, tree lines, and smaller, unconnected stands reached up to 224 ha, i.e., 12.6% of the total area in 1941 (Figures 8 and 9). According to Table 3, this LUC area decreased by about 3% (165.2 ha) by 2019.

The existence of woodland and shrubs areas in the studied area's landscape structure has a positive impact on mitigating the negative impact of plowing on the local landscape and its ecological stability. The extent of this LUC has been growing since the last quarter of the 19th century, shown by Cai and Cclo coefficients in Table 4.

The location and area of intensive LUCs (arable land, permanent cultures, settlements) in the years 1782–1785 to 2019 were the result of the interaction of historical factors and social driving forces. The Slovak population began to cultivate the wasteland on the edge of the Dorog Basin bordered by the slopes of Piliš Mts. between 1701 and 1709.

Analogous contexts of natural and social driving forces of land-use were also found in the Slovak exclaves of Cápar (hung. Szápár), Čerňa (hung. Bakonycsernye), and Jášd' (hung. Jásd) in Hungary [80] or Borumlak (rom. Borumlaca) and Varzaľ (rom. Vărzari) in Romania [14].

Analyses confirmed that in 1782-1785 the meadows and pastures (permanent grasslands) spread mainly on the floodplain of the Čív Brook (site Velká and Malá pažíc, Dolná pažíc, Dlhé luki). Their area reached less than 158 ha (approximately 9%). By 1841, this LUC area more than doubled to 364 ha, 20.6%. This significant increase is probably related to the grassing of eroded fields, which arose in places with sandy soils (e.g., Homoki) or in the wet valley bottoms (Figure 8). The emergence of new pastures was not directly related to the development of animal production in the village. The area of permanent grasslands in the studied area has been without significant changes from the eighties of the 19th century until the present (2019) (Table 3, Figure 9). A relatively small number of cattle grazed the pastures; in 1911, it was only 13 pcs/100 inhabitants [104] (p. 153). Some areas of meadows and pastures (or vineyards) were built up (Homoki) in the second half of the 20th century. This decrease was induced by the transformation of arable land into permanent grasslands at the Tatarské salaše hill and elsewhere (Figure 8). Some parts of arable land (e.g., Na vršku hill and sites Pustovňík, Ot Santova) are temporarily used as grasslands, which is potentially related to the European Union's agricultural policy.

The area of grasslands remained relatively stable during the observed periods. Development of Cai and Cclo coefficients in Table 4 shows that meadows and pastures, along with other, less intensive LUCs (forest, woodland and shrubs), were (and are) important in improving the ecological stability of the landscape.

The arable land occupied almost 57% of the area surveyed in the first period (1782–1785). In addition to localities with a high groundwater level (e.g., the floodplain of the Čív Brook) and the afforested slopes of the Piliš Mts. with a steeper inclination, the fields managed by the three-field system spread practically over the entire area (Figure 8). Their extent was approximately 1005 ha. By 1841, the area of arable land had decreased by 16% (718.9 ha). This phenomenon is primarily related to the grassing of eroded fields on loess and windy sands. According to map 2, the selected areas of arable land on the warm substrates of the basin uplands were transformed into vineyards (e.g., Homoki, Holé vrški, Od Lamváru sites). The increase in the area of fields in 1882 (764.3 ha) did not have a more permanent character, because there was a gradual reduction in the extent of arable land in the studied area in the next time horizons (1941 and 2019), which was not significantly affected by collectivization (Table 3, Figure 9). The development of plant production was limited by natural driving forces, and natural conditions, especially light soils, slope inclination, and wetting of selected localities.

For this reason, Slovak farmers from Čív, there was a lease (renting) of fields in the nearby village of Jászfalu, and Pilisjászfalu which inhabited the Hungarian [93] (p. 77). A significant reduction of arable land areas in or near urban areas began in the 1970s. The reduction was also related to the expansion of the settlement itself, and the spatial expansion of agricultural and industrial area on the southwestern edge of the territory (Figure 8). Following Hungary's accession to the E.U., the fields in the east and south-east of the study area were converted to grasslands.

The large extent of the LUC in question negatively affects the local landscape's originality and ecological stability. The Cai and Cclo coefficients show a considerable anthropogenic load, especially during the first period (1782–1785). The area of arable land decreased in the following periods, which also had a positive effect on the coefficients (Table 4).

The home gardens were created together with the development of the village's urban area (Figure 8). In 1782–1785, they occupied 11.6 ha. The area of gardens increased to 15.2 ha by 1841. There was an increase in this LUC (17.9 ha) in the third period (1882). The significant decrease in the area of home gardens in 1941 does not correspond to reality but is related to the generalization of the underlying map's content. The largest area of gardens was in 2019 (78.7 ha), representing more than 6.5 times increase compared to the first time horizon (Table 3, Figure 9).

The vineyards in the Starí vinohrad and Vršek hills occupied an area of 69.4 ha (Figure 8) in 1782–1785. This relatively large area proves the developed viticulture in Čív. During the 19th century, there was an expansion of the existing and creation of new vineyards on the sunny slopes of the basin hills, protected from the wind by the forest stands (sites Homoki, Od Verešvára, and others). The reason was the quality of the wine, which motivated the local population to transform the sloping, less productive fields on light soils such as regosol or brown soil. In 1841, the area of the analyzed LUC was 106.9 ha (6.0%). In 1882, vineyards with an area of 114.4 ha covered about 6.5% of the area studied. At the beginning of the 20th century, vineyards were also located at the Tatarské salaše hill [91] (p. 18). The given LUC is absent on the map from 1941 (Figure 8); the reason is the scale, and the generalized content of the cartographic base. However, the field research confirmed that viticulture continued to develop in the village also during this period.

In 2019, the vineyards occupied only 1.9 ha, i.e., 0.1% of the studied area (Figure 8). Unused, abandoned vineyards were absorbed by the village urban area's development, or they were transformed into social fallows with grasslands and other LUCs. The development of vineyard areas is shown in Table 3 and Figure 9.

Orchards with vineyard huts and cellars were not mapped in the studied area until 1841 (Figure 8). Their extent almost doubled to 18.4 ha by 1882 from the original 5.4 ha (Figure 8). The increase in their area is also characteristic for 1941 when they covered an area of 15.2 ha (Figure 8). However, the acreage of orchards with viticultural architecture decreased to 4.3 ha in the last time horizon (2019) (Figure 8). The reason is the transformation of selected areas into extensive grasslands and residential development (Figures 8 and 9, Table 3).

The areas of permanent crops influenced the degree of anthropogenic influence of the local landscape together with the selected LUCs, as reflected in the development of the Cai coefficient in Table 4.

The LUC vineyards and orchards (including vineyard huts and cellars) represent characteristic elements of the studied area's landscape structure. Specific areas were used as grasslands and building plots after 1989 (Homoki site). Vineyards and orchards with vineyard huts at Starí vinohrad hill also lost their significance at the end of the 20th century. However, the archetype of the vineyard landscape with a landscape-creating and recreational function has been preserved.

The wetland on the floodplain of the Čív Brook was mapped in the area only in 1882 and 1941 (Figure 8). The data from Table 3 indicate a stable area of this LUC, which was around 95 ha. Interestingly, the wetland was not found at these sites on cartographic data from other periods, which, however, does not correspond to reality—e.g., wet grasslands (Figure 8) spread here in the years 1782–1785 and 1841. It was possible to use some areas of the former wetland as fields and gardens, or areas of non-forest woody vegetation formed by hydrophilic woody plants (willow, poplar, and others) after the land consolidation in the second half of the 20th century. This character of land-use was maintained until 2019 (Figure 8).

The wetland area (together with forests, grasslands, and other extensive LUCs) mitigated the intensity of the anthropogenic influence during the two periods (1882 and 1941), which is confirmed by the values of Cai in Table 3.

The areal extent of settlements increased during the specific periods following the growth of Čív's population (Figure 10).

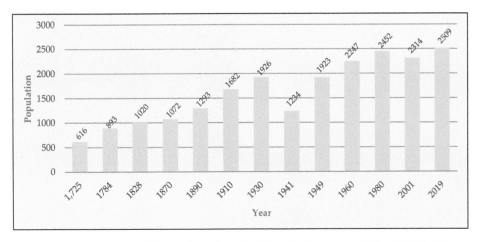

Figure 10. Development of Čív population form the 18th to the 21st century E. Fügedi [93] (p. 77), Piliscsév népessége (1870–2019).

In 1782-1785, the settlement spread over an area of 22.6 ha. The village had the character of a compact settlement with an irregular and seemingly chaotic arrangement of streets with houses that marked dry or drier locations in the basin (Figure 8). The LUC area and spatial disposition did not change significantly even during the two following periods (1841, 1882), which is confirmed by the data in Table 3, Figure 8.

The economic effect of job opportunities in the coal mines of town Dorog (and additional jobs in quarries and lime production) and the alternating system of arable land management manifested itself by an increase in natural growth, which increased the population of the village. Immigration from the surrounding settlements also played a role. These factors caused the built-up area extent to increase more than 2.5 times to 70.7 ha by 1941. According to Figure 8, the construction of new houses has affected fields, grasslands, and vineyards near the southern edge of the urban area. In 2019, the area of the monitored LUC exceeded 100 ha. Mainly the sloping locations with vineyards (Homoki site), fields, and grasslands (Dlhé, Dlhé lúki sites) were used as development areas for a set of new buildings in Čív or Piliscsév (Hungarian name of the village since 1954), or the existing built-up areas in the village became more concentrated.

The oldest cemetery in the village is located at Vršek hill near the Calvary. An elevated position was the motive for the cemetery location, together with a suitable substrate (dry loess) and the proximity of human settlements with a church.

The cemetery, with an area of 0.6 ha, has appeared in the landscape structure of the studied area since 1882. It was located in the southern part of the built-up area, today's kindergarten. In 1941 it was mapped on an area of 1.3 hectares. The new cemetery at Vinohrady (in Homoki site) had an area of about 1 ha in 2019. A military cemetery for members of the Red Army, victims of World War II, is situated at the kindergarten, at the crossroads of Béke Street and Vörösvári Street.

The areal extent of the agro-industrial site on the southwestern edge of the studied area reached 17.4 ha in 2019 (Table 3). Kilns for lime burning stood initially in these places [89] (p. 68), which is also confirmed by the map marker on the cartographic base from 1941.

The areal extent of settlements during particular periods (in 2019 also the area of the agro-industrial site) influenced the values of the Cai coefficient (Table 4). The relatively small areal extent of positive LUCs (forest, grasslands, woodland, and shrubs) could not significantly compensate for the intense anthropogenic pressure on the landscape of the studied area. The improvement of the situation in 2019 is mainly related to the decline in crop production on less efficient areas of agricultural land.

3.4. Sustainable Ecotourism of Rural Historical (Čív) Landscape

At present, the village is not a tourist destination, but research results indicate that the cultural landscape around Slovak exclave Čív (historic buildings and the Dorog basin itself) has the potential for development of ecotourism (individual tourism, or tourism for small groups, such as families with children). In terms of the development of recreational activities, the historical landscape has the potential for cultural tourism (folk traditions, municipal museum), hiking, cycling, and agrotourism.

Based on the research of the study area, we can also determine the position of the cultural landscape of the Dorog basin as the primary source for the development of rural cultural tourism, in which ecotourism finds its most significant application.

The cultural landscape of Slovak exclave Čív provides the potential for the development of cultural rural tourism, which can take the form of ethic tourism, religious tourism, adventure tourism, gastronomic tourism, dark tourism, tourism of living history due to its specific landscape elements, and inhabitants (Figure 11).

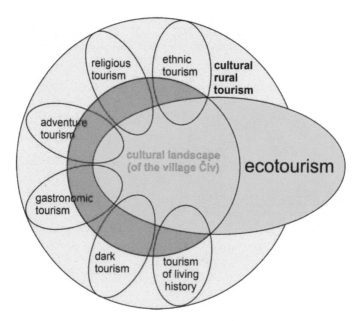

Figure 11. The position of cultural landscape of the village of Čív (in the Dorog Basin) as a primary source of development of individual types of tourism with an emphasis on ecotourism. Source: authors' research

In order for modern ecotourism to be sustainable in a long-term stable cultural rural landscape, it must be formed by closed-cyclical components (criteria) in interaction with the local community, i.e., in the case of our research, by a rural settlement—the municipality of Čív. It must also be environmentally friendly to the surrounding landscape, and it must be in accordance with the protection and preservation of natural and cultural components of the landscape (Figure 12). In order for ecotourism to be sustainable in a historic rural landscape, we must consider them both as interconnected vessels. If the (sustainable) historical rural landscape is stabilized, ecotourism will also be sustainable, and vice versa, if the historic landscape is unsustainable, ecotourism will also be unsustainable.

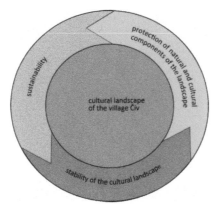

Figure 12. Sustainable ecotourism in the cultural landscape is formed by cyclical components (criteria) in interaction with the local community (rural settlement-municipality Čív). (Modified and supplemented based on the authors' research according to Trifumov, Soim [44]).

In order for the country to maintain its long-term potential for sustainable tourism, it is necessary to manage its current state in terms of proposals that we have compiled on the basis of research results and present them at the end of the paper. The proposals emphasize the four pillars on which sustainable ecotourism (ecological, economic, social, and cultural) is built.

4. Discussion

The historical rural landscape as a source for the development of sustainable ecotourism in the Carpathian Mountains of Central Europe represents a substantial potential for this European region, which has not yet been fully exploited.

Wilderness (natural landscape), which forms the essential space for ecotourism development, no longer exists in Central Europe (Carpathian region). It disappeared due to the influence of human society and anthropogenic transformations, which have acquired an intense character in this area since the Middle Ages. The last written reports about the existence of the primeval forests in the Carpathian Basin are from the turn of the 12th and 13th centuries. It was a relatively sparsely populated region at the beginning of the 13th century, as evidenced by the report preserved in the chronicle called Gesta Hungarorum, but better known as the Anonymous Chronicle, written by an anonymous notary who probably served under King Belo III. (1173–1196). Its content is also a contemporary description of the landscape as it existed in this period in the Carpathian Mountains of Central Europe. The chronicle mentions an extensive royal forest, or a primeval forest [105]. Other documents also confirm the presence of large forested areas in this geographical area at the turn of the 12th and 13th centuries, in which settlements are gradually beginning to densify [106]. During this period, we can look for the emergence of a cultural landscape in the Carpathian Mountains, which currently occupies almost the entire territory. The above-mentioned national parks are also an example.

It is the stable cultural landscape, as a "natural" heritage, that is becoming an increasingly sought-after environment for tourism, or rather ecotourism, in this geographical area. Our research confirms the potential and suitability of a cultural landscape for ecotourism of the Slovak exclave Čív. By comparing selected indicators (pros and cons) influencing the development of ecotourism in a cultural landscape resulting from our research and based on the evaluation of literary sources related to the wilderness (as the primary space for the development of ecotourism), we found that a cultural landscape is more suitable (more attractive) for the development of ecotourism than wilderness in many cases (Table 5). We can point out the characteristics of the landscape, such as its vastness, naturalness, and natural diversity, which significantly dominate in the wilderness. However, on the contrary, the cultural landscape has the advantage for today's ecotourist that it is relatively more comfortable and much faster to access. The remoteness of the wilderness can have a significant negative impact on time and transport, representing an increased financial cost.

Table 5. Basic characteristics of the environment (wilderness-cultural landscape) suitable for the development of ecotourism from the point of view of modern (contemporary) ecotourist (basic specifics resulting from the research of authors in the cultural landscape of the Dorog Basin). The green color indicates better conditions for ecotourism in one environment compared to another.

Wilderness	Cultural Landscape
the vastness of the landscape	the vastness of the landscape
natural landscape	natural landscape
relatively short distance for the client	relatively short distance for the client
natural diversity	natural diversity
cultural diversity	cultural diversity
the originality of local communities	the originality of local communities
health risks	health risks
ambulance availability	ambulance availability
access to the Internet and social networks in terms of security	access to the Internet and social networks in terms of security
elements of adrenaline tourism	elements of adrenaline tourism
the uniqueness of the visual experience	the uniqueness of the visual experience
hiking trails and local roads	hiking trails and local roads
Touristic facilities	Touristic facilities
The need for tourist guides	The need for tourist guides
General knowledge of potential ecotourists about the area	General knowledge of potential ecotourists about the area
Physical fitness of ecotourists	Physical fitness of ecotourists
Estimated financial costs	Estimated financial costs

Cultural diversity, whether of the landscape itself or the cultural (technical) monuments, is significantly more attractive for today's tourists than sterile monotonous (large) cities. The reduction of health risks and the relative availability of health care, which, on the other hand, can reduce the adrenaline attractiveness of this environment, are closely related to the advantages of the cultural landscape mentioned above. Almost complete coverage of areas with an Internet signal in Central Europe's cultural landscape enables fast, high-quality and cheap communication with "civilization". The cultural landscape also provides "comfort" for modern ecotourists with a relatively dense structure of roads, hiking trails, and other necessary infrastructure.

It is also debatable, which will require further research, whether the rural landscape with its communities is attractive enough for ecotourists compared to a natural community outside of civilization in a remote primeval forest-wilderness.

5. Conclusions

The development of selected and researched LUCs and the identified archetypes of the landscape reflects the specifics of the so-called land-use driving forces. The effective use of research results and their transfer into social practice can be realized through cultural and educational activities aimed at supporting the identity of the local population [107,108]. The Minority Slovak local government can saturate this role in cooperation with the municipality's management.

The research results also can address the issues and problems of harmonization and revitalization of the natural and socio-economic subsystems of the local landscape. The methodology used and the results obtained can be applied in basic research in the study of similar areas. Findings on changes in the local landscape, their aspects, and contexts can be used, e.g., to saturate social practice in the creation of integrated landscape management, which we present in the form of framework proposals. Their goal is to harmonize the development of anthropogenic activities while preserving the cultural and historical potential of the studied area and its sustainability through ecotourism. In order for

Sustainability **2020**, *12*, 6048

the rural cultural landscape to maintain agreed stability with the possibility of sustainable use in ecotourism, it would be appropriate to put the following proposals resulting from the research results into practice:

Proposal no. 1: It is necessary to support the biodiversity (as one of the elements of ecotourism) in the basin by increasing the area of permanent grasslands through delimitation of the temporarily grassed arable land, revitalization of wetlands on the floodplain of Čív Brook, and non-forest woody vegetation, to preserve the local landscape's ecological stability.

Proposal no. 2: Ensuring the social and cultural diversity of the local landscape through the restoration and maintenance of the traditional way of life of the "original" Slovak inhabitants of the village. In terms of ecotourism development, it is primarily a matter of gaining an authentic experience of learning about the traditional tangible and intangible culture and local cuisine.

Proposal no. 3: Ensuring the active economic involvement of the local community in the development of ecotourism in the form of offering local services and products (especially the production of traditional souvenirs, preparation of local cuisine, preparation of traditional cultural and social events, and the provision of guide services).

These proposals can stabilize and preserve the heritage of the cultural landscape as the primary source for the development of ecotourism, which would lead to mutual interaction and sustainability not only of the cultural landscape and its tangible and intangible elements itself but also of ecotourism.

Author Contributions: Conceptualization, P.H., B.G. and P.C.; methodology, P.C., P.H. and B.G.; software, M.Ž.; validation, M.Ž., B.G. and P.H.; formal analysis, P.C., B.G. and P.H.; investigation, B.G. and P.H.; resources, P.C., P.H., B.G. and M.Ž.; data curation, P.C., and M.Ž.; writing—original draft preparation, P.C., P.H., B.G. and M.Ž.; writing—review and editing, P.H., B.G. and M.Ž.; visualization, P.C., P.H., B.G. and M.Z.; supervision, P.C., P.H.; project administration, B.G.; funding acquisition, P.C. All authors have read and agree to the published version of the manuscript.

Funding: This research and the APC was funded by SLOVAK RESEARCH AND DEVELOPMENT AGENCY, grant number APVV-18-0196 "Vedomosti Nitrianskej stolice M. Bela (interpretácia a aplikácia)", and grant number APVV-18-0185 "Transformácia využívania kultúrnej krajiny Slovenska a predikcia jej ďalšieho vývoja" and also by CULTURAL AND EDUCATIONAL GRANT AGENCY, grant number 005UCM/4-2019 "Prírodné pomery Nitrianskej stolice v 18. storočí pohľadom M. Bela (vysokoškolská učebnica)".

Conflicts of Interest: The authors declare no conflict of interest. The funders had no role in the design of the study; in the collection, analyses, or interpretation of data; in the writing of the manuscript, or in the decision to publish the results.

References

1. Fescenko, A.; Nikodemus, O.; Brūmelis, G. Past and Contemporary Changes in Forest Cover and Forest Continuity in Relation to Solis (Suthern Latvia). *Pol. J. Ecol.* **2014**, *62*, 625–638. [CrossRef]

2. Olah, B.; Boltižiar, M.; Gallay, I. Transformation of the Slovak cultural landscape since the 18th century and its recent trends. *J. Landsc. Ecol.* **2009**, *2*, 41–55. [CrossRef]

3. Fialová, D.; Chromý, P.; Marada, M. Historicko geografická analýza změn funkčního využití břehů Vltavy (v období od přelomu 18. a 19. století do současnosti). *Hist. Geogr.* **2007**, *34*, 307–317.

4. Trpáková, I. *Krajina ve Světle Starých Pramenů*; Lesnická práce: Praha, Czech Republic, 2013; p. 248.

5. Wang, W.; Zhang, C.H.; Allen, J.; Li, W.; Boyer, M.; Segerson, K.; Silander, J. Analysis and prediction of land use changes related to invasive species and major driving forces in the State of Connecticut. *Land* **2016**, *5*, 25. [CrossRef]

6. Boltižiar, M.; Chrastina, P.; Trojan, J. Vývoj využívania kultúrnej krajiny slovenskej enklávy Šára v Maďarsku (1696-2011). *Geogr. Inf.* **2016**, *20*, 24–37.

7. Chrastina, P. Krajina Veľkého Bánhedeša a jej Premeny. In *Acta Nitriensiae 10*; Gadušová, Z., Ed.; FF UKF: Nitra, Slovakia, 2008; pp. 74–94.

8. Chrastina, P. Pivnica: Krajina—človek—kultúra slovenskej enklávy v srbskej Báčke. In *Svedectvá Slovenského Dolnozemského Bytia: Aspekty zo Slovenskej Dolnozemskej Kultúrnej Histórie a Kultúrnej Antropológie*; Ambruš, I.M., Ed.; Vyd. I. Krasko: Nadlak, Romania, 2012; pp. 187–201.

9. Chrastina, P. Zmeny využívania krajiny Békešskej Čaby. In *Kapitoly z Minulosti a Súčasnosti Slovákov v Békešskej Čabe*; Kmeť, M., Tušková, T., Uhrinová, A., Eds.; Magyarországi Szlovákok Kutatóintézete: Békešská Čaba, Hungary, 2018; pp. 378–401.

10. Chrastina, P.; Boltižiar, M. Butín: Krajina—človek—kultúra slovenskej enklávy v rumunskom Banáte. *Studia Hist. Nitriensia* **2008**, *14*, 165–193.

11. Chrastina, P.; Boltižiar, M. Senváclav: Krajina—človek—kultúra slovenskej enklávy vo Vyšegrádskych vrchoch. *Studia Hist. Nitriensia* **2010**, *15*, 53–86.

12. Chrastina, P.; Trojan, J.; Župčán, L.; Tuska, T.; Hlásznik, P.P. Land use ako nástroj revitalizácie krajiny na príklade slovenskej exklávy Tardoš (Maďarsko). *Geogr. Cassoviensis* **2019**, *13*, 121–140.

13. Chrastina, P.; Trojan, J.; Valášek, P. Historický "land use" Tardoša. In *Slovenské Inšpirácie z Tardoša. 35. Interdisciplinárny výskumný tábor Výskumného ústavu Slovákov v Maďarsku*; Tušková, T., Žiláková, M., Eds.; VÚSM: Békešská Čaba, Hungary, 2018; pp. 65–88.

14. Chrastina, P.; Křováková, K.; Brůna, V. Zmeny krajiny v rumunskom Bihore (na príklade slovenskej enklávy Borumlak a Varzaľ). *Hist. Geogr.* **2007**, *32*, 371–398.

15. Drumm, A.; Moore, A. *Ecotourism Development—A Manual for Conservation Planners and Managers Volume 1: An Introduction to Ecotourism Planning*, 2nd ed.; The Nature Conservancy: Arlington, VA, USA, 2005; p. 96.

16. Ziffer, K. *Ecotourism: The Uneasy Alliance*; Conservation International: Washington, DC, USA, 1989; p. 36.

17. Brandon, K. *Ecotourism and Conservation: A Review of Key Issues*; The World Bank: Washington, DC, USA, 1996; p. 80.

18. Honey, M. *Ecotourism and Sustainable Development. Who owns Paradise?* Island Press: Washington, DC, USA, 1999; p. 405.

19. Weaver, D.B. Magnitude of Ecotourism in Costa Rica and Kenya. *Ann. Tour. Res.* **1999**, *26*, 792–816. [CrossRef]

20. Weaver, D.B. *Ecotourism*; John Wiley & Sons Inc: New York, NY, USA, 2008; p. 360.

21. Kurek, W. *Turystyka*; Wydawnictwo Naukowe PWN: Warszawa, Poland, 2008; p. 540.

22. Rahman, A. Application of GIS in Ecotourism Development: A Case Study in Sundarbans, Bangladesh. Master's Thesis, Mid-Sweden University Master of Arts, Human Geography Focusing on Tourism, Sundsvall, Sweden, 2010; p. 79.

23. The International Ecotourism Society. Available online: https://ecotourism.org/what-is-ecotourism (accessed on 10 May 2020).

24. Global Ecotourism Network. Available online: https://www.globalecotourismnetwork.org (accessed on 10 May 2020).

25. Cambridge Dictionary. Available online: https://dictionary.cambridge.org/ (accessed on 10 May 2020).

26. Sharpley, R. Ecotourism: A Consumption Perspective. *J. Ecotour* **2006**, *5*, 7–22. [CrossRef]

27. Blamey, R.K. *Principles of Ecotourism, the Encyclopedia of Ecotourism*; CABI Publishing: New York, NY, USA, 2001; pp. 5–22.

28. Honey, M. *Ecotourism and Sustainable Development: Who Own Paradise*, 2nd ed.; Island Press: Washington, DC, USA, 2008.

29. Cheia, G. Ecotourism: Definition and concepts. *J. Tour.* **2013**, *15*, 56–60.

30. Fennell, D. A content analysis of ecotourism definitions. *Curr. Issues Tour.* **2001**, *4*, 403–421. [CrossRef]

31. Barkauskiene, K.; Snieska, V. Ecotourism as an Integral Part of Sustainable Tourism Development. *Econ. Manag.* **2013**, *18*, 449–456. [CrossRef]

32. Tsaur, S.; Lin, Y.; Lin, J. Evaluating Ecotourism Sustainability from the Integrated Perspective of Resource, Community and Tourism. *Tour. Manag.* **2006**, *27*, 640–653. [CrossRef]

33. Wall, G. Is ecotourism sustainable? *Environ. Manag.* **1997**, *21*, 483–491. [CrossRef]

34. Kim, M.; Xie, Y.; Cirella, G.T. Sustainable Transformative Economy: Community-Based Ecotourism. *Sustainability* **2019**, *11*, 4977. [CrossRef]

35. Donohoe, H.M. Defining culturally sensitive ecotourism: A Delphi consensus. *Curr. Issues Tour.* **2011**, *14*, 27–45. [CrossRef]

36. Luchman, H. Cultural Landscape Preservation and Ecotourism Development in Blambangan Biosphere Reserve, East Java. In *Landscape Ecology for Sustainable Society*; Hong, S.K., Nakagoshi, N., Eds.; Springer: Cham, Switzerland, 2018; pp. 341–358.

37. Luchman, H.; Kim, J.; Hong, S. Cultural Landscape and Ecotourism in Bali Island, Indonesia. *J. Ecol. Field Biol.* **2009**, *32*, 1–8.

38. Poulot, D. Identity as Self-Discovery: The Ecomuseum in France. In *Museum Culture: Histories, Discourses, Spectacles*; Sherman, D., Rogoff, I., Eds.; Routledge: London, UK, 1994; pp. 66–84.

39. Schlichtherle, H. *'Der archäologische Moorlehrpfad im Südlichen Federseeried', in Urgeschichte Erleben Fährer zum Federseemuseum mit Archäologischem Freigelände und Moorlehrpfad*; Verlagsbüro Wais & Partner: Stuttgart, Germany, 2008; pp. 47–71.

40. Schlichtherle, H.; Strobel, M. *Archaeology and Protection of Nature in the Federsee Bog*; Landesdenkmalamt Baden-Württemberg: Stuttgart, Germany, 1999.

41. Russell, A. Sustaining Reflections on the Saimaa Lakes. *Muut. Matk.* **2001**, *2*, 60–63.

42. Wallace, G. Getting a Buzz from Winter Tourism on the Saimaa Lakes. *Muut. Matk.* **2002**, *2*, 68–71.

43. Kranjčević, J.; Šaban, S. Tourism in the cultural landscape. An attempt to create new value in the example of the Medvednica nature park, Croatia. In *Tourism—Current and Future Challenges for Urban Development*; Institut za turizam: Zagreb, Croatia, 2009; p. 13.

44. Trofimov, V.; Soimu, O. Ecotourism concept in the light of cultural diversity and regional development. *Agric. Econ. Rural Dev.* **2011**, *8*, 117–125.

45. Dzhandzhugazova, E.A.; Ilina, E.L.; Latkin, A.N.; Davydovich, A.R.; Valedinskaya, E.N. Ecotourism Programs in the Context of the Perception of Natural and Cultural Landscapes (on the Example of the Kizhi Museum Reserve). *Ekoloji* **2018**, *27*, 377–382.

46. Campbell, L. Ecotourism in rural developing communities. *Ann. Tour. Res.* **1999**, *26*, 534–553. [CrossRef]

47. Frochot, I. A benefit segmentation of tourists in rural areas: A Scottish perspective. *Tour. Manag.* **2005**, *26*, 335–346. [CrossRef]

48. Kiper, T.; Özdemir, G.; Sağlam, C. Enviromental, socio-cultural and economical effects of ecotourism perceived by the local people in the northwestern Turkey: Kiyiköy case. *Sci. Res. Essays* **2011**, *6*, 4009–4020.

49. Okan, T.; Köse, N.; Arifoğlu, E.; Köse, C. Assessing Ecotourism Potential of Traditional Wooden Architecture in Rural Areas: The Case of Papart Valley. *Sustainability* **2016**, *8*, 974. [CrossRef]

50. Bunruamkaew, K.; Murayama, Y. Land Use and Natural Resources Planning for Sustainable Ecotourism Using GIS in Surat Thani, Thailand. *Sustainability* **2012**, *4*, 412–429. [CrossRef]

51. Clifton, j.; Benson, A. Planning for Sustainable Ecotourism: The Case for Research Ecotourism in Developing Country Destinations. *J. Sustain. Tour.* **2006**, *14*, 238–254. [CrossRef]

52. Semple, W. Traditional architecture in Tibet: Linking issues of environmental and cultural sustainability. *Mt. Res. Dev.* **2005**, *25*, 15–19. [CrossRef]

53. Angelica, M.; Zambrano, A.; Broadbent, E.N.; Durham, W.H. Social and environmental effects of ecotourism in the Osa Peninsula of Costa Rica: The Lapa Rios case. *J. Ecotourism* **2010**, *9*, 62–83.

54. Pénzes, E. *Országos ökoturizmus Fejlesztési Stratégia*; Pannon Egyetem Turizmus Tanszék: Budapest, Hungary, 2008; p. 170.

55. Michalkó, G. *Turizmológia: Elméleti alapok*; Akadémiai Kiadó: Budapest, Hungary, 2012; 266p.

56. Matlovičová, K.; Klamár, R.; Mika, M. *Turistika a Jej Formy*; Fakulta humanitných a prírodných vied PU v Prešove: Prešov, Slovakia, 2015; p. 549.

57. Gregorová, B. The issue of pilgrimage tourism from the point of view of geography. *Acta Geoturistica* **2019**, *10*, 1–9.

58. Ryglová, K. *Cestovní Ruch*; KEY Publishing s.r.o.: Ostrava, Česká Republika, 2009; p. 187.

59. *Zaręba, Dominika 2010: Ekoturystyka*; Wydawnictwo Naukowe PWN: Warszawa, Poland, 2010; p. 180.

60. Kolejka, J. Czech experience with land use and land cover change research. In *Land Use/Land Cover Changes in the Period of Globalization: Proceedings of the IGU-LUCC International Conference*; Bičík, I., Chromý, P., Jančák, V., Janů, H., Eds.; Charles University: Prague, Czech Republic, 2002; pp. 144–152.

61. Gobin, A.; Campling, P.; Feyen, J. Logistic modelling to derive agricultural land use determinants: Case study from southeastern Nigeria. *Agric. Ecosyst. Environ.* **2002**, *89*, 213–228. [CrossRef]

62. Kuplich, T.M.; Freitas, C.C.; Soares, J.V. The study of ERS-1 SAR and Landsat TM synergism for land use classification. *Int. J. Remote Sens.* **2000**, *21*, 2101–2111. [CrossRef]

63. Bezák, P.; Izakovičová, Z.; Miklós, L. *Reprezentatívne Typy Krajiny Slovenska*; Ústav krajinnej Ekologié Slovenskej akadémie Vied: Bratislava, Slovakia, 2010.

64. Köhler, R.; Olschofsky, K.; Gerard, F. *Land Cover Change in Europe from the 1950'ies to 2000*; Institute for Worldforestry, University of Hamburg: Hamburg, Germany, 2004.
65. Frisnyák, S. Magyarország tájai. In *Magyarország Földrajza*; Frisnyák, S., Ed.; Tankönyvkiadó: Budapest, Hungary, 1988; pp. 145–150.
66. Süli-Zakar, I. Dunántúli-közephegység. Dunazug-hegyvidék. In *Magyarország földrajza*; Frisnyák, S., Ed.; Tankönyvkiadó: Budapest, Hungary, 1988; pp. 238–243.
67. Šomšák, L. *Flóra a Fauna v Rastlinných Spoločenstvách Strednej Európy (Aplikovaná Biocenológia)*; PríF UK: Bratislava, Slovakia, 1998; p. 308.
68. Dražil, T. Teplomilné ponticko-panónske dubové lesy na spraši a piesku. In *Katalóg biotopov Slovenska*; Stanová, V., Valachovič, M., Eds.; Daphne: Bratislava, Slovakia, 2002; pp. 93–94.
69. Krištofičová, E. *1997 Prostriedky Hodnotenia Knižničných a Vedeckoinformačných Procesov*; CVTI: Bratislava, Slovakia, 1997; p. 157.
70. Carrizo-Sainero, G. Toward a Concept of Bibliometrics. *J. Span. Res. Inf. Sci.* **2000**, *1*, 1–6.
71. Ondrišová, M. *Bibliometria*; STIMUL: Bratislava, Slovakia, 2011; p. 134.
72. Hartshorne, R. The Concept of Geography as a Science of Space, from Kant and Humboldt to Hettner. *Ann. Assoc. Am. Geogr.* **1958**, *48*, 97–108. [CrossRef]
73. Demek, J. *Úvod do štúdia Teoretickej Geografie*; SPN: Bratislava, Slovakia, 1987; p. 241.
74. Ivanička, K. *Základy Teórie a Metodológie Socioekonomickej Geografie*; SPN: Bratislava, Slovakia, 1983; p. 432.
75. Butlin, R.A.; Dodgshon, R.A. *An Historical Geography of Europe*; Claredon Press: Oxford, UK, 1998; p. 373.
76. Chrastina, P. Krajina ako jeden zo styčných fenoménov prírodných a spoločenských vied. *Acta Hist. Nitriensia* **2001**, *4*, 333–344.
77. Chrastina, P. *Vývoj Využívania Krajiny Trenčlanskej Kotliny a jej Horskej Obruby*; UKF: Nitra, Slovakia, 2009; p. 285.
78. Rábik, V.; Labanc, P.; Tibenský, M. *Historická Geografia*; Filozofická fakulta Trnavskej univerzity v Trnave: Trnava, Slovakias, 2013; p. 82.
79. Semotanová, E. *Historická Geografie českých zemí*; Historický ústav AV ČR: Praha, Czech Republic, 2002; p. 279.
80. Chrastina, P.; Boltižiar, M. Historicko-kultúrn Ogeografické črty obcí Čápar, Čerňa a Jášť. In *Kultúrne Tradície Slovákov v Oblasti Bakonského Lesa*; Šusteková, I., Ed.; FF UKF: Nitra, Slovakia, 2008; pp. 7–33.
81. Mapire. Available online: https://mapire.eu/en/ (accessed on 15 May 2020).
82. Google Earth. Available online: https://earth.google.com/web/ (accessed on 1 May 2020).
83. Timár, G.; Lévai, P.; Molnár, G.; Varga, J. A második világháború német katonai térképeinek koordinátarendszere. *Geodézia És Kartográfia* **2004**, *56*, 25–35.
84. Timár, G.; Molnár, G.; Székely, B.; Biszak, S.; Varga, J.; Jankó, A. *Digitized Maps of the Habsburg Empire—The Map Sheets of the Second Military Survey and Their Georeferenced Version*; Arcanum: Budapest, Hungary, 2006; p. 59.
85. Timár, G.; Biszak, S. Digitizing and georeferencing of the historical cadastral maps (1856–60) of Hungary. In Proceedings of the 5th International Workshop on Digital Approaches in Cartographic Heritage, Vienna, Austria, 22–24 February 2010.
86. Blišťan, P. Geographic information system for major mining area Dubník-opalmines. In *Proceedings of the SGEM 2013: 13th International Multidisciplinary Scientific GeoConference: Informatics, Geoinformatics and Remote Sensing*; STEF92 Technology Ltd.: Albena, Bulgaria, 2013; pp. 729–736.
87. Blišťan, P. Some possibilities of using geographic information systems in analysis of the potential of destination Slovensky raj (Slovakia) in tourism. In *Advances and trends in geodesy, cartography and Geoinformatics: Proceedings of the 10 International Scientific and Professional Conference on Geodesy, Cartography and Geoinformatics*; CRC Press: Leiden, The Netherland, 2018; pp. 147–152.
88. Blišťan, P.; Šoltésová, M.; Kršák, B.; Sidor, C.; Štrba, Ľ. Modelling of geological phenomena in GIS (Geographical information system). *Metalurgija* **2008**, *47*, 278.
89. Mihalovič, A. *Zemepisné Mená Čívu (Jazykovedná štúdia)*; Tankönyvkiadó: Budapest, Hungary, 1987; 147p.
90. Osváth, A. *Komárom és Esztergom KözígazgatáSilag Egyelőre Egyesített Vármegyék Multja és Jelene*; A Magyar Vármegyék és Városok Multja és Jelene: Budapest, Hungary, 1938; p. 936.
91. Reiszig, E. Esztergom vármegyei köszégei. In *Esztergom Vármegye (Magyarország Vármegyéi és Városai)*; Borovszky, S., Ed.; Országos Monografia Társaság: Budapest, Hungary, 1908; pp. 6–39.
92. Csombor, E. Piliscsév története. In *300 rokov v Piliši: Štúdie z Minulosti a Prítomnosti Čívu. (300 év a Pilisben: Tanulmányok Piliscsév múltjából és Jelenéből*; Divičanová, A., Ed.; Piliscsévi Szlovák Kisebbségi Önkormányzat: Čív-Piliscsév, Hungary, 2002; pp. 15–29.

93. Fügedi, E. Príspevky k dejinám osídlenia niektorých slovenských obcí na území dnešného Maďarska: Čív (Csév). In *Atlas Slovenských nárečí v Maďarsku: Atlas der Slowakischen Mundarten in Ungarn*; Király, P., Ed.; VÚSM: Budapest, Hungary, 1993; p. 77.

94. Udvari, I. *A Mária Terézia-Féle úrbérrendezés Szlovák nyelvű Dokumentumai: Slovenské Dokumenty Urbárskej Regulácie Márie Terézie. Adatok a Szlovák nép Gazdaság- és Társadalomtörténetéhez (Szepességi Ruszin Falvak Népélete Mária Terézia Korában)*; Bessenyei György Kiadó: Nyíregyháza, Hungary, 1991; p. 370.

95. Bičík, I.; Jeleček, L.; Kabrda, J.; Kupková, L.; Lipský, Z.; Mareš, P.; Šefrna, L.; Winklerová, J. *Vývoj využití ploch v Česku*; Edice Geographica: Praha, Czech Republic, 2010; p. 251.

96. Mather, A. The reversal of land-use trends: The beginning of the reforestation of Europe. In *Land Use/Land Cover Changes in the Period of Globalization: Proceedings of the IGU-LUCC International Conference*; Bičík, I., Chromý, P., Jančák, V., Janů, H., Eds.; Charles University: Prague, Czech Republic, 2002; pp. 23–30.

97. Lettner, C.H.; Wrbka, T. Historical development of the cultural landscape at the nothern border of the Eastern Alps: General trends and regional peculiarities. In *Proceedings: Workshop on Landscape History*; Balász, P., Konkoly-Gyuró, E., Eds.; University of West Hungary Press: Sopron, Hungary, 2011; pp. 109–121.

98. Skokanová, H. Long-term changes in the landscape structure in three border areas, of the Czech Republic. In *Proceedings: Workshop on Landscape History*; Balász, P., Konkoly-Gyuró, E., Eds.; University of West Hungary Press: Sopron, Hungary, 2011; pp. 152–160.

99. Krausmann, F.; Naberl, H.; Schulz, N.B.; Erb, K.; Darge, E.; Gaube, V. Land-use change and socioeconomic metabolism in Austria, Part I: Driving forces of land-use change: 1950–1995. *Land Use Policy* **2003**, *20*, 1–20. [CrossRef]

100. Hanuškin, J.; Lacika, J. Vybrané environmentálne súvislosti zmien historickej laznickej krajiny (na príklade obce Hrušov okres Veľký Krtíš). *Geogr. Časopis* **2018**, *70*, 57–77.

101. Trojan, J.; Chrastina, P.; Boltižiar, M. Case study area Veľký Bánhedeš/Nagybánhegyes: Land use changes from the past to 2015. In *Land Use/Cover Changes in Selected Regions in the World*; International Geographical Union: Asahikawa, Japan, 2018; pp. 53–60.

102. Kupková, L. Data o krajině včera a dnes. 160 let v tváři české kulturní krajiny. *GEOInfo* **2001**, *7*, 16–19.

103. Žigrai, F. Forming of the Cultural Landscape of Liptov in the Past and Today. *Acta Geogr. Univ. Comen. Econ.-Geogr.* **1971**, *10*, 137–155.

104. Auerhan, J. *Čechoslováci v Jugoslávii, v Rumunsku, v Maďarsku a v Bulharsku*; Melantrich: Praha, Czech Republic, 1921; p. 203.

105. Múcska, V. *Kronika Anonymného Notára Kráľa Bela: Gesta Hungarorum*; Vydavateľstvo Rak: Budmerice, Czech Republic, 2000; p. 159.

106. Musil, F. Gesta Hungarprum a historicko-zemepisný obraz Slovenska. *Hist. Časopis* **2004**, *52*, 433–450.

107. Králik, R.; Lenovský, L.; Pavlíková, M. A few comments on identity and culture of one ethnic minority in central Europe. *Eur. J. Sci. Theol.* **2018**, *14*, 63–76.

108. Lenovský, L. Identity as an Instrument for Interpreting the Socio-cultural Reality. *Eur. J. Sci. Theol.* **2015**, *11*, 171–184.

Article

The Analysis of Urban Fluvial Landscapes in the Centre of Spain, Their Characterization, Values and Interventions

Pedro Molina Holgado [1], Lara Jendrzyczkowski Rieth [2], Ana-Belén Berrocal Menárguez [3,*] and Fernando Allende Álvarez [1]

[1] Departamento de Geografía, Universidad Autónoma de Madrid, 28049 Madrid, Spain;
pedro.molina@uam.es (P.M.H.); fernando.allende@uam.es (F.A.Á.)

[2] Programa de Pos Graduação em Engenharia Civil: Construção e Infraestrutura, Universidade Federal do Rio Grande do Sul, Porto Alegre-RS 90040-060, Brazil; larart.arq@gmail.com

[3] Departamento de Ingeniería del Transporte, Territorio y Urbanismo, ETSI Caminos, Canales y Puertos, Universidad Politécnica de Madrid, 28040 Madrid, Spain

* Correspondence: anabelen.berrocal@upm.es

Received: 5 May 2020; Accepted: 29 May 2020; Published: 7 June 2020

Abstract: River areas are undoubtedly among the most valuable territorial areas in Europe, not only in terms of their eco-landscape and use but also, culturally. However, there is currently a sharp reduction in the extension and increase of deterioration of riverbanks around the world. A substantial part of losses and deterioration are associated with the artificialization of the territories, derived mainly from a less than respectful urbanization around these landscapes. Urban and peri-urban riverbanks are landscapes in expansion due to the continuous growth of built-up spaces. Therefore, they should be areas of preferential consideration, especially in territories with a marked tendency to dryness, like the centre of the Iberian Peninsula. This article aims to contribute to our understanding of these spaces through the study of four distinct cases in the centre of the peninsula, in particular: the river Manzanares running through the city of Madrid, the river Tagus in Toledo and running through Talavera de la Reina, and the river Henares in Guadalajara. Three of the four urban water courses analyzed are zones of special interest for waterfowl: they sustain a winter population that varies between 745 and 1529 birds and they provide a home to some globally threatened species. The density of the riparian birds is also very high during winter, these values oscillating between a mean of 141.16 and 240.12 birds/10 ha. It should be noted that the diversity of this group of birds in the four regions studies is also high (H > 2.4 nats). The article also examines the interventions and the urban planning criteria applied to these urban and peri-urban river spaces, inferring the need to reassess urban planning in river areas to ensure it is compatible with their operation, values and possible uses.

Keywords: urban rivers; urban planning; cultural landscape; natural landscape; waterfowl; riparian birds

1. Introduction

River landscapes are among those most affected by human intervention around the world [1–4], especially in historically longstanding urban areas where their alteration dates back many centuries [5]. All over the planet, rivers and their associated landscapes have been affected by intense fragmentation processes, which cause significant simplification and a loss of connectivity. This circumstance limits the river's ability to flow uninterrupted and it affects many of their fundamental processes and functions. Accordingly, it produces a rapid decline in biodiversity [6] and in essential ecosystem services [7]. Moreover, the occupation of areas near riverbanks by infrastructures often intensifies these processes [8].

River areas are very valuable areas in terms of their use and ecology, as well as culturally [9–12]. Indeed, riverbanks are considered to be the most diverse, dynamic, and complex terrestrial habitats on earth [13], and to be true accumulators of history. Like all landscapes, riverbanks are laden with objective and subjective value, derived from the elements they contain and from the cultural references they have accumulated [14]. These values are typical of the most natural spaces that are well-preserved and only sporadically affected by human pressure. However, they are also characteristic of other less well valued river areas, such as those located in areas that have received intense intervention and that have been modified most significantly by human activity, the greatest example of which are probably urban areas.

Urban river spaces, like other green areas and free spaces within cities, sustain species of fauna and flora typical of non-urbanized environments [15], sometimes even rare or globally threatened species [16]. The presence of these elements often reflects the complexity of these spaces, as well as bearing witness to the persistence of natural remnants [17–19] and the density of the vegetation [20]. In terms of biodiversity, riverbanks are areas of particular interest, although except for some groups of vertebrates like birds [21–24], their potential has not been widely explored [25,26]. However, river ecosystems are also very vulnerable to invasion by non-indigenous species [27], the presence of such exotic species generating problems for the conservation of biodiversity [28].

Although rivers and their banks are areas of special value, their proximity or position within the built-up space generates multiple problems and tensions [29], which are generally associated with a reduction in their area and the loss of specific values (environmental, territorial, landscape, and cultural). In particular, the impact of human activities produces a significant homogenization and a reduction of riparian vegetation [30], as well as reducing their possibilities to serve as spaces for public use. Their conservation is justified for diverse reasons, including the preservation of local biodiversity, the creation of transitional territories and areas of extra-urban habitats, or to understand and facilitate responses to climate change [31].

The marked reduction in the extension of riverbanks is a widespread global process and it is evidence of their deterioration. There have been very intense losses of this type of landscape in Europe, with an estimated decrease of 12% (7508 ha) during the period 1987–2000. In Spain, this has affected more than 2 million hectares, which represents approximately 4% of the Spanish territory [32]. A substantial part of these losses are associated with the artificialisation of these territories, mainly derived from disrespectful urbanization of these landscapes [33]. In this sense, urbanization currently represents the most relevant territorial action on the planet. In Europe, where 74% of the population lives in urban areas [34], the extent of urbanization is quite dominant. Indeed, Spain has experienced significant urban concentration in the 20th century [35], generating extensive and dense metropolitan areas such as the administrative region of Madrid. This metropolitan area is the third largest in the European Union in terms of the number of inhabitants, after Paris and London [36,37].

The continuous expansion of urban areas ultimately increases the extension of urban riverbanks. These spaces are in many cases areas of risk that are difficult to integrate into the urban fabric due to their configuration, characteristics, and hydrological and geomorphological nature. They are also commonly areas of marginal use, often occupied by less-favoured social groups, living in substandard houses on the river banks, and they are essentially residual spaces in part due to the threat and risk posed by periods when the rivers are in full flow. In Spain, they have been, and still are, territories separated from urban spaces in our minds, often existing at the limits of those urban spaces and with no specific public use. Thus, they are poorly integrated in the best-case scenario, and often completely abandoned. These circumstances have led to the non-formal occupation of flood plains in urban and peri-urban areas [38], causing a marginal and often private use of these public spaces. In Spain, the limited connection between the cities that border the rivers that flow through them has, in most cases, generated strong social detachment [39]. This phenomenon is almost certainly associated with the perceived risk traditionally posed by proximity to river courses, which has traditionally manifested itself in neglect and degradation of the fluvial territory, and its poor integration into the urban fabric. To mitigate the impact of this situation, it would be useful to integrate societal considerations into the management models of these spaces, as occurs in some river reserves in Europe and the United States [40].

The tensions and threats that affect these spaces are also related to the profound alteration of river systems, most notably, the poor water quality, the loss of water resources, the morphological alteration of rivers, and modifications to the hydrological and hydraulic characteristics of the river course. Water is the driving force upon which the operation of river systems depends, and the changes to the hydrological and hydraulic behaviour of river courses in central Spain have been quite intense [41]. One example is the River Tagus that passes through the city of Toledo, where the average annual flow contribution at the Toledo gauging station fails to reach 40% of the estimated contribution in a natural regime (3256.93×10^6 hm^3) [32], a value that decreases to 29.1% in the latest available data for the hydrological year 2008–2009 [42].

The reduced flood risk in the centre of Spain has favored occupation of the flood plains associated with river courses in urban areas, as well as the alteration of canals. In general, the environmental and landscape value of these areas has generally diminished [43]. However, river spaces are also areas of clear opportunity that must be properly considered, understood, and managed in urban and territorial planning, both locally and regionally. This is particularly relevant in terms of public use, intervention, landscape recovery, and conservation of environmental diversity. In Spain, fluvial territories in general, and the urban river territory in particular, are largely public land integrated into the Public Hydrological Domain (which governs the public water supply).

Despite the intense occupation of these areas, there are often semi-natural remnants lying between infrastructures that are little occupied. Riverbanks are also key areas for the conservation of environmental and landscape diversity, as indicated by the European Biodiversity Strategy [44] or the European Landscape Convention [45], with significant environmental value and potential and, in general, with a strong potential for regeneration, even in the case of heavily intervened banks. They are undoubtedly key spaces for the recovery of ecological, landscape, and territorial connectivity, in the sense specified in the EU Strategy on Green Infrastructure [46], and they are easy to recover through low-cost interventions. In Spain, urban riverbanks, like all riverbanks, fall under a specific regulatory framework for water courses and fluvial territories [47], which clearly provides legal protection for many actions, especially those carried out under the governance of the Public Hydrological Domain.

2. Materials and Methods

This study analyzes and assesses the treatment given to stretches of four Spanish rivers in four projects that specifically centred on those fluvial territories. An environmental characterization of the study areas was initially performed, considering the value of each area for waterfowl and riparian birds during the winter 2019–2020, and also the general structure of the landscape and the presence of habitats included in Annex I of the Directive 92/43. This characterization allows us to assess the impact of the projects undertaken by applying objective criteria, and to estimate their potential impact while they are still in the design phase.

2.1. Study Area

The four stretches of urban river (Figure 1, Table 1) selected for study were situated in the Tagus Hydrographic Basin on the banks of the rivers Manzanares, Henares, and Tagus, lying in the provinces of Madrid and Castilla-La Mancha (Central Spain). These four river sections were chosen based on our interest in analyzing comparable urban river spaces in similar or assimilable landscape contexts, which are also representative of the characteristic of the urban river landscapes in central Spain (Figure 2). All the cases analyzed are situated in the same climatic (Continental Mediterranean) [48] and morphostructural context (Tertiary Sedimentary Tagus Basin or Madrid Basin) [49], passing through large (Madrid) or intermediate cities (Guadalajara, Toledo, and Talavera) that are representative of the different urban and demographic configurations and dynamics in Central Spain [50]. The rivers analyzed are courses of different hydrological entities that are subject to disparate regulation at their headwaters, as already indicated, and all are integrated into the Tagus Hydrographic Basin.

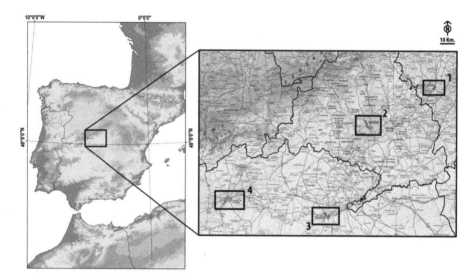

Figure 1. Location of study areas. 1. Henares (Guadalajara); 2. Manzanares (Madrid); 3. Tagus (Toledo); and 4. Tagus (Talavera de la Reina).

Table 1. Basic characteristics of the river stretches analyzed. Source: demographic information [51], hydrological data [43].

River	Length of Stretch Analyzed (km)	City	Population 2019	Region	Annual 10^6 m^3 (1987–2008)
Manzanares	6.86	Madrid	3,266,126	Madrid	66.92
Henares	1.47	Guadalajara	85,871	Castilla-La Mancha	291.13
Tagus (Toledo)	4.56	Talavera	84,417		1338.26
Tagus (Talavera)	6.36	Toledo	84,873		1126.05

Figure 2. The rivers analyzed. Upper left, (**a**) *Populus alba* forest on the Henares river (5/12/2017). Upper right (**b**), the Manzanares river channel at the Segovia Bridge (14/1/2020). Lower left (**c**), the Tagus river in Toledo at the level of the "Artilleros" dam (16/1/2018). Lower right (**d**), a mixed forest of *Populus alba* and *Salix alba* on the Tagus river in Talavera de la Reina (25/2/2020).

2.1.1. River Manzanares in Madrid

The stretch of the river Manzanares chosen for study was a 6.86 km urban section channelled over 5.57 km and limited by a breakwater over 1.29 km after intense intervention from 1946 [52]. Associated with this canalisation, 9 small dams were built that transformed the river into a succession of bound water canals, thereby eliminating sections of rapid-flow. After the opening of the dams, the canalisation of the river course has become the main reason for the loss of its natural aspect, as well as provoking alterations to its hydrological and hydraulic characteristics. The closest gauging station is located 7 km upstream from the starting point of the river stretch studied, where an average annual contribution of 66.92×10^6 m^3 was recorded (1987–2008). The last gauging station is situated 19.6 km downstream of the southern limit of the study area, where the Manzanares reaches 410.12×10^6 m^3 after receiving only 4.9×10^6 m^3 from the "Culebro" stream or arroyo. This increase, six times higher than the contribution of the river in the city of Madrid, is due to water discharge from the city and the sanitation system in its immediate metropolitan area. In the section studied, the river lacks forest and riparian vegetation on its banks as it borders urbanized or landscaped areas. There are also nineteen footbridges and four historical bridges: the Queen Victoria (1908–1909), King's (1815–1816), Segovia (1574–1577), and Toledo

(1715–1722) bridges [53]. After the completion of the "Madrid Río" landscaping project, the whole section became an area of intense recreational use.

2.1.2. River Henares in Guadalajara

A 1.47 km urban stretch of the river Henares in the city of Guadalajara was studied. This section was very well preserved until 2019, protected on its entire left bank by a continuous embankment that separates the free zone of the river from the urban territory. Both riverbanks are occupied by an authentic *Populus alba* forest, a habitat included in Annex I of Directive 92/43/EC. There are no structures spanning this section, which today is an urban park. Unlike the section studied, the river sections located outside the limits of the Guadalajara municipality, upstream and downstream, are included in the ZEC Ribera de Henares ES4240003 [54], all of which share similar values. The average annual contribution at the closest gauging station reaches 29.12×10^6 m^3.

2.1.3. The River Tagus in Toledo

The urban river section studied was a 6.36 km stretch, open in sedimentary materials, and partly embedded in granitoids, leukogranites, and pegmatites (Carboniferous-Paleozoic) [55]. The entire riverbank maintains a narrow and discontinuous canopy of *Populus alba*, *Populus nigra*, *Salix alba*, *Tamarix africana*, and *Tamarix canariensis*, a habitat included in Annex I of Directive 92/43/EC, as well as dense masses of marsh vegetation (*Phragmites australis* and *Typha domingensis*). There are remains of 14 water mills preserved along this section, some originating from the 9th century [56], and seven weirs that reduce the course of the river to a slow flow. There are river islands of great interest downstream of three of the weirs, one a refuge to an interesting heron colony (*Ardea cinerea*, *Bubulbus ibis*, and *Nycticorax nycticorax*). There are five bridges along this stretch of the river, including the historical bridge of "Alcántara" (of Roman origin but rebuilt in the 10th century) and the bridge of San Martín, for which an exact date of construction has not been established but that was cited as long ago as 1165 [57]. The average annual hydrological contribution in the section is 1126.05×10^6 m^3 for the period 1987–2008. The entire stretch analyzed is a space used intensely by the local population for recreational activities and also, by a significant number of visitors since Toledo is a World Heritage City [58] (913,796 overnight stays in 2019) [59].

2.1.4. The River Tagus in Talavera de la Reina

The section studied here is a 4.56 km urban and peri-urban stretch that houses four weirs, which reduce the flow of the river to predominantly slow. The urban façade occupies 2.95 km of the right bank and 565.2 m of the left bank. There is intense recreational activity on both margins, a historic pedestrian bridge (Puente Viejo "the Old Bridge", XIII century) [60], and three bridges for vehicles along this section. The section houses two large islands (47.65 ha, 66.47 ha), on which 35% of their surface area is covered by alluvial forests and mosaic forests, dominated by *Populus alba*, *Populus nigra*, *Salix alba*, *Tamarix africana*, and *Tamarix canariensis*, and structured into three large groves (18.34 ha, 14.98 ha, and 6.8 ha). The remaining 65% (74 ha) is covered by open woodland (tree density ≤25%), pastures of diverse nature, and masses of marsh vegetation. The banks of the bankfull channel maintain a meagre arboreal–arborescent canopy made up of the aforementioned species.

2.2. Data Collection from the Natural Environment

The analysis of the natural environment provides data from the wintering waterfowl and forest–marsh communities, as well as the basic structure of the vegetation. Monthly censuses were carried out between December 2019 and February 2020 to study the bird communities. The censuses of forest–marsh birds was carried out using "census itineraries" [61], with a main counting strip of 25 + 25 m, the results of which are expressed as global numbers (abundance, a), number of species (richness, r), and birds/10 ha (density, d). Waterfowl censuses are global counts performed in accordance with the "field protocol for water bird counting" [62], expressed as the global figures (abundance a: total number of waterfowl; richness r: number of species) or as birds per km of river (density d: birds/km). The length of the stretches examined did not extend over the entire river sections analyzed, except for the river Manzanares, whereas the waterfowl censuses did cover the entirety of the sections (Table 2). The diversity of the communities is calculated according to the Shannon index [63] (H), using a Napierian logarithm (expressed in nats). The values of abundance and density in the different zones were compared using a Kruskall–Wallis Test (K). To only compare two series of values, a Kolmogorov–Smirnov (D) Test was used in the case of independent samples and a Student *t*-test (t) for paired simples. The Statgraphics Centurion 18© software and Microsoft Excel 2016© were used to manage the data and to perform the aforementioned tests.

Table 2. Characteristics of the census sections.

River	Length of the Waterfowl Section Analyzed (km)	Length of Riparian Bird Section Analyzed (km)	Riparian Bird Area Analysis (ha)
Manzanares	6.86	6.86	34.31
Henares	1.47	1.47	7.35
Tagus (Toledo)	6.36	4.23	21.15
Tagus (Talavera)	4.56	1.18	5.90
Total	19.25	13.74	68.71

For each species, its Status and SPEC Category in Europe is indicated, and the categories recognized in the study area were [64]:

European Population Status Category [65]:

- Vulnerable (VU), European population meets any of the IUCN Red List criteria for VU.
- Near Threatened (NT), European population close to meeting the IUCN Red List criteria for VU.
- Declining, European population has declined by ≥20% since the 1970s (when the Birds in Europe series began), and has continued to decline since 2001.
- SPEC Category (Species of European Conservation Concern):
- SPEC 1, European species of global conservation concern, i.e., classified as Critically Endangered, Endangered, Vulnerable, or Near Threatened at global level [66].
- SPEC 2, Species whose global population is concentrated in Europe, and which is classified as Regionally Extinct, Critically Endangered, Endangered, Vulnerable, Near Threatened, Declining, Depleted, or Rare at European level [67].
- SPEC 3, Species whose global population is not concentrated in Europe, but which is classified as Regionally Extinct, Critically Endangered, Endangered, Vulnerable, Near Threatened, Declining, Depleted, or Rare at European level [67].

The species included in Annex I of the Directive 2009/147/EC (Birds Directive) are also cited [68], "Those that shall be the subject of special conservation measures concerning their habitat in order to ensure their survival and reproduction in their area of distribution." (Article 1.4). To determine the structure of the vegetation and the landscapes, a basic analysis was carried out to identify 16 categories.

- Water areas.

  - Sandbanks and water cover.
  - Rockfill/sandbanks with scattered trees.
  - Marsh vegetation: dense masses of *Typha domingensis*, *Phragmites australis*, at times with *Schoenoplectus lacustris*, often accompanied by hygrithrophilous species (*Rumex cristatus*, *Rumex crispus*, *Polygonum persicaria*, and *Polygonum lapathifolium*).
  - Sandbanks with marsh vegetation.
  - Wooded areas: arboreal–arborescent groves and canopies with at least 40% coverage, dominated by *Populus alba*, *Populus nigra*, *Salix alba*, and *Tamarix canariensis*, sometimes with

      Tamarix africana, Fraxinus angustifolia, Salix purpurea, and *Salix salviifolia.* Allochtonous or hybrid elements often appear (*Platanus × hybrida, Robinia pseudoacacia, Gleditsia triacanthos, Ailanthus altissima, Ulmus pumila, Populus × canadensis,* and *Eleagnus angustifolia*).

- Wastelands: areas occupied by herbaceous vegetation, from hygrophilous to mesophilic, sometimes nitrophilous, with or without wooded areas (coverage ≤10%). Among many other communities they include grassland-rushes of *Scirpoides holoschoenus,* pastures of *Brachypodium phoenicoides-Elymus pungens,* hygrithrophilous pastures (*Rumex cristatus, Rumex crispus, Polygonum persicaria,* and *Polygonum lapathifolium*), and many ruderal communities dominated by Gramineae, Brassicaceae, or Asteraceae.
- Rocky areas.
- Mosaic of wooded areas and marsh vegetation.
- Mosaic of wooded areas and wasteland: a mix of dense wooded areas and wastelands.
- Mosaic of scattered trees and wasteland: mixture of low coverage wooded areas (40–10%) and wastelands.
- Mosaic of scattered trees and marsh vegetation: mixture of low coverage wooded areas (40–10%) and marsh vegetation.
- Mosaic of scattered trees and rocky areas.
- Garden areas.
- Artificial land.

- Gravel pits: (extraction of aggregates and sand).

      These areas were defined by a buffer of 200 meters on each side of the central axis of the channel, which in the case of a canalized channel was adjusted to the canalised banks (i.e., river Manzanares). Within this buffer the land use was photointerpreted, generally involving the area of the floodplain nearest to the main channel. This process was carried out on the 1:5000 orthophotographs obtained as part of the Spanish National Plan of Aerial Orthophotography by the National Geographic Institute, available in ECW file format from 2012. The information was coded on the basis of the estimated proportion of the covered space dedicated to each use, differentiating a total of 16 categories, where single-use or mosaic polygons were also contemplated. The ArcGis 10.5 © software was used to perform digitization, coding, and topological validation, and to produce the shapefile format for the layer of results obtained (Figures A1–A4).

      The habitats included in Annex I of Directive 92/42/EC [69] were also considered for each study area.

2.3. Project Reviews

      In the case of the work carried out between 2016 and 2019 on the river Manzanares in Madrid, the plan for Naturalization of the Manzanares as it passes through the city of Madrid was examined [70], while some complementary field work was also carried out. In the case of Toledo, the initial document of the Environmental Impact Assessment for the project to integrate the river Tagus into the city of Toledo was examined [71]. The information analyzed in the study of the River Henares in Guadalajara came from the city's Strategy for Sustainable Urban Development [72], which was complemented by field work as the project is currently being carried out. Finally, for the actions proposed in Talavera de la Reina, the report and plans that form part of the technical proposal that adjudicated the project following the international call for ideas were consulted [73]. In addition, abundant auxiliary information was evaluated, such as news reports in the press, debate forums, consultations with the European Parliament, etc. [74–76]. A synthesis of the collected data is available in Appendix B.

3. Results

3.1. Analysis of the Cases

In May 2016, the Madrid City Council began to implement the measures included in the "Plan for Renaturalization of the River Manzanares as it passes through the city of Madrid." This was a plan based on the proposal that the group "Ecologists in Action" sent to the City Hall in January of that same year. The measures undertaken did not involve large construction work and no new structures were built (Table A1). The techniques employed in the canals were based on bioengineering principles, which involve using natural materials or those that are biodegradable in the short or mid-term.

The Integrated Strategy for Sustainable Urban Development of the municipality of Guadalajara was approved in December 2016, and it included a set of actions aimed at, "Recovering the land occupied by the right bank of the River Henares from the neighborhood of 'Los Manantiales' to the 'Finca de Castillejos' [72]." Under this undefined objective, "Recovery" work on the right bank of the river Henares, between the "Arab bridge" and the "Julián Basteiro bridge" was proposed and executed. Among the work undertaken, urbanization and gardening actions were implemented under the auspices of the Public Hydrological Domain between October 2018 and May 2019 (Table A2). This work, with a total cost of €1,438,277.90, was 80% co-financed by the European Regional Development Fund. According to various publications in the media, the action implemented is far from that indicated in the application submitted to the European Union for co-financing, in which the main objective was the environmental recovery of the river Henares and the increase in its biodiversity. In this sense, on 25 October 2018, the Equo European parliamentarian, Florent Marcellesi, registered two questions in the European Parliament related to the legitimacy of the actions undertaken with the co-financing of ERDF funds. In December of the same year, the Parliament responded by stating that, "According to the Spanish authorities, the project implemented by the City Council received authorization from the Tagus Hydrographic Confederation, yet it does not currently meet the requirements necessary to obtain EU funding". At the time this response was registered, the works were fully underway, and the site had put up a sign indicating the European co-financing received.

The notification by the Tagus Hydrographic Confederation of the results of the "International Call for Ideas and Projects aimed at integrating the river Tagus into the city of Toledo," presided over by a panel of judges, took place in December 2009 (BOE no. 289, 1 December, 2009). However, it was not until 2014 that the project chosen was released into the public domain [71]. The project proposed an extensive, intense, and ambitious program of actions along 43 km of the river's course as it runs through the municipality of Toledo. However, it is in the urban section of the city of Toledo where the project is most intense, in part due to the profound modifications to the existing configuration of the river's course, considering the implementation of important interventions, such as earth movements and the construction of concrete structures (Table A3).

The final criterion to define the appropriate actions was to protect this space against floods over a 100-year return period, with reference to other considerations like the volume of earth filling required or the degree of transformation of the current environment. Likewise, the landscape relevance in the intervention was prioritized over its integration into the environment, as witnessed by the choice of the geometry of the river walk, which was built using concrete flagstones with an angular broken geometry that will intentionally be perceived as a clearly artificial and autonomous element of the landscape [71]

In January 2018, the Tagus Hydrographic Confederation unveiled the winning proposal in the "International Call for Ideas and Projects aimed at recovering the banks of the rivers Tagus and Alberche in the municipality of Talavera de la Reina." It was the so-called "When the river sounds" ("Cuando el río suena") idea, submitted by a temporary union of three architectural and engineering consultancy companies [73]. The action proposed covers an extended stretch of the river, from the mouth of the river Alberche to the natural site known as the "Charca del Cura", passing through the urban section of the city of Talavera. The actions are quite diverse in nature and scope, depending on the proximity to the urban nucleus, although in general terms they involve a permeation of flow towards well-preserved river spaces that are currently difficult to access (Figure 3, Table A4).

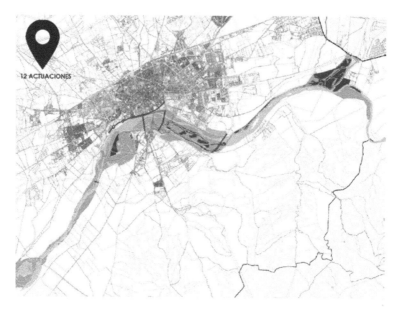

Figure 3. Interventions proposed in the planned rehabilitation project for the riverbanks and islands of the river Tagus in Talavera de la Reina [73].

3.2. Basic Structure of the Landscape

When the 16 landscape categories identified were considered (Appendix A), the river landscape at the four study areas did not show significant differences ($K = 4.57$; $p > 0.05$). The landscape of the Tagus river bank in Talavera was the most diverse (H = 1.95), followed by that of the river Henares (H = 1.83), of the river Tagus in Toledo (H = 1.59), and finally, that of the river Manzanares (H = 0.83). Nevertheless, all have low values of diversity.

Due to its structure and composition, the landscape of the river Manzanares as a whole differs from the other river landscapes analyzed, with the proximity between the two sections of the river Tagus representing the maximum. In the river Henares, landscaped and artificial surfaces occupy a large area (49%), along with wooded areas (20%). The wooded areas, in their different categories, are precisely the larger areas in the two sections of the river Tagus analyzed. The river Manzanares is quite different to the other river courses in terms of landscape. This is largely due to the dominance of the categories, "Sandbank and water cover" (>60%) and "Sandbanks with marsh vegetation" (>25%), that developed on the canalised river bed after the opening of the nine dams that exist in the section analyzed.

3.3. Waterfowl and Riparian Birds

The three study areas are not equally important to waterfowl, as reflected in the significant differences between the mean values of abundance (K = 36.20; $p < 0.005$) and density (K = 31.09; $p < 0.005$). The Manzanares river maintains the largest (a = 1528.00 ± 523.51) and most dense (d = 222.7 ± 94.94) wintering population of waterfowl, followed by the river Tagus in Talavera (a = 1131 ± 859.53; d = 178.85 ± 86.10) and in Toledo (a = 745.33 ± 358.04; d = 178.29 ± 56.12). The river Henares was always the river course with the lowest waterfowl values, with very low global figures below 20 birds. As expected, there were no significant differences in the wintering populations at the two stretches of the Tagus analyzed (D = 0.25; $p = 0.2710$).

There was a marked variability in the density and abundance values (see Tables A5–A7). This is because the values for the month of February, quite late in winter, were included in the analysis. The highest global figures were always obtained in January: 1915 birds for the Manzanzares, 1110 birds for the Tagus in Talavera, and 915 birds for the Tagus in Toledo. The values for the river Henares are irrelevant as they never exceeded 55 birds a month.

The mean richness values were high for the river Tagus in Talavera (r = 19.67 ± 1.53) and Toledo (r = 19.00 ± 2.65), they were in the mid-range for the river Manzanares (r = 10.67 ± 2.52), and very low for the river Henares (r < 4). This distribution approximates very well to the H diversity index, although in this case the river Tagus is the river registering the highest values (H = 2.06 ± 0.24), which are very close to those of the river Tagus in Talavera (H = 2.03 ± 0.14). There is a very strong positive correlation between richness and diversity (r_s = 0.99; p < 0.05), and a moderate positive correlation between density and diversity (r_s = 0.59; p < 0.05). For the river Manzanares, the high density (birds/km) of *Chroicocephalus ridibundus* (d = 99.95 ± 60.03), *Anas platyrhynchos* (d = 40.38 ± 4.40), *Gallinula chloropus* (d = 20.85 ± 6.35), and *Larus fuscus* (d = 54.86 ± 43.69) was particularly notable. For the river Tagus in Toledo, *Fulica atra* (d = 17.92 ± 14.38), *Anas platyrhynchos* (d = 17.61 ± 11.87), *Chroicocephalus ridibundus* (d = 16.77 ± 14.95), and *Phalacrocorax carbo* (d = 16.93 ± 6.49) were dominant. For the river Tagus in Talavera, the dominant species were *Fulica atra* (d = 45.10 ± 31.49), *Phalacrocorax carbo* (d = 34.87 ± 14.67), *Anas platyrhynchos* (d = 26.32 ± 16.86), and *Chroicocephalus ridibundus* (d = 60,67 sd > $\overline{X}$).

The natural and semi-natural spaces of the four fluvial territories studied host important populations of forest and marsh birds. Unlike waterfowl, there were no significant differences in the abundances of these species (K = 1331; p > 0.05). The maximum density values (birds/10 ha) were registered on the banks of the river Tagus in Talavera (d = 240.12 ± 80.88), followed by those on the banks of the river Manzanares (d = 172.64 ± 27.11), the river Henares (d = 151.02 ± 27.18), and the river Tagus in Toledo (d = 141.16 ± 27.11). The richness values range from a maximum of 28.33 species (Toledo) to a minimum of 21 species (Henares). In terms of diversity, the maximum was recorded in Talavera (H = 2.9) and the minimum in the river Manzanares (H = 2.44). These values were on the whole strongly positively correlated with the structural diversity in the areas studied (r_s = 0.87; p < 0.05). The dominant species on the banks of the river Henares were *Columba palumbus* (d = 31.29 ± 8.92), *Turdus merula* (d = 11.34 ± 3.42), *Erithacus rubecula* (d = 11.34 ± 1.57), and *Aegithalos caudatus* (d = 11.79 ± 5.66). Those on the banks of the river Manzanares were *Columba livia domestica* (d = 52.75 ± 11.73), *Passer domesticus* (d = 27.79 ± 8.67), *Sturnus unicolor* (d = 9.72 ± 6.35), and *Pica pica* (d = 7.58 ± 4.31). On the banks of the river Tagus in Toledo, *Columba livia domestica* (d = 20.79 ± 5.73), *Columba palumbus* (d = 16.38 ± 2.43), *Passer domesticus* (d = 13.07 ± 0.98), and *Phylloscopus collybita* (d = 11.97 ± 1.44). On the Tagus river banks in Talavera, *Passer montanus* (d = 30.41 ± 9.96), *Columba palumbus* (d = 20.87 ± 11.91), *Passer domesticus* (d = 19.68 ± 1.79), and *Sturnus unicolor* (d = 19.08 ± 16.91). In general, these are species typical of humanized environments, except for those on the banks of the Henares where forestry elements dominate.

3.4. Sensitive Species and Habitats Included in Annex I of the Habitat Directive

Sensitive species, according to the Annex I of the Birds Directive, represented 17.01% of the 21 species identified in the four study areas, which fall within the unfavorable or threatened SPEC category in Europe (Table 3). In addition, seven habitats were recognized as corresponding to those in Annex I of the Habitat Directive, although only three were cited in the official cartography of the Spanish Inventory of Terrestrial Habitats [77] (Table 4).

Table 3. Protected, threatened, or declining bird species in the study areas during 2019–2020 wintering: (1) Annex I Bird Directive; (2) Status in Europe—VU Vulnerable, D Declining, and NT Near Threatened; and (3) SPEC (Species of European Conservation Concern).

	Annex I (1)	European Status (2)	SPEC (3)	River
		Waterfowl		
Actitis hypoleucos		D	3	TTO, TTA
Alcedo atthis	*	VU	3	MNZ, TTO, TTA
Aythya ferina		VU	1	TTA
Ciconia ciconia	*			MNZ, TTO, TTA
Circus aeroginosus	*			TTO, TTA
Egretta garzetta	*			MNZ, TTO, TTA
Fulica atra		NT	3	TTO, TTA
Gallinago gallinago		D	3	MNZ, TTO, TTA
Ixobrychus minutus	*		3	TTO TTA
Nycticorax nycticorax			3	TTO, TTA
Porphyrio porphyrio	*			TTO, TTA
Porzana porzana	*			MNZ
		Other Species		
Delichon urbicum		D	2	TTA
Hirundo rustica		D		TTA
Passer domesticus		D	3	HNR-MNZ-TTO-TTA
Passer montanus			3	HNR-MNZ-TTO-TTA
Picus sharpei			1	HNR-MNZ-TTO-TTA
Serinus serinus		D	2	HNR-MNZ-TTO-TTA
Sturnus vulgaris			3	MNZ
Troglodytes troglodytes	*			HNR-MNZ-TTA
Turdus iliacus			1	TTA

Table 4. Habitat of Annex I of the Habitat Directive and communities: [EUc] Habitat code of the European Union [69], (Sc) Spanish habitat code [77]. HRN, Henares river; MNZ, Manzanares river; TTO, Tagus river in Toledo; and TTA, Tagus river in Talavera (*) [78].

[EUc 3250] Constantly flowing Mediterranean rivers with *Glaucium flavum*
　　(Sc 225011) *Andryaletum ragusinae* Br.-Bl. & O. Bolòs 1958: HNR, TTO, TTA.

[EUc 3280] Constantly flowing Mediterranean rivers with *Paspalo-Agrostidion* species and hanging curtains of *Salix* and *Populus alba*
　　(Sc 228012) *Ranunculo scelerati-Paspaletum paspalodis* Rivas Goday 1964 corr. Peinado, Bartolomé, Martínez-Parras & Ollala 1988: HNR, MNZ, TTO, TTA.

[EUc 7210] Calcareous fens with *Cladium mariscus* and species of the *Caricion davallianae*
　　(Sc 621123) Reedbeds with *Schoenoplectus lacustris*: HNR, TTO (*), TTA.

[EUc 92A0] *Salix alba, Salix* sp. *and Populus alba galleries*
　　(Sc 82A056) *Salicetum salviifoliae* Oberdorfer and Tüxen in Tüxen and Oberdorfer 1958: MNZ.
　　(Sc) *Salicion salviifoliae* Rivas-Martínez, T.E. Díaz, F. Prieto, Loidi & Penas 1984: HNR.

[EUc 92D0] Mediterranean riparian galleries (*Nerio-Tamariceteae*) and south-west Iberian Peninsula riparian galleries (*Securinegion tinctoriae*)
　　(Sc 82D013) *Tamarix canariensis, Tamarix gallica* and *Tamarix africana* Grove: HNR*, TTO (*), TTA.

[EUc 6420] Mediterranean tall humid grasslands of the *Molinio-Holoschoenion*
　　(Sc 542015) *Holoschoenetum vulgaris* Br.-Bl. ex Tchou 1948: HNR, TTO, TTA.
　　(Sc 54201P) *Trifolio resupinati-Holoschoenetum* Rivas Goday 1964: MZN.

[EUc 6430] Hydrophilous tall herb fringe communities of plains and of the montane to alpine level
　　(Sc 543135) *Myrrhoidi nodosae-Alliarietum petiolatae* Rivas-Martínez & Mayor ex V. Fuente 1986. HNR, TTO, TTA.
　　(Sc 543110) *Convolvulion sepium* Tüxen 1947: HRN, TTO, TTA.

4. Discussion

The recovery of river spaces in urban areas is one of the most useful tools for the conservation of biodiversity [79], particularly due to their intrinsic value and importance as a functional connector between residual natural elements. However, the treatment given to urban riverbanks, frequently proposed in terms of "restoration", does not usually consider the intrinsic value of these areas, their characteristics or their singularities. While the environmental and socio-economic objectives of any such intervention should be mutually beneficial [38], they are often disconnected.

Three of the four cases considered here are good examples of this, as they cause or will cause a high degree of artificialization and an objective loss of value in the affected river territories. Like other far-reaching interventions [80,81], these have completely ignored the local river landscape, its value, dynamics, and its ecoterritorial and landscape context. Indeed, as indicated in various studies [8–84], the concept of "river restoration" is used to refer to projects of a quite diverse nature, and not always focused on their protection and improvement.

The same is true for the projects proposed for the banks of the river Henares or those of the river Tagus in Toledo and Talavera de la Reina. It is not possible that, "Respect for the natural values of the river, its fauna and flora, with the aim of safeguarding existing ecosystems and bringing them closer to their status prior to the negative man-made alterations" [85] is among the project's objectives in the case of the river Tagus in Talavera. Indeed, a substantial part of the actions considered will cause the destruction of one of the most valuable river landscapes in the centre of the Iberian Peninsula if implemented under the terms established in the proposal. Thus, as indicated previously, it is fundamental that these projects, whose principles and effects are contrary to the recovery and integration of river areas, are not formally presented as such.

The most expensive projects usually incorporate a good number of elements linked to structures that make the riverbanks and flood plains resilient and waterproof. This is the case for the "Salón fluvial", proposed for the banks of the river Tagus in Toledo, which is nothing but concrete terraces, like tiered seating, that artificially modifies the original morphology of that stretch of the river considerably. On the other hand, the four projects analyzed have the common and legitimate objective of permitting the use and enjoyment of riparian spaces by citizens. However, these projects make the conservation of space subject to this consideration, without differentiating valuable spaces that should remain inaccessible or with limited accessibility. In particular, the river Tagus projects in Talavera and Toledo serve to "thematise" the river space, with a variety of construction proposals that will stimulate activities that induce a greater influx of people, including a hostel and a restaurant in a flood zone in the case of the river Tagus in Talavera. The opening of multiple trails and of dog parks, in hitherto well-preserved river spaces, increases the risk of their degradation due to the problems derived from affluence, such as noise, garbage, and direct predation in the case of pets running around leash-free, etc.

The projects studied did not analyse or consider the value of river spaces in terms of diversity. How is it possible that projects of high environmental impact are implemented without previously determining the value of the territory affected? The sections of the river Tagus analyzed in Toledo and Talavera are areas of great interest for wintering water birds, with average densities above 140 and 240 birds/km, respectively, and with high values of diversity (H > 2). The value of the community parameters for passerine species and related communities is also very high, especially the diversity values (H > 2.8). Both riverbanks also maintain a good number of habitats included in Annex I of the Habitats Directive, as well as rare, threatened, or sharply declining species in Europe (Figure 4), in particular *Circus aeroginosus*, *Nycticorax nycticorax*, *Ixobrychus minutus*, *Passer montanus* or *Picus sharpei*.

Figure 4. Some sensitive species. Above, *Lymnocryptes minimus* (**a**) (river Manzanares, 5/12/2019), *Alcedo atthis* (**b**) (river Manzanares, 26/12/2019), *Nycticorax nycticorax* (**c**) (river Tagus in Toledo, 27/2/2020). Below, *Picus sharpei* (**d**) (river Manzanares, 28/2/2020), *Passer montanus* (**e**) (river Manzanares, 26/12/2019).

In the specific case of the project developed on the banks of the river Henares (Figure 5), we were able to verify the extent of the damage caused by the implementation of the project supposedly aimed at achieving the "Recovery of the right bank of the Henares river, between the Arab bridge and the Julián Besteiro bridge", 80% co-financed, to the tune of €1,150,622, by the European Regional Development Fund under the "Operational Program for Sustainable Growth 2014–2020" [86]. Indeed, the average density values for the forest–marsh bird communities in the winter of 2019–2020 were significantly lower than those in the winter of 2013–2014, with a 45% decrease of the average values in the winter of 2019–2020. Diversity values also fell, although they were only 2.5% lower in 2019–2020 than in 2013–2014. In terms of waterfowl, despite the limited interest of this group of birds in this stretch of the river, the 60.4% drop in the figures for 2019–2020 relative to those registered in 2013–2014 is particularly striking.

(a)   (b)

Figure 5. Bank of the river Henares before (**a**) (10/5/2014) and after (**b**) (29/10/2019) the implementation of the project over a small area of activity.

Any project whose objective is the recovery of river spaces in urban or peri-urban areas must initially locate the remnants of any existing natural landscape. The priority should be to protect these areas in order to try to extend these habitats [87], not least as they are source areas and host habitats of great value to wildlife [88]. Indeed, the rapid recovery of the river Manzanares in the city of Madrid can be largely attributed to the existence of upstream river spaces currently undergoing restoration that maintain extensive natural remnants [89]. Furthermore, the planning of this space was based on criteria of diversity, coherence, and continuity, which are basic elements to guarantee the quality of the landscape [90]. As such, a space of great value in terms of biodiversity has been created within an intensely urbanized area ex novo, and following a behavior observed in other river areas [22,91,92].

Of the four projects analyzed, only the re-naturalization of the river Manzanares meets one of the basic requirements of the Water Framework Directive: to situate environmental concerns at the centre of any interventions [93]. Of all the cases studied, this is the one that started from a more precarious environmental situation and with limitations in terms of the scope of the actions undertaken, particularly due to the physical constriction of the river space and its disconnection with its former banks and its hyporheic medium, the latter due to the burying and waterproofing of the circular M-30 urban motorway. In absolute terms, even after the re-naturalization intervention, the urban section of the river Manzanares exhibits environmental values lower than those of the other sections, which can be attributed to the limitations described above, limitations that hinder the full development of the river's natural behavior. However, the intervention has explored and taken advantage of the margins for improvement provided by this stretch of the river, resulting in an exponential increase in riparian quality. All this has been achieved at low cost and of the four actions analyzed, this action has exhibited the lowest cost per linear meter.

At present, the average wintering values of waterfowl exceed 1500 individuals and the average density of passerine and related species is 220 birds/10 ha. Both are dominant generalist species, but they also include rare, threatened, or declining common species such as the *Porzana porzana*, *Lymnocryptes minimus* or *Passer montanus* species. This is without doubt a good example of how an urban sector of a river of no value has become a valuable space within a city for both the birds in the river area and for ecological diversity in general, contributing to the permeation of a high-density urban space and promoting regional connectivity. Moreover, the restoration measures applied were low impact, causing a re-naturalization of a river course highly valued by society. In this respect, it is interesting to note that rivers where only mild restoration measures have been implemented and that maintain clear features of their natural state are socially more valued than artificialized rivers [94]. Furthermore, the bond established between the resident population and the restored space has a positive effect on the valuation of the new river landscape, as reported previously [95]. This may have occurred because thanks to its re-naturalization, this space provides new possibilities of use, as has been the case with other rivers located in high-density urban areas [96].

Another major problem with most interventions in river spaces is the consideration of these spaces as independent sectors [97]. Like all river spaces, urban river spaces must be considered as part of a complex system of interconnected units that have to be managed at both the river and drainage network level [98]. Furthermore, urban and peri-urban river landscapes can be internalised and understood as "Third Landscapes" [99]: residual spaces of high value in terms of environmental and landscape diversity, the evolution—and disappearance—of which is usually associated with urban planning and development. For this reason, it is important to highlight the need to consider principles, values, criteria, objectives, and actions that differ considerably from those usually contemplated. That is, the essence of urban and peri-urban river landscapes must be founded on an understanding of its value, dynamics, and possibilities of use. A major challenge for the conservation of urban green spaces—including river spaces—is perhaps understanding how they can be developed, remodelled or restored whilst favoring natural processes, and conserving functional ecosystems [21]. That would enhance models of intervention that often do not consider environmental improvements [100].

In the current climatic scenario, with a clear tendency towards aridity, there is a clear increase in the value of river areas as "environmental corridors". The importance of river corridors as priority elements within the "European Green Infrastructures" must also be taken into account [101]. The European proposal points out the need to improve the connectivity between natural areas, using river corridors to counteract the effects of fragmentation of these territories. In particular, there is a growing importance placed on rivers in urban and peri-urban areas to maintain biodiversity due to the steady increase in the extension of these spaces [102–104].

5. Conclusions

Before undertaking any actions involving urban river courses, it is essential to understand the environmental and landscape interest of the spaces under consideration. This is especially important when these interventions highlight sustainability among their objectives, and they establish "activities of environmental recovery or integration". At three of the sites analyzed in this study interventions were presented that will cause important modifications, yet the environmental interest of the areas in which they will be carried out was not analyzed. Nor was their capacity to integrate into the urban landscape evaluated. These are expensive projects, with costs lying between 900,000 and €2 million per linear km. Nevertheless, it is noteworthy that of the cases analyzed, the project that best integrated the river into the city, respecting and potentiating its natural value and interest as a landscape, is that which cost the least money: €162,000 per linear km. Specifically, we refer to the project to recover the natural habitat of the river Manzanares in Madrid. This project set out to convert this urban fluvial stretch into an area of interest for wintering aquatic birds. It also hosts some globally threatened species and it maintains habitats that are included in Annex I of the Habitats Directive. In addition, it is also a highly valued social and leisure area for the local population.

Author Contributions: P.M.H. and A.-B.B.M.: devised the ideas for this study, collected and analyzed the data, and wrote the original manuscript; L.J.R. contributed to developing the original idea of the study and reviewed the manuscript; and F.A.Á. assisted in the data collection and analysis, and reviewed the associated literature. All authors have read and agreed to the published version of the manuscript.

Funding: This research was funded by Erasmus+ Project RailtoLand (Grant agreement: 2019-1-ES01-KA203-065554 - Erasmus+ Programme of the Europe Union).

Acknowledgments: We thank Tomás Velasco and Íñigo Vicente for their participation in the waterfowl census.

Conflicts of Interest: The authors have no conflict of interests to declare.

Appendix A

Figure A1. Landscape units for the river Henares.

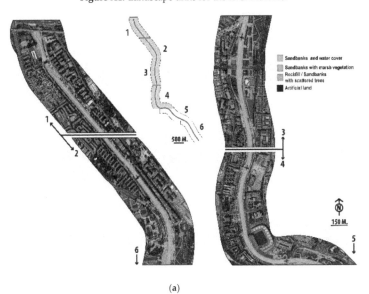

(a)

Figure A2. *Cont.*

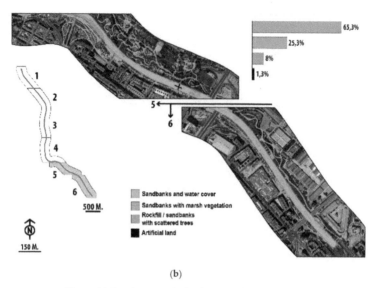

(b)

Figure A2. Landscape units for the river Manzanares (a,b).

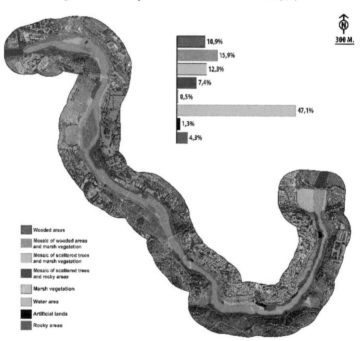

Figure A3. Landscape units for the river Tagus in Toledo.

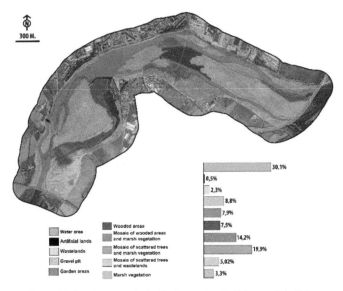

Figure A4. Landscape units for the Tagus river in Talavera de la Reina.

Appendix B

Table A1. Analysis of the actions on the river Manzanares.

City	Madrid
Length	7.5 km; of which only 6.86 km are accessible
Character	Urban
Total cost	€1,216,054
Unit cost (per linear meter)	€162,140.53/m
Action status	Implemented May 2016–May 2019
Promoter	Madrid City Council, following a suggestion by Ecologists in Action
Funding	Municipal budgets

Overall aim of the action

The plan aims to recover and preserve, to the extent possible, the functions of the ecological corridor that the river Manzanares defines as it passes through Madrid relative to the upper and lower sections, and to manifest its value within the urban environment, in line with the guidelines set out in the relevant European legislation (Water Framework Directive) and the National Strategy for River Restoration.

Actions with an impact on the hydrological regime and their effects

Actions:

- Opening of all the gates that regulate the water flow along the section stipulated.

Effects:

- Partial recovery of the natural sedimentation and erosion processes, as well as those of natural colonization of riparian species of flora and fauna.
- Improvement of the longitudinal continuity of flow.

Actions and Effects on the morphology of the channel and its bank

Actions:

- Partial removal of the breakwaters at the ends of the section and the conditioning of its embankment.
- Preparation of the banks employing bioengineering techniques.
- Creation of deflectors using bioengineering techniques to create a meandering river course in the central section.

Effects:

- Improvement of riparian environments.
- Improvement of the lateral connectivity of the channel with its banks.
- Landscape integration.

Table A1. *Cont.*

Actions and Effects on vegetation and fauna

Actions:

- Revegetation of breakwaters.
- Planting of autochthonous species of riparian trees and shrubs (more than 5000 specimens).
- Elimination of exotic species.
- Installation of nesting boxes.
- Release of stripe-necked terrapin (*Mauremys leprosa*).

Effects:

- Increase in biodiversity as a consequence of the improvements of the riparian habitat for many species of fauna, especially birds, but also for other relevant elements like otters (*Lutra lutra*).

Table A2. Analysis of the actions on the river Henares in the city of Guadalajara.

City	Guadalajara
Length	1.47 km
Character	Urban
Total cost	€1,438,277.90
Unit cost (per linear meter)	€978,420.34/m
Action status	Implemented October 2018–May 2019
Promoter	Guadalajara City Council
Funding	European Regional Development Fund

Overall aim of the action

The overall objectives included in the Strategy for the Sustainable Urban Development of Guadalajara were the:

- Creation of new spaces for leisure and recreation.
- Aesthetic improvement of the river.
- Improvement of pedestrian safety.
- Preservation of environmental conditions.
- Establishment of a new park alongside the existing one in the area known as "La Chopera".

Actions with an impact on the hydrological regime and their effects

Actions: Non-existent
Effects: -

Actions and Effects on the morphology of the channel and its bank

Actions:

- Construction of viewpoints by remodelling the terrain of the banks and advancing it towards the channel.
- Actions to compact and waterproof a trail, which will have a width of 2.5 m along most of the route and that runs through a Public Hydrological Domain (P.H.D.).

Effects:

- Artificialisation of the riverbank.
- Increased pedestrian traffic in the P.H.D.
- Compaction and waterproofing of part of the riverbank, which by definition is a permeable area.
- Direct effects on the riverside vegetation due to land clearing and filling in the levelling works.

Actions and effects on vegetation and fauna

Actions:

- Clearing and felling of large poplar trees on the riverbank to permit the construction of roads, parks and viewpoints.
- Installation of illumination.

Effects:

- Increase of traffic and use in the P.H.D.
- Increase in nocturnal illumination affecting the riparian fauna.
- Artificialisation of the riverbank.

Table A3. Analysis of the actions on the Tagus river in the city of Toledo.

City	Toledo
Length	43 km; of which 6.36 are in urban and peri-urban sections
Character	Urban, peri-urban and rural
Total cost	€90,648,611.94
Unit cost (per linear meter)	€2,108,107.25 (although the actions focus on the urban section)
Action status	Call for tenders pending
Promoter	Tagus Hydrographic Confederation, with the collaboration of the Toledo City Council
Funding	General State Budget

Overall aim of the action

According to the call for tenders, the overall objective was the harmonization of the urban environment of Toledo with the river environment, recovering the riverbanks and its adjacent spaces, as well as improving pedestrian connections between the city and the river.

Actions with an impact on the hydrological regime and their effects

Actions:

- Rehabilitation of historical hydraulic infrastructures.

Effects:

- Increased longitudinal fragmentation of the river.
- Possible modification of the hydraulic regime of the river channel based on the management of the infrastructure and on the final provisions laid down in the rehabilitation project.

Actions and Effects on the morphology of the channel and its bank

Actions:

- "Salón fluvial" in an urban section of 4 ha, involving terracing with concrete walls filled with earth.
- Construction of longitudinal mounds on both banks along the entire section.
- Opening of compacted or paved roads in riparian areas and through rarely frequented slopes.

Effects:

- Artificialisation of the river course and its bank. Waterproofing of the bank affected by the action named the "Salón fluvial".
- Loss of lateral connectivity between the river channel and its surrounding areas.
- Alteration of the natural dynamics of river flooding, avoiding the dispersion of nutrients and energy to its flood plains, increasing the flow speed and therefore, the risk of downstream flooding.
- Increase of pedestrian traffic and use in the P.H.D.
- Loss of landscape quality.

Actions and effects on vegetation and fauna

Actions:

- Night lighting.

Effects:

- Increase in nocturnal illumination affecting the riparian fauna.

Table A4. Analysis of the actions on the Tagus river in Talavera de la Reina.

City	Talavera de la Reina
Length	14 km; of which 4.56 are in urban sections
Character	Urban, peri-urban and rural
Total cost	€14,033,300.00
Unit cost (per linear meter)	€1,002,378.57/m
Action status	Call for tenders pending
Promoter	Tagus Hydrographic Confederation, with the collaboration of the Talavera de la Reina City Council
Funding	General State Budget

Overall aim of the action

- To respect the natural values of the river, its fauna and flora, with the aim of safeguarding existing ecosystems and bringing them closer to their status prior to negative man-made alterations.
- To recognise the historical value of the heritage and constructions in order to avoid their deterioration through functional rehabilitation, either for their original uses or for other uses that guarantee their adequate conservation.
- To perform interventions on the banks and islands aimed at promoting the use and the enjoyment of the river, making recreational and cultural activities compatible with its natural life. A proposal for preventive measures to ensure that river does not pose a natural barrier to communications and the functioning of the city, and to mitigate the negative effects of these activities on the existing ecosystems.
- Control the flooding of various streams that are tributaries of the river Tagus.

Actions with an impact on the hydrological regime and their effects

Actions:

- Functional rehabilitation of weirs and canals associated with hydroelectric plants.

Effects:

- Increased longitudinal fragmentation of the river.
- Possible modification of the hydraulic regime of the river channel based on infrastructure management and the final provisions established in the rehabilitation project.

Actions and Effects on the morphology of the channel and its bank

Actions:

- Expansion of the sections of "La Parra" and "Cornicabral" tributary arroyos.
- Opening of paths facilitating access along the entire area of action, even to Mill Island.
- Creation of an artificial lagoon in the park used as a nursery.

Effects:

- Profound alteration of the structure and composition of the streams affected.
- Artificialisation of the river space.
- Compaction of the soil in the Public Hydrological Domain.
- Increase in the influx of passers-by and opening of large isolated spaces that have been well preserved to date.
- Alteration of the local phreatic level as a consequence of the opening of an artificial lagoon.

Actions and effects on vegetation and fauna

Actions:

- Clearing and felling of spontaneous vegetation in the urban front.
- Clearing and felling of vegetation in the "La Parra", "Cornicabral" and "La Portiña" streams.
- Selective plantation of autochthonous species.
- Access for domestic animals (canine park) and access to highly frequented spaces (restaurants; hostel; etc.).
- Installation of recreational fishing zones in protected and hitherto non-accessible areas, such as the "Charca del Cura".

Effects:

- Artificialisation of large well-preserved spaces, such as the "Isla Grande", where the construction an urban park is planned.
- Increased pressure on relevant elements of flora and fauna, as a consequence of the influx of people and predatory animals.
- Risk of introduction of exotic species linked to recreational fishing

Appendix C

Table A5. Average Waterfowl Abundance (± SD): HRN, River Henares; MNZ, River Manzanares; TTO, River Tagus in Toledo; and TTA, River Tagus in Talavera.

	HNR	MNZ	TTO	TTA
Actitis hypoleucos			11 ± 7.55	7.67 (*)
Aix galericulata	0.33 (*)			
Alcedo atthis		2.00 (*)		2.33 (*)
Alopochen aegyptiaca		19.33 ± 7.09	0.33 (*)	
Anas clypeata			1.33 (*)	5.00 (*)
Anas crecca			16.33 ± 16.01	23.33 ± 11.72
Anas platyrhynchos	24.00 ± 20.95	277.00 ± 30.20	112.00 ± 75.50	120.00 ± 76.86
Ardea cinerea	1.00 ± 1.00	0.67 (*)	39.67 (*)	15.00 ± 7.00
Aythya ferina				1.00 (*)
Bubulcus ibis			5.33 (*)	2.33 (*)
Cairina moschata		0.67 ± 0.58		
Charadrius dubius				0.33 (*)
Chroicocephalus ridibundus		685.67 ± 411.78	106.67 ± 95.10	276.67 (*)
Ciconia ciconia		1.33 (*)	1.67 ± 0.58	4.00 ± 3.61
Circus aeroginosus			0.67 ± 0.58	0.67 ± 0.58
Egretta garzetta		8.67 ± 1.15	2.00 ± 1.73	2.33 ± 1.53
Fulica atra			114.00 ± 91.43	205.67 ± 143.60
Gallinago gallinago		1.67 ± 0.58	37.33 (*)	22.67 (*)
Gallinula chloropus	1.67 ± 0.58	143.00 ± 43.55	43.33 ± 37.65	53.67 (*)
Ixobrychus minutus			3.00 ± 1.73	3.67 ± 1.15
Larus fuscus		376.33 ± 299.74	101.00 ± 92.83	163.33 (*)
Lymnocryptes minimus		2.00 ± 0.00		
Mareca strepera			25.00 (*)	36.33 (*)
Netta rufina				0.33 (*)
Nycticorax nycticorax			7.33 ± 5.51	7.00 ± 6.56
Podiceps cristatus			1.33 ± 1.15	1.67 ± 0.58
Porphyrio porphyrio			0.33 (*)	3.00 ± 2.65
Porzana porzana				
Phalacrocorax carbo	1.00 (*)	9.00 ± 1.00	107.67 ± 41.31	159.00 ± 66.90
Rallus aquaticus			1.50 ± 0.71	5.50 ± 3.54
Tachybaptus ruficollis			6.00 ± 5.29	8.67 ± 4.93
Tringa ochropus			1.67 ± 1.15	2.50 ± 0.71
Medium Abundance	28 ± 24.27	1528.00 ± 523.51	745.33 ± 358.04	1131 ± 859.53

(*) Standard deviation (SD) higher than the average values ($\overline{X}$).

Table A6. Average density (± SD) of Waterfowl (birds/km): HRN, River Henares; MNZ, River Manzanares; TTO, River Tagus in Toledo; and TTA, River Tagus in Talavera.

	HNR	MNZ	TTO	TTA
Actitis hypoleucos			1.73 ± 1.19	1.68 (*)
Aix galericulata	0.45 ± 0.393			
Alcedo atthis		0.29 (*)		0.5 (*)
Alopochen aegyptiaca		2.82 ± 1.03	0.05 (*)	
Anas clypeata			0.21 (*)	1.10 (*)
Anas crecca			2.57 ± 2.52	5.12 ± 2.57
Anas platyrhynchos	22.7 ± 16.1	40.38 ± 4.40	17.61 ± 11.87	26.32 ± 16.86
Ardea cinerea	1.13 ± 0.393	0.10 (*)	6.24 (*)	3.29 ± 1.54
Aythya ferina				0.22 (*)
Bubulcus ibis			0.84 (*)	0.51 (*)
Cairina moschata		0.10 ± 0.08		
Charadrius dubius				0.07 (*)
Chroicocephalus ridibundus		99.95 ± 60.03	16.77 ± 14.95	60.67 (*)
Ciconia ciconia		0.19 ± 0.34	0.26±0.09	0.88 ± 0.79
Circus aeroginosus			0.10 ± 0.09	0.15 ± 0.13
Egretta garzetta		1.26 ± 0.17	0.31 ± 0.27	0.51 ± 0.33
Fulica atra			17.92 ± 14.38	45.10 ± 31.49
Gallinago gallinago		0.24 ± 0.08	5.87 (*)	4.97 (*)
Gallinula chloropus	1.13 ± 0.393	20.85 ±6.35	6.81 ± 5.92	11.77 (*)
Ixobrychus minutus			0.47 ± 0.27	0.80 ± 0.25
Larus fuscus		54.86 ± 43.69	15.88 ± 14.60	11.84 ± 10.54
Lymnocryptes minimus		0.29 ± 0.00		
Mareca strepera			3.93 (*)	7.97 (*)
Netta rufina				0.07 (*)
Nycticorax nycticorax			1.15 ± 0.87	1.54 ± 1.44
Podiceps cristatus			0.21 ± 0.18	0.37 ± 0.13
Porphyrio porphyrio			0.05 (*)	0.66 ± 0.58
Porzana porzana		0.10 ± 0.17		
Phalacrocorax carbo	1.36 ± 1.18	1.31 ± 0.15	16.93 ± 6.49	34.87 ± 14.67
Rallus aquaticus			0.16 ± 0.16	0.80 (*)
Tachybaptus ruficollis			0.94 ± 0.83	1.90 ± 1.08
Tringa ochropus			0.26 ± 0.18	0.37 ± 0.33
Global Average Density (birds/km)	19 ± 16.5	222.74 ± 94.93	117.29 ± 56.12	177.85 ± 86.10
Global Average Richness	3.67 ± 1.53	10.67 ± 2.52	19.00 ± 2.65	19.67 ± 1.53
Diversity	0.55 ± 0.17	1.37 ± 0.12	2.06 ± 0.14	2.04 ± 0.13

(*) Standard deviation (SD) higher than the average values ($\overline{X}$).

Table A7. Average density values (± SD) of Riparian birds (birds/10ha): HRN, River Henares; MNZ, River Manzanares; TTO, River Tagus in Toledo; and TTA, River Tagus in Talavera.

	HNR	MNZ	TTO	TTA
Aegithalos caudatus	11.79 ± 5.66	0.78 (*)	4.88 ± 1.19	16.10 ± 4.73
Carduelis carduelis		1.46 (*)	4.25 ± 2.95	10.14 ± 2.73
Certhia brachydactyla	4.5 ± 2.08	1.17 ± 0.77	0.95 ± 0.00	5.96 ± 2.07
Cettia cetti	9.07 ± 3.42	5.63 ± 1.94	7.88 ± 2.89	4.77 ± 5.47
Chloris chloris			1.89 ± 1.89	3.58 ± 3.58
Cisticola juncidis			0.47 ± 0.00	
Coccothraustes coccothraustes			0.32 (*)	11.33 ± 3.72
Columba livia		52.75 ± 11.73	20.79 ± 5.73	
Columba oenas	6.80 (*)			
Columba palumbus	31.29 ± 8.92	4.76 (*)	16.38 ± 2.43	20.87 ± 11.91
Corvus monedula		0.39 ± 0.34		
Cyanistes caeruleus	6.35 ± 5.50	3.21 ± 0.87	10.24 ± 0.27	12.52 ± 4.73
Delichon urbicum			0.79 (*)	2.39 (*)
Dendrocopos major	1.81 ± 0.79	0.29 (*)	0.16 (*)	1.19 (*)
Dryobates minor				1.79 ± 0.00
Emberiza cirlus	0.68 (*)			
Emberiza schoenichlus		0.10 (*)		
Erithacus rubecula	11.34 ± 1.57	3.30 ± 1.71	2.84 ± 1.70	7.75 ± 2.73
Fringilla coelebs	1.36 ± 1.36	0.10 (*)	2.99 ± 1.36	6.56 ± 2.73
Hirundo rustica				3.46 ± 1.65
Motacilla alba	1.81 ± 0.79	7.48 ± 3.38	0.95 ± 0.00	2.39 ± 1.03
Motacilla cinerea	0.45 (*)	1.65 ± 0.89	0.47 ± 0.47	2.39 ± 1.03
Myiopsitta monachus		4.08 ± 1.05		1.79 (*)
Parus major	4.99 ± 0.79	0.68 (*)	5.20 ± 1.25	10.14 ± 4.13
Passer domesticus	0.9 (*)	27.79 ± 8.67	13.07 ± 0.98	19.68 ± 1.79
Passer hispaniolensis	0.45 (*)		2.52 ± 1.44	
Passer montanus	4.54 (*)	11.17 ± 3.98	7.88 ± 5.44	30.41 ± 9.96
Periparus ater	0.91 (*)	2.14 ±1.50		
Phoenicurus ochruros		1.46 ± 1.34	0.95 ± 0.00	1.79 ± 0.00
Phylloscopus collybita	7.26 ± 5.50	7.48 ± 4.38	11.97 ± 1.44	10.14 ± 2.73
Pica pica	7.71 ± 6.28	7.58 ± 4.31	2.05 ± 0.27	2.39 ± 1.03
Picus sharpei	0.45 (*)	1.07 ± 1.02	0.47 ± 0.00	0.60 (*)
Prunella modularis	0.91 (*)	0.58 (*)	0.16 (*)	
Psittacula krameri		0.29 ± 0.29		
Ptyonoprogne ruspestris		4.66 (*)		
Regulus ignicapillus		0.10 (*)		
Remiz pendulinus	1.36 (*)		0.32 (*)	
Serinus serinus	6.35 (*)	3.40 (*)	5.67 ±1.89	13.12 ± 7.23

Table A7. *Cont.*

	HNR	MNZ	TTO	TTA
Spinus spinus		0.19 (*)		
Streptopelia decaocto	3.17 ± 1.57		0.16 (*)	3.58 ± 3.58
Sturnus unicolor	6.80 ± 2.72	9.72 ± 6.35	2.68 ± 0.72	19.08 ± 16.91
Sturnus vulgaris		0.10 (*)		
Sylvia atricapilla	0.45 (*)	0.39 ± 0.17	3.78 ± 1.25	1.79 ± 1.79
Sylvia melanocephala		0.19 (*)	0.79 ± 0.27	
Troglodytes troglodytes	3.17 ± 0.79	0.29 ± 0.29		0.60 (*)
Turdus iliacus	0.45 (*)			
Turdus merula	11.34 ± 3.42	6.22 ± 3.37	5.99 ± 0.55	12.52 ± 1.79
Turdus philomelos	2.72 ± 2.36		1.26 ± 0.27	1.19 (*)
Upupa epops			0.16 (*)	0.60 (*)
Global Average Density (birds/10ha)	151.02 ± 27.18	172.64 ± 27.11	141.16 ± 8.74	240.12 ± 80.88
Global Average Richness	21 ± 1.73	27 ± 0.47	28.33 ± 2.49	24.67 ± 2.49
Diversity	2.65 ± 0.48	2.44 ± 0.13	2.86 ± 0.09	2.90 ± 0.09

(*) Standard deviation (SD) higher than the average values ($\overline{X}$).

References

1. Nilsson, C.; Reidy, C.A.; Dynesius, M.; Revenga, C. Fragmentation and flow regulation of the world's large river systems. *Science* **2005**, *308*, 405. [CrossRef]
2. Paul, M.J.; Meyer, J.L. Streams in the urban landscape. *Annu. Rev. Ecol. Syst.* **2001**, *32*, 333–365. [CrossRef]
3. Tockner, K.; Stanford, J.A. Riverine flood plains: Present state and future trends. *Environ. Conserv.* **2002**, *29*, 308–330. [CrossRef]
4. Walsh, C.J.; Roy, A.H.; Feminella, J.W.; Cottingham, P.D.; Groffman, P.M.; Morgan, R.P. The urban stream syndrome Current knowledge and the search for a cure. *J. N. Am. Benthol. Soc.* **2005**, *24*, 706–723. [CrossRef]
5. Petts, G.E.; Heathcote, J.; Martin, D. *Urban Rivers: Our Inheritance and Future*; IWA Publishing/Environment Agency: London, UK, 2002; p. 117.
6. Wozniak, M.; Leuven, R.S.E.W.; Lenders, H.J.R.; Chmielewski, T.J.; Geerling, G.W.; Smits, A.J.M. Assessing landscape change and biodiversity values of the Middle Vistula river valley, Poland, using BIO-SAFE. *Landsc. Urban. Plan.* **2009**, *92*, 210–219. [CrossRef]
7. Grill, G.; Lehner, B.; Thieme, M.; Geenen, B.; Tickner, D.; Antonelli, F.; Zarfl, C. Mapping the world's free-flowing rivers. *Nature* **2019**, *569*, 215–221. [CrossRef] [PubMed]
8. Blanton, P.; Marcus, W.A. Railroads, roads and lateral disconnection in the river landscapes of the continental United States. *Geomorphology* **2009**, *112*, 212–227. [CrossRef]
9. Berrocal-Menárguez, A.B.; Molina-Holgado, P. El valor de los paisajes fluviales: Su consideración en la planificación y la normativa. *Planur-E* **2009**, *6*, 4.
10. Everard, M.; Moggridge, H.L. Rediscovering the value of urban rivers. *Urban. Ecosyst.* **2012**, *15*, 293–314. [CrossRef]
11. Gurnell, A.; Lee, M.; Souch, C. Urban rivers: Hydrology, geomorphology, ecology and opportunities for change. *Geogr. Compass* **2007**, *1*, 1118–1137. [CrossRef]
12. Ribas, A. Los paisajes del agua como paisajes culturales: Conceptos, métodos y experiencias prácticas para su interpretación y visualización. *Apogeo* **2007**, *32*, 39–48.
13. Naiman, R.J.; Decamps, H.; Pollock, M. The Role of Riparian Corridors in Maintaining Regional. Biodiversity. *Ecol. Appl.* **1993**, *3*, 209–212. [CrossRef]
14. Martínez de Pisón, E. Valores e identidades. In *El Paisaje: Valores E Identidades*; Martínez de Pisón, E., Ortega, N., Eds.; Universidad Autónoma de Madrid-Fundación Duques de Soria: Madrid, Spain, 2010; pp. 11–46.

15. Cornelis, J.; Hermy, M. Biodiversity relationships in urban and suburban parks in Flanders. *Landsc. Urban. Plan.* **2004**, *69*, 385–401. [CrossRef]

16. Ives, C.D.; Lentini, P.E.; Threlfall, C.G.; Karen Ikin, K.; Danielle, F.; Shanahan, D.F.; Garrard, G.E.; Bekessy, S.A.; Fuller, R.A.; Laura Mumaw, L.; et al. Cities are hotspots for threatened species. *Glob. Ecol. Biogeogr.* **2016**, *25*, 117–126. [CrossRef]

17. Haase, D. Urban Ecology of Shrinking Cities: An Unrecognized Opportunity? *Nat. Cult.* **2008**, *3*, 1–8. [CrossRef]

18. Crane, P.; Kinzig, A. Nature in the Metropolis. *Science* **2005**, *308*, 1225. [CrossRef] [PubMed]

19. Millard, A. Semi-natural vegetation and its relationship to designated urban green space at the landscape scale in Leeds, UK. *Landsc. Ecol.* **2008**, *23*, 1231–1241. [CrossRef]

20. Pennington, D.N.; Blair, R. Habitat selection of breeding riparian birds in an urban environment: Untangling the relative importance of biophysical elements and spatial scale. *Divers. Distrib.* **2011**, *17*, 506–518. [CrossRef]

21. Blair, R.B. Land-use and avian species diversity along an urban gradient. *Ecol. Appl.* **1996**, *6*, 506–519. [CrossRef]

22. Dallimer, M.; Rouqette, J.R.; Skinner, A.M.; Armsworth, P.R.; Maltby, L.M.; Warren, P.H.; Gaston, K.J. Contrasting patterns in species richness of birds, butterflies and plants along riparian corridors in an urban landscape. *Divers. Distrib.* **2012**, *18*, 742–753. [CrossRef]

23. Rodewald, A.D.; Bakermans, M.H. What is the appropriate paradigm for riparian forest conservation? *Biol. Conserv.* **2006**, *128*, 193–200. [CrossRef]

24. Rottenborn, S.C. Predicting the impacts of urbanization on riparian bird communities. *Biol. Conserv.* **1999**, *88*, 289–299. [CrossRef]

25. Suri, J.; Anderson, P.M.; Charles-Dominique, T.; Hellard, E.; Cumming, G.S. More than just a corridor: A suburban river catchment enhances bird functional diversity. *Landsc. Urban. Plan.* **2017**, *157*, 331–342. [CrossRef]

26. Andrade, R.; Bateman, H.L.; Frank, J.; Allen, D. Waterbird community composition, abundance, and diversity along an urban gradient. *Landsc. Urban. Plan.* **2018**, *170*, 103–111. [CrossRef]

27. Säumel, I.; Kowarik, I. Urban rivers as dispersal corridors for primarily wind-dispersed invasive tree species. *Landsc. Urban. Plan.* **2010**, *94*, 244–249. [CrossRef]

28. Aguiar, F.C.F.; Ferrereira, M.T. Plant invasions in the rivers of the Iberian Peninsula, south-western Europe: A review. *Plant. Biosyst.* **2013**, *147*, 1107–1119. [CrossRef]

29. Bravard, J.P. La gestión de los ríos en el medio urbano: Tendencias francesas. In *Ríos Y Ciudades*; De la Call, P., Pellicer, F., Eds.; Institución Fernando El Católico: Zaragoza, Spain, 2002; 400p.

30. Inoue, M.; Nakagoshi, N. The effects of human impact on spatial structure of the riparian vegetation along the Ashida river, Japan. *Landsc. Urban. Plan.* **2001**, *53*, 111–121. [CrossRef]

31. Dearborn, C.C.; Kark, S. Motivations for Conserving Urban Biodiversity. *Conserv. Biol.* **2010**, *24*, 432–440. [CrossRef]

32. Molina, P.; Berrocal, A. Dinámica fluvial, propiedad de la tierra y conservación del paisaje de ribera en el entorno de Aranjuez (Madrid, Toledo). *Estud. Geográficos* **2013**, *64*, 495–522. [CrossRef]

33. González, M.A.; de la Lastra, I.; Rodríguez, I. La urbanización y su efecto en los ríos. In *Mesa de Trabajo Estrategia Nacional de Restauración de Ríos y Riberas*; Ministerio de Medio Ambiente y Universidad Politécnica de Madrid: Madrid, Spain, 2007; 45p.

34. United Nations, Department of Economic and Social Affairs. Population Division. In *World Urbanization Prospects: The 2018 Revision (ST/ESA/SER.A/420)*; United Nations: New York, NY, USA, 2019.

35. Bayona, J.; Pujadas, I. Las grandes áreas metropolitanas en España: Del crecimiento y la expansión residencial al estancamiento poblacional. *Doc. D'anàlisi Geogràfica* **2020**, *66*, 27–55. [CrossRef]

36. Eurostat. *Eurostat Regional Yearbook*; European Union: Luxembourg, 2019.

37. Major Metropolitan Areas in Europe. Available online: http://www.newgeography.com/content/003879-major-metropolitan-areas-europe (accessed on 9 April 2020).

38. Cengiz, B. Urban river landscapes. In *Advances in Landscape Architecture*; Ozyavuz, M., Ed.; InTech: Rijeka, Croatia, Yugoslavia, 2013.

39. Vadillo, A.; Molina, P. Los paisajes del río Pisuerga en la ciudad de Valladolid: Evolución, sostenibilidad y participación ciudadana. In *Colloque International Paysages de la Vie Quotidienne. Regards Croisés Entre la Recherche et L'action*; CEMEGRAF: Perpignan, France, 2011; 39p.

40. Hale, B.W.; Adams, M.S. Ecosystem management and the conservation of river -floodplain systems. *Landsc. Urban. Plan.* **2007**, *80*, 23–33. [CrossRef]

41. Molina, P.; Sanz, C.; Mata-Olmo, R. *Los Paisajes del Tajo*; Ministerio de Agricultura, Pesca y Alimentación: Madrid, Spain, 2010; p. 358.

42. Anuario de Aforos 2015–2016. Available online: http://ceh-flumen64.cedex.es/anuarioaforos/default.asp (accessed on 20 March 2020).

43. Molina-Holgado, P. Análisis y Comparación de la Vegetación de las Riberas de los Ríos Ebro, Tajo Y Jarama. Ph.D. Thesis, Universidad Autónoma de Madrid, Madrid, Spain, 2003.

44. Biodiversity Strategy. Available online: https://ec.europa.eu/environment/nature/biodiversity/strategy/index_en.htm (accessed on 10 December 2019).

45. The European Landscape Convention. Available online: https://www.coe.int/en/web/landscape (accessed on 10 December 2019).

46. The EU Strategy on Green Infrastructure. Available online: https://ec.europa.eu/environment/nature/ecosystems/strategy/index_en.htm (accessed on 10 April 2020).

47. Real Decreto Legislativo 1/2001, de 20 de Julio, por el que se Aprueba el Texto Refundido de la Ley de Aguas. Available online: https://www.boe.es/buscar/act.php?id=BOE-A-2001-14276 (accessed on 5 September 2019).

48. De Castro, M.; Martín-Vide, J.; Alonso, S. El clima de España: Pasado, presente y escenarios de clima para el siglo XXI. In *Evaluación Preliminar de los Impactos en España por Efecto del Cambio Climático*; Moreno-Rodríguez, J.M., Ed.; Ministerio de Medio Ambiente: Madrid, Spain, 2005.

49. Alonso-Zarza, A.M. Cuenca del Tajo. In *Geología de España*; Vera, J., Ed.; Instituto Geológico y Minero de España: Madrid, Spain, 2004.

50. Solís, E.; Ureña, J.M.; Ruiz-Apilánez, B. Transformación del sistema urbano-territorial en la región central de la España peninsular: La emergencia de la región metropolitana policéntrica madrileña. *Scr. Nova.* **2012**, *16*. Available online: https://www.ub.edu/geocrit/sn/sn-420.htm (accessed on 10 December 2019).

51. Instituto Nacional de Estadística. Cifras Oficiales de Población Resultantes de la Revisión del Padrón Municipal A 1 de Enero 2019. Available online: https://www.ine.es/dynt3/inebase/es/index.htm?padre=517&capsel=525 (accessed on 10 December 2019).

52. De Coca, J.; Fernández, F. La renovación del Manzanares: Transformaciones y reciclaje. Paisajes urbanos. *Proy. Prog. Arquit.* **2011**, *4*, 88–105.

53. Belinches, A. *Aquitectura de Madrid*; Fundación COAM: Madrid, Spain, 2003.

54. Zona de Especial Conservación Ribera de Henares ES4240003. Available online: https://www.castillalamancha.es/gobierno/agrimedambydesrur/estructura/dgapfyen/rednatura2000/zecES4240003 (accessed on 25 February 2020).

55. Instituto Geológico Y Minero de España. Navegador de Información Espacial. Available online: http://info.igme.es/visorweb (accessed on 24 February 2020).

56. Méndez-Cabezas, M. *Los Molinos de Agua de la Provincia de Toledo*; Diputación Provincial de Toledo: Toledo, Spain, 1989.

57. Masa, F. Provincia de Toledo. In *Guía de Castilla-La Mancha. Patronio Histórico*; Lara, P., Masa, F., Eds.; Servicio de Publicaciones de la Junta de Comunidades de Castilla-La Mancha: Toledo, Spain, 1992.

58. World Heritage List. Historical City of Toledo. Available online: https://whc.unesco.org/en/list/379 (accessed on 26 February 2020).

59. Instituto Nacional de Estadística. Viajeros Y Pernoctaciones por Puntos Turísticos. Toledo. Available online: https://www.ine.es/jaxiT3/Datos.htm?t=2078#!tabs-tabla (accessed on 26 February 2020).

60. Pacheco, C. Obras públicas en Talavera de la Reina: Los puentes medievales. Aproximación histórica y arqueológica. *Espac. Tiempo Y Forma* **2001**, *14*, 163–191.

61. Bibby, C.J.; Burguess, N.D.; Hill, D.A.; Mustoe, S.H. *Bird Census Techniques*; Academic Press: London, UK, 2000.

62. Delany, S. Guidance on Waterbird Monitoring Methodology: Field Protocol for Waterbird Counting. Wetlands International: Wageningen, The Netherland, 2010.

63. Magurran, A.E. *Ecological Diversity and Its Measurement*; Coom Helm: London, UK, 1988.

64. Staneva, A.; Burfield, I. *European Birds of Conservation Concern. Populations, Trends and National Responsibilities*; Bird Life International: Cambridge, UK, 2017.
65. IUCN. *IUCN Red List Categories and Criteria*, 2nd ed.; IUCN: Gland, Switzerland; Cambridge, UK, 2012.
66. Red List for Birds, Bird Life International. Available online: http://datazone.birdlife.org/spe-cies/search (accessed on 3 April 2020).
67. European Red List of Birds, Bird Life International. Office for Official Publications of the European Communities: Luxembourg. Available online: http://datazone.birdlife.org/info/euroredlist (accessed on 3 April 2020).
68. Bird Species of Annex I of the Birds Directive. Available online: https://ec.europa.eu/environment/nature/conservation/wildbirds/threatened/index_en.htm (accessed on 4 April 2020).
69. The Habitat Directive. Available online: https://ec.europa.eu/environment/nature/legislation/habitatsdirective/index_en.htm (accessed on 4 April 2020).
70. Plan de Naturalización del Manzanares a su paso por la Ciudad de Madrid. Available online: https://www.esmadrid.com/sites/default/files/dossier_plan_naturalizacion_manzanares.pdf (accessed on 7 April 2020).
71. Documento Inicial Para el Procedimiento de Evaluación de Impacto Ambiental del Proyecto para la Integración del río Tajo en la Ciudad de Toledo. Available online: https://realacademiatoledo.es/wp-content/uploads/2014/05/noticias_2014_proyectotajo2.pdf (accessed on 7 April 2020).
72. Estrategia de Desarrollo Urbano Sostenible de la Ciudad de Guadalajara. Available online: https://www.guadalajara.es/recursos/doc/portal/2017/09/19/estrategia-de-desarrollo-urbano-sostenible-integrado-2014-2020.pdf (accessed on 7 April 2020).
73. Memoria Y Presentación de la Propuesta Técnica Ganadora del Concurso Internacional de Ideas Para la Integración de los ríos Tajo Y Alberche en la Ciudad de Talavera de la Reina. Available online: http://www.chtajo.es/Servicios/Contratacion/Documents/talavera/MEMORIA%20GLOBAL.pdf (accessed on 7 April 2020).
74. News in Relationship of Henares´ Project. Available online: https://www.eldiario.es/clm/destrozo-Henares-ambiental-Ayuntamiento-Guadalajara_0_868263921.html (accessed on 7 April 2020).
75. Consultations to the European Parliament, E-005479-18. Available online: https://www.europarl.europa.eu/doceo/document/E-8-2018-005479_ES.html (accessed on 7 April 2020).
76. News Article in El País. La Inesperada Recuperación Medioambiental del Manzanares. Available online: https://elpais.com/ccaa/2018/09/14/madrid/1536928384_530297.html (accessed on 7 April 2020).
77. Morillo, C. *Atlas Y Manual de los Hábitats Terrestres de España*; Ministerio de Medio Ambiente: Madrid, Spain, 2003.
78. Hábitats de Interés Comunitario del Anexo I de la Directiva 92/43/CEE. Available online: https://www.miteco.gob.es/es/biodiversidad/servicios/banco-datos-naturaleza/informacion (accessed on 25 March 2020).
79. Franz, K.W.; Romanowski, J.; Saavedra, D. Effects of prospective landscape changes on species viability in Segre River valley, NE Spain. *Landsc. Urban. Plan.* **2011**, *100*, 242–250. [CrossRef]
80. Perini, K. Strategies and Techniques. In *Sustainability and River Restoration: Green and Blue Infrastructure*; Perini, K., Sabbion, P., Eds.; John Wiley & Sons Ltd.: Madrid Río, Spain, 2016; pp. 117–126.
81. Garrido, G. Madrid Río, o el retorno de la urbe a la geografía del Manzanares. *Revista PH* **2017**, *91*, 100–117. [CrossRef]
82. Chin, A.; Gregory, K. From research to application: Management implications from studies of urban river channel adjustment. *Geogr. Compass* **2009**, *3*, 297–328. [CrossRef]
83. Downs, P.; Gregory, K. *River Channel Management: Towards Sustainable Catchment Hydrosystems*; Routledge: Cambridge, UK, 2014.
84. Díaz-Orueta, U. Megaproyectos urbanos y modelo de ciudad. El ejemplo de Madrid Río. Cuaderno urbano. *Espac. Cult. Soc.* **2015**, *19*, 179–200.
85. Proyecto de Integración de los Ríos Tajo y Alberche en Talavera de la Reina. Available online: http://www.chtajo.es/Servicios/Contratacion/Paginas/Talavera2.aspx (accessed on 10 March 2020).
86. Crecimiento Sostenible FEDER 2014-20 PO. Available online: https://www.idae.es/uploads/documentos/documentos_PO_CrecimientoSostenible_FEDER_2014-2020_cb50c638.pdf (accessed on 10 April 2020).
87. Baschak, L.A.; Brown, R.D. An ecological framework for the planning, design and management of urban river greenways. *Landsc. Urban. Plan.* **1995**, *33*, 211–225. [CrossRef]

Sustainability **2020**, *12*, 4661

88. Fujihara, M.; Kikuchi, T. Changes in the landscape structure of the Nagara River Basin, central Japan. *Landsc. Urban. Plan.* **2005**, *70*, 271–281. [CrossRef]

89. Magdaleno, F.; Fernández-Yuste, J.A.; Martínez-Santa-María, C.; Sánchez, F.J.; Aparicio, M. De Madrid al cielo a través del Manzanares: Restauración del río en el entorno de El Pardo. In Proceedings of the 6° Congreso Forestal Español, Vitoria-Gasteiz, Spain, 10–14 June 2013.

90. Kuiper, J. Landscape quality based upon diversity, coherence and continuity Landscape planning at different planning-levels in the River area of The Netherlands. *Landsc. Urban. Plan.* **1998**, *43*, 91–104. [CrossRef]

91. Donnelly, R.; Marzluff, J.M. Importance of reserve size and landscape context to urban bird conservation. *Conserv. Biol.* **2004**, *18*, 733–745. [CrossRef]

92. McKinney, R.A.; Raposa, K.B.; Cournoyer, R.M. Wetlands as habitat in urbanising landscapes: Patterns of bird abundance and occupancy. *Landsc. Urban. Plan.* **2011**, *100*, 144–152. [CrossRef]

93. Hering, D.; Borja, A.; Carstensen, J.; Carvalho, L.; Elliott, M.; Feld, C.K.; Heiskanen, A.-S.; Johnson, R.K.; Moe, J.; Pont, D.; et al. The European Water Framework Directive at the age of 10: A critical review of the achievements with recommendations for the future. *Sci. Total Environ.* **2010**, *408*, 4007–4019. [CrossRef] [PubMed]

94. Junker, B.; Buchecker, M. Aesthetic preferences versus ecological objectives in river restorations. *Landsc. Urban. Plan.* **2008**, *85*, 141–154. [CrossRef]

95. Verbrugge, L.; Born, R.V.D. The role of place attachment in public perceptions of a re-landscaping intervention in the river Waal (The Netherlands). *Landsc. Urban. Plan.* **2018**, *177*, 241. [CrossRef]

96. Chou, R.J. Achieving Successful River Restoration in Dense Urban Areas: Lessons from Taiwan. *Sustainability* **2016**, *8*, 1159. [CrossRef]

97. Shi, S.; Kondolf, M.L.D. Urban River Transformation and the Landscape Garden City Movement in China. *Sustainability* **2018**, *10*, 4103. [CrossRef]

98. Landon, N.; Piégay, H.; Bravard, J.P. The Drome river incision (France): From assessment to management. *Landsc. Urban. Plan.* **1998**, *43*, 119–131. [CrossRef]

99. Clément, G. *Manifiesto por el Tercer Paisaje*; Gustavo Gili: Barcelona, Spain, 2018.

100. Francis, A. Urban rivers: Novel ecosystems, new challenges. *Wires Water* **2013**, *1*, 19–29. [CrossRef]

101. Green Infrastructure: Better Living through Nature-Based Solutions. Available online: https://www.eea.europa.eu/articles/green-infrastructure-better-living-through (accessed on 10 March 2020).

102. Fernández-Juricic, E.; Jokimäki, J. A habitat island approach to conserving birds in urban landscapes: Case studies from southern and northern Europe. *Biodivers. Conserv.* **2001**, *10*, 2023–2043. [CrossRef]

103. Urueña, J.M. La ordenación de los espacios fluviales en las ciudades. In *Ríos y Ciudades*; De la Call, P., Pellicer, F., Eds.; Institución Fernando El Católico: Zaragoza, Spain, 2002; 400p.

104. Spirn, A.W. *O Jardim de Granito: A Natureza no Desenho da Cidade*; Editora da Universidade de São Paulo: São Paulo, Brazil, 1995; p. 345.

Article

Decolonizing Pathways to Sustainability: Lessons Learned from Three Inuit Communities in NunatuKavut, Canada

Amy Hudson * and Kelly Vodden

Environmental Policy Institute, Grenfell Campus, Memorial University of Newfoundland,
Corner Brook, NL A2H 5G5, Canada; kvodden@grenfell.mun.ca
* Correspondence: ahudson@mun.ca

Received: 28 April 2020; Accepted: 25 May 2020; Published: 28 May 2020

Abstract: Community led planning is necessary for Inuit to self-determine on their lands and to ensure the preservation of cultural landscapes and the sustainability of social-ecological systems that they are a part of. The sustainability efforts of three Inuit communities in Labrador during a Community Governance and Sustainability Initiative were guided by a decolonized and strength-based planning framework, including the values of Inuit in this study. This paper demonstrates that Inuit led planning efforts can strengthen community sustainability planning interests and potential. We situate the experiences of NunatuKavut Inuit within, and contribute to, the existing body of scholarly decolonization and sustainability literature. For many Indigenous people, including Inuit, decolonization is connected to inherent rights to self-determination. The findings suggest that decolonizing efforts must be understood and actualized within an Indigenous led research and sustainability planning paradigm that facilitates autonomous decision making and that is place based. Further, this study illustrates five predominant results regarding Inuit in planning for community sustainability that support sustainable self-determination. These include: inter and cross community sharing; identification of community strengths; strengthened community capacity; re-connection to community and culture; and the possibility for identification of sustainability goals to begin implementation through community led governance and planning processes.

Keywords: Inuit; sustainability; decolonization; self-determination; community planning

---

1. Introduction

Sustainability planning is necessary for community and cultural survival in remote Indigenous regions, like those in NunatuKavut (coastal Labrador). There is increasing recognition within the sustainability science literature of the need for place-based sustainability goals in Arctic communities that align with Arctic needs, based on the fact that these needs may in fact differ from global responses and efforts [1]. The literature reveals that both Indigenous and sustainability sciences contribute to the sustainability of "resilient landscapes", and to our understanding of them [2,3] (p. 1). This recognition further validates the need to work with Indigenous peoples in planning, by doing planning and sustainability scholarship differently. Sustainability science has been disconnected from Indigenous science and this has meant that Indigenous rights and knowledge have not been adequately engaged or privileged by Western scientific enquiry [3]. The participation of Indigenous peoples in planning processes have also been notably marginalized in Canada and around the world [4], with outside planning actors participating in the dispossession and marginalization of Indigenous peoples in the planning process [5]. This is despite the fact that "Indigenous peoples possess deep connections to place and knowledge of the land upon which they have lived for thousands of years" [6] (p. 428)

123

and that planning is a vital aspect of governance, including Indigenous forms of governance that have also endured marginalization resulting from colonization [7]. Planners must be cognizant of this colonial history as "state-based planning has provided the conceptual and practical apparatus for institutionalizing marginalization" [8] (p. 643).

Sustainability work in rural and remote Indigenous communities offers important contributions to the sustainability science knowledge base. Recent collaborative, community-based research in the area of renewable energy in Labrador, for example, demonstrates that the voice of Inuit and their active participation in decision making is an integral part of process and outcome, building on the strengths and knowledge of Inuit themselves while reinforcing their role as decision makers and experts on their lands [9]. Land-use planning in the Nunatsiavut region of Labrador offers further insight into Indigenous planning in Labrador and the North. The land use plan of the Nunatsiavut government has been designed to "respond, first and foremost, to Inuit environmental, social, cultural, and economic interest" [10] (p. 438). Earlier research related to the process of mine development in Voisey's Bay, Labrador cited the apparent success of agreements reached between Indigenous and non-Indigenous parties that was based on "sustainability centered decision making" [11] (p. 343). Yet, O'Faircheallaigh [12] illustrates the tensions and complexities involved in the Voisey's development. The Province of Newfoundland (at the time), committed to advancing the development of the mine as expeditiously as possible, left the Innu and Inuit (the latter group represented by the LIA-Labrador Inuit Association) emphatic about their inclusion and participation in negotiations and reaching satisfactory agreements. The Innu were opposed to development early on but felt (along with the Inuit represented by LIA) that they had no choice but to seek inclusion as the development was set to proceed [12]. Moreover, Archibald and Crnkovich [13] point to a lack of Inuit women's representation and voice in the Voisey's Bay development, adding that analysis into the differential impacts on Inuit women were lacking in this development.

Indigenous planning has been broadly defined as a process whereby Indigenous people make their own decisions on their lands, and drawing upon the knowledge, values and principles within themselves to "define and progress their present and future social, cultural, environmental and economic aspirations" [8] (p. 642). To date, planning in practice has yielded limited opportunities to share and exercise principles and practices of Indigenous planning, particularly in the context of sovereign nations [7]. Indigenous planning has been identified as an approach that respects Indigenous sovereignty and worldviews [14], requiring sustainability planning approaches in Indigenous communities that are cognizant of inherent and sovereign rights to land and culture.

Indigenous peoples assert jurisdiction over their lands and within their communities in various ways (e.g., land claims, advocacy, agreements with the state, planning efforts). Most Inuit groups in Canada have settled land claims agreements with the state [15]. Inuit in NunatuKavut have not yet settled a final land claim agreement. However, they have a long history of asserting their rights on their land. Most recently, Canada has accepted the NunatuKavut Community Council (NCC), a governing organization that represents the Indigenous rights of NunatuKavut Inuit, into a Recognition of Indigenous Rights and Self-Determination (RIRSD) process to negotiate on matters of mutual interest between NunatuKavut Inuit and Canada [15]. Today, NunatuKavut Inuit continue to assert their rights on their land to ensure the future of their people and communities. Community-led sustainability planning during a Community Governance and Sustainability Initiative (CGSI) in NunatuKavut should be understood within a rights-based paradigm.

The CGSI (described in more detail below), was piloted in three select Inuit communities in NunatuKavut during 2017 and 2018 to facilitate opportunities for those communities to think about the future from the perspective of sustainability, grounded in their rights as Inuit belonging to their ancestral lands, and to plan accordingly. Baxter and Purcell [16] define Integrated Community Sustainability Planning (ICSP) as "a high-level overarching document for a community that is informed by sustainability principles and guides the community into the future" (p. 35). ICSPs are one example of a model of sustainability planning that have been employed across Canada, including the province of Newfoundland and Labrador (NL) [17]. This paper presents an alternative Indigenous sustainability planning perspective and approach, particularly one that is grounded in the efforts of Inuit in NunatuKavut through a community led, decolonized and strength-based planning framework. This study builds upon normative ideas of community sustainability planning, like ICSP, at the same time as privileging Inuit knowledge, expertise and values that are vital to the planning process within Inuit territories.

Throughout this paper, we draw upon and situate Inuit planning within the overarching concept of decolonization, while building on the work of Indigenous scholars who have informed our analysis such as Jeff Corntassel [18], Pam Palmater [19], Linda Smith [20], and Shawn Wilson [21]. In NunatuKavut, where Inuit are planning for sustainable communities and futures, planning efforts invoke a necessary and simultaneous process of self-decolonization. The decolonizing of the self is integral to a larger order of decolonization and to anti-colonial sustainability efforts that connect both theory and practice. The concept of "sustainable self-determination," a term coined by Indigenous scholar Jeff Corntassel [18], is useful for understanding Inuit planning in NunatuKavut as a pathway to decolonized self-determination. In the context of NunatuKavut Inuit, we argue that Inuit led, decolonized and strength-based planning, can strengthen community sustainability planning interest and overall potential. The results of this process give rise to sustainable self-determination that contribute to the preservation of cultural landscapes and the sustainability of social-ecological systems that make up Inuit society.

1.1. Decolonization and Sustainable Self-Determination

Community sustainability planning approaches designed and developed by and for Indigenous peoples are integral to Indigenous self-determination efforts. Indigenous governance practices and methods, including planning efforts, can be conducive to the creation of societies that are more sustainable [22]. Recent research with First Nations in Saskatchewan, for example, point to the success of Indigenous planning when the approach results in trust relationships between the First Nation community, other participants and university researchers and community capacity is strengthened [23]. The ability of communities to self-determine in ways that reflect Indigenous ways of knowing and being is in part, contingent upon Indigenous autonomy and control of decision making about the future. Yet, Indigenous community planning and approaches to planning have often been marginalized by external decision makers [4]. Externally controlled community development and planning processes are indicative of colonial ideas and mentalities that undermine Indigenous knowledge and expertise in favor of Western European knowledge in deciding matters for the future of Inuit and their lands. Therefore, any approach to decolonized community planning must be cognizant of historic and modern impacts of colonization.

Indigenous scholar, lawyer and advocate Pamela Palmater defines colonization as a process by which "a state or colony attempts to dispossess and subjugate the original Indigenous peoples of the land," [19] (p. 3) and she maintains that colonization, in this form, has not ended for Indigenous peoples. Corntassel [24] portrays colonization as a dysfunctional force that disconnects peoples from their home, land and culture. He maintains that Indigenous resurgence is about connecting to home, land and culture, a central feature of decolonization.

Decolonization has been defined and drawn upon by academia, institutions and governments. Leading Indigenous scholars like Linda Smith [20] and Margaret Kovach [25] have engaged decolonization discourse, enlightening a world that resonates for many Indigenous peoples and offering insights into how to think about and do research differently. Conceptually and practically, decolonization is a necessary and integral step towards acknowledging and confronting the legacy of colonization (past and ongoing). Decolonizing work is an ever evolving, dynamic and site-specific process. Decolonization and decolonized planning can be further linked to Corntassel's key concept of sustainable self-determination, with a view towards privileging and bringing attention to Inuit efforts to self-determine that may otherwise go unnoticed by outside decision makers or planners.

We engage decolonization as a process that sets the foundation for everyday acts of resurgence, including Indigenous-led planning. Corntassel [24] recalled pathways to decolonization that are and can be realized through Indigenous led self-determination efforts. Learning from Fanon [26], we are alert to the reality that decolonization implies a commitment to embracing differing worldviews and perspectives, and the tensions that are inherent in this process. This entails moving beyond European norms and ways of thinking. Decolonization must be a unique and context specific process that includes individual and collective acts of resurgence, revitalization and determination contingent upon time and place, in Indigenous peoples' pursuit of self-determination. We argue that a decolonial approach to community sustainability planning in NunatuKavut is integral to ensuring that the sustainability goals identified and the planning process itself is embedded in a vision for the future that is self-determined by Inuit in their time and place and reflective of Inuit values and ways of knowing and being. In this way planning can, in turn, further sustainable self-determination and create the pathways to decolonization observed and called for by Corntassel and others.

1.2. Grounding Decolonization: Recognizing the Role of Indigenous Peoples and Their Communities

The participation of planning actors in the "dispossession, oppression and marginalization of Indigenous peoples has implications for the field" [5] (p. 403). Recognizing colonial realities allows for the challenging of western, well intentioned, and persistent assumptions imbued in planning that seek to "better the world" [5] (p. 403). Indigenous claims to self-determination, land restitution, etc., make the need to challenge planning assumptions evident and timely. When Indigenous people question ongoing normative assumptions and practices by privileging their own ways of knowing and being, opportunities arise to plan for a future that is shaped by their own worldview(s). The ability to inform planning approaches from one's own space (values, goals, etc.), as opposed to outside perceptions of what is good or necessary, is optimal for decolonizing planning processes that are Indigenous designed and led.

In many cases, Indigenous peoples, communities, nations and governments continue to work towards building a future and a path that is reflective of their values, perspectives and worldviews, despite ongoing colonial interference. Indigenous peoples have been finding opportunities to revitalize as nations, while making small movements towards reclamation—whether that be of culture, language, education, political society, etc. [20,24]. We contend that acts of resistance and resurgence in these forms are a necessary part of the process of decolonization and are necessarily linked to community planning, yet they often go unrecognized as a source of knowledge or expertise integral to planning work by outsiders. Additionally, these acts are rarely upheld or highlighted as integral and tangible decolonizing work, particularly by states and/or institutions who often set the standard for how reconciliation and/or decolonization is to be approached in Canada and within institutions (i.e., academia). This provides evidence that as a society we are still unwilling to really learn or accept the knowledge and expertise of Indigenous peoples in their place and as autonomous rights holders on their lands. Realities like these are well established and have been demonstrated over time as the courts have consistently failed to consider Indigenous people's perspectives in law and legal analysis [27,28]. This too has implications for the field of Indigenous sustainability planning.

The idea that the state and its government know best is an age-old way of thinking and doing and is perpetuated in relations with Indigenous peoples, and even in times of good will and positive intention. Eisenberg, Webber, and Coulthard [29] maintain that Indigenous peoples and communities themselves are the sole agents with the power to recognize and give expression to the knowledge that make up who they are. When Indigenous peoples, organizations, and communities take on the arduous tasks of reclamation through tangible and practical everyday acts on their lands and in their communities, they are in fact pursuing and leading decolonizing work that lends toward self-determination.

A strength-based approach to community sustainability planning, that rested on the values, hopes and goals of Inuit in this study, guided the approach of the CGSI. This work exists as an example of a community based and community driven approach to decolonization, grounded in and guided by connection to home, values and individual and collective determination to ensure the survival and preservation of community and culture. In what follows, we describe and interpret acts of resurgence, revitalization and sustainable self-determination in three Inuit communities within community sustainability planning efforts as part of, and emblematic of, a larger process of decolonization.

2. Methodology

This research was guided by Indigenous and qualitative research methodologies. Indigenous research methodology is integral to understanding and making space for sustainable self-determination in Indigenous communities. The ability to share, learn and listen through stories is fundamental to understanding Indigenous worldviews and perspectives and storytelling is an integral and valued method and approach [30]. This research seeks to ensure that the voice and knowledge of Inuit are privileged and drive the findings of this paper. A culturally relevant research paradigm (as employed in this research), ensures that Indigenous methods are validated and used [21], contributing to decolonization and supporting the assertion of rights and sovereignty. Research within this paradigm remains cognizant of a history of colonially rooted research practices (including a tradition that privileges research practices that are value neutral), while remaining committed to research that seeks to better the well-being of Indigenous peoples as per their ways of being and knowing [20]. Booth and Muir [31] understand Indigenous planning as an attempt to "recognize the unique and specific legal, political, historical, cultural and social circumstances in which the world's Indigenous peoples find themselves" (p. 422). It can be argued that this is also the case for the Inuit of NunatuKavut and their representative governing organization the NunatuKavut Community Council (NCC), as they seek to enhance capacity and knowledge for planning that is specific to their needs, interests, and historical and modern realities and as they engage in culturally relevant planning to advance self-determination efforts. This research initiated and facilitated community capacity strengthening efforts so that community members and leaders are better equipped to effectively engage in the planning of their communities for the future and validated in doing so.

2.1. Community Governance and Sustainability Initiative (CGSI): A Framework for Designing and Implementing Community Led and Responsive Research and Planning Practices

There is a growing interest in planning that is adaptable to uncertain conditions and realities [32]. Adaptability is a central feature of Inuit societies. Cognizant of the social and political history of the Inuit communities in NunatuKavut, and moreover, a legacy of research on and within Indigenous communities broadly, the overall approach to this research was to work with NunatuKavut Inuit and to locate positive attributes of their communities, and to privilege Inuit worldviews and perspectives in the process. We collaboratively identified approaches and ways of doing based on what has worked well in the past, locating expertise and assets within communities themselves, all to further strengthen and benefit from the adaptive capacities required to vision and plan for a positive and vibrant future that is relevant to Inuit themselves.

We examined contributions in NunatuKavut in the areas of self-determination, decolonization, resurgence and rights that are Indigenous led and inspired, building upon scholarly literature in discussions surrounding decolonization and sustainability. The worlds of academia and Inuit community life have come together in this project to support the creation of space and opportunities for community sustainability planning. These opportunities have implications for the preservation of culture and communities in NunatuKavut, and for the methodology used in this research.

Respectful community engagement was guided by the work of leading Indigenous scholars in the field like Smith, Wilson and Kovach, along with Hudson's connection to her home community and to NunatuKavut generally. This approach to community engagement helped to ensure that the research study was informed by the community in both purpose and methods. We also drew from the expertise, knowledge and guidance of three NunatuKavut communities: Black Tickle, Norman Bay and St. Lewis (Appendix A, Table A1). This research was community led and driven and the research methods support this end. Hart [33] writes of research that is "structured within an epistemology that includes a subjectively based process for knowledge development and a reliance on Elders and individuals who have or are developing this insight" (p. 9). Hudson's own experiences, as a result of growing up in and belonging to one of the pilot communities of this study and her work with the NCC, further embedded and ensured accountability to this research approach.

Strength-based decision making and planning was introduced as the framework for our discussions. This assisted in situating Inuit participants as knowledge holders and experts on matters that impact them and on their lands. This strength-based approach is particularly fundamental to decolonized sustainability planning in NunatuKavut. Deficit based research has often been conducted in Indigenous communities, failing to acknowledge and respect Indigenous knowledge and expertise [34]. The use of strength-based planning allowed for Inuit worldviews, values and perspectives to lead and guide the planning process. Planning with and by Indigenous peoples in this way has elsewhere resulted in positive outcomes across a range of areas like culture, identity-building, healing, etc. [35]. In this study, dialogue around strength-based thinking was integral to envisioning a sustainable future. It is noteworthy that females pre-dominantly led the sustainability work and all three community sustainability coordinators (described below) were female. In remote communities such as these, there is often a tendency to focus on what has not been working in communities, or how governments or other governing bodies are not working, without looking at the potential and individual and collective agency that already exists within communities. Strength based discussions, asset mapping and visioning exercises assisted communities in maneuvering around this paradigm to get to a place of planning without the baggage of what has gone wrong in the past, which stands in the way of planning a desired future. Planning from a place of strength that privileges local Inuit knowledge is also key to the pursuit of sustainable self-determination.

As a way to initiate the CGSI a regional workshop was held in Happy Valley-Goose Bay (HVGB), Spring 2017. This gathering brought together the three pilot communities, including three representatives from each of the communities. We worked with community participants and engaged in various awareness, skills and capacity building exercises. They included: (a) strength-based decision making and planning; (b) community visioning exercises; (c) community asset mapping; (d) community engagement; and (f) proposal writing.

Following the initial gathering in HVGB, pilot community participants applied and furthered the lessons that they had learned once they returned home to their community (e.g., asset mapping). As research lead, Hudson identified an external funding opportunity to further the community sustainability planning work. This allowed NCC to employ a community sustainability coordinator in each of the three communities for a period of seven months. Throughout the scope of this work, and working directly with Hudson, community sustainability coordinators were able to solidify sustainability committees in their respective communities and then co-led the committees in a range of activities and areas relevant and localized to each community. Hudson oversaw the work of the coordinators as NCC lead and as a part of this study. The coordinators furthered asset mapping exercises, participated in and co- led visioning exercises and activities (feast, cultural events, community games, etc.), wrote proposals, and engaged in networking opportunities with stakeholders.

2.2. Recruitment and Data Collection

Interactive workshops, gatherings and community meetings supported both collaboration and consensus building discussions and provided the space and environment to engage participants throughout 2017 and into 2018. These workshops, meetings and gatherings were predominantly held in the study communities, with the exception of two larger gatherings that brought together all three communities to learn and share in a larger setting in HVGB. Recruitment strategies within communities relied on local knowledge and expertise from community members and the NCC. Other NCC partners, past and present, with experience and knowledge of NCC governance and land claims, were also invited to participate. Participants were contacted in various ways depending on the data collection strategy (i.e., email, public notices, in person, email). In order to achieve the goals of the project across three communities, it was necessary to employ a multi-dimensional approach to community outreach and engagement, and the project lent itself to learning and refining best practices, in working with the three communities.

Qualitative data collection methods included one on one interviews, focus groups, and surveys. Participants were recruited by email, telephone and word of mouth for each of these methods. Four one on one interviews were conducted in the communities (one from Black Tickle, two from St. Lewis, one from Norman Bay). Additionally, two external interviews were conducted with individuals who have been participatory to NCC's land claim and research journey over the past two decades. See Table A2 for a detailed list of activities undertaken with participants from each of the three pilot communities. Interviews occurred simultaneously with other forms of data collection. We chose interviews as a data collection method given the centrality of interviewing to qualitative methodology. However, it was clear that action-oriented data collection that directly engaged participants in gatherings (like those described above) and settings designed to share and learn from one another, were much more conducive to collecting rich data and in engaging participants throughout the research. In some instances, such as the two gatherings in HVGB, stakeholders were invited by email to participate, listen and respond to community interests and goals. Some of the stakeholders in attendance included representatives from funding agencies (e.g., Atlantic Canada Opportunities Agency), business advisors from Nunacor (NCC's business arm), and academics in related fields at Memorial University.

The two larger, centralized gatherings, also referred to as workshops, were held in HVGB and brought representatives from all three study communities together. Recruitment for these two gatherings was done by contacting the local governing structure by telephone and email in each of the study communities (municipality, local service district, recreation committee). It was appropriate to work with the local governing boards to not only seek their interest in the project, but to identify recruits to attend the gatherings in HVGB. The second gathering, recruited in much the same manner, also hosted a focus group discussion with participants from all three communities. The dynamics of these gatherings were comfortable, supportive, open and transparent. Existing best practices in engagement by NCC in the past also assisted in implementing spaces that were conducive to sharing and dialogue. Community gatherings ranged in size and were influenced by community population size, with 25+ people attending in Black Tickle at a full day youth and community event, approximately six people in Norman Bay and 40+ people in St Lewis at a community feast and youth/family event. The community feast in St. Lewis resulted in 43 written submissions by community members detailing what they value most about life in St. Lewis.

There were four focus groups in total (one in each individual pilot community and one collective focus group at the second sustainability gathering in HVGB-described above). There were seven participants in the focus group in Black Tickle, two in Norman Bay, six in St. Lewis and ten in the HVGB workshop. Participants attended and engaged in two workshops in Happy Valley Goose Bay with ten participants in each workshop. Survey respondents totaled 26 in Norman Bay and St. Lewis. The surveys sought to elicit information about the age, gender, and connection community members felt towards their home. The surveys were not initiated or completed in Black Tickle as the community is all of Hudson's relations. While surveys assist in gathering relevant information for analysis, in this context the use of a survey in Hudson's home community felt too impersonal. Hudson knows each individual personally and shares ancestral ties and modern-day kinship and social networks with them.

Further data were collected through collaborative community development efforts (planning and ideas sharing), and a manual to guide community planners/coordinators was compiled by the sustainability coordinators in this study. The development of this manual was informed by work in each of the pilot communities through a process of reflection and community engagement. In addition, written submissions from individual community members about what they value most about their community were collected and compiled separately into community booklets. There were 12, 12 and

26 individual submissions respectively, numbering 50 submissions in total. Participants were recruited by advertisement, telephone and word of mouth.

2.3. Data Analysis

One on one interviews and focus groups were audio recorded and transcribed. Notes were taken and reflected upon in instances where audio recording did not take place. Prominent themes from all sources of data were identified and interpreted. Due to the Indigenous storytelling nature of data collection, the interpretation of data sets was validated during conversations, focus groups, and gatherings with participants. This ensured that participants had ample opportunity to reflect, discuss, share what they meant, and what they saw as important for the future. The community led and driven approach of this research meant that participant stories (i.e., submissions on what they love about community, asset mapping, visioning), reflect the voices of communities in this study and explicitly reinforce connection to community. Thus, community voice and direction underscore the results and discussion that follows and will be central to any future efforts that result from planning for sustainability in NunatuKavut.

3. Results: Planning for Sustainability in NunatuKavut

Five predominant results regarding Inuit planning, through the Community Governance and Sustainability Initiative (CGSI), materialized from this study, identified in Table 1 below. A discussion of each of these key results follows. These results illustrate how Inuit led community planning materialized in this study. These results offer an alternative approach to conducting Inuit community led sustainability planning that is guided by a decolonized and strength-based framework. In doing so, we respond to the above described call by Johnson et al. [3], Ugarte [5], McGregor [6] and others to engage and privilege Indigenous rights and knowledge and participation by Indigenous peoples in planning processes.

Table 1. Key Results.

1. Inter and cross community sharing integral to community planning
2. Community strengths identified
3. Strengthened community capacity
4. Re-connection to community and culture during the planning process
5. Sustainability goals identified and implementation begun

The results reflect the multifaceted engagement of participants, and their contributions to this study, and are embedded and interpreted from a place of strength, autonomy and Inuit rights. In sum, the results point to a reality whereby commitment and connection to community is paramount and where knowledge and expertise has been borne from generations of living on and with the land and this knowledge is paramount to continued community planning and ultimately survival.

3.1. Inter and Cross Community Sharing Integral to Community Planning

Storytelling and knowledge passed down through generations are integral to the continuity and survival of Inuit societies, and in community sustainability planning efforts. The exchange of knowledge and expertise between Inuit and as it relates to their collective and individual experiences living on and with the land, within their respective communities and in the region as a whole, is an integral method within a decolonized and strength based planning framework. This is particularly relevant given the many accounts of how Indigenous peoples have been marginalized by external planners in planning processes on Indigenous lands [4]. Therefore, this approach seeks to privilege the voice of Inuit in planning a future on their own terms, and from their own perspectives. This also assists in motivating and empowering community members to reject a history of outsider knows best, inherent in mainstream Western sustainability planning, and to reclaim agency on their lands. Previous in depth research with NunatuKavut Inuit demonstrates the important role of storytelling and local knowledge and expertise to family and community survival [33]. Participant feedback about participation in the sustainability workshops revealed that participants saw value in coming together, across communities, to share and learn from one another. Community members gained encouragement to move forward in their own communities as a result of this co-learning and sharing. Community participants thought deeply about the values, assets, and overall strengths of their respective communities and how their communities were similar and dissimilar in NunatuKavut, as well as how they could support one another and learn from one another moving forward. One of the participants commented:

"During these workshops I've learned with my community how to try and embrace the negative in our community and turn it into a positive. I've experienced other communities address issues that are similar to ours that I didn't know existed … Just overall this experience have been amazing and so insightful".

Demonstrating further the importance of relationship building to this work, another community participant described the key benefits she gained from participating in the process. She stated: "The connections and relationships/bonds I made. The confidence to return to my community with knowledge I didn't know before".

Sharing and co-learning was key to the success of this work. While communities often work alone to achieve their goals (lack of resources and time to collaborate and remote geography, contribute to this reality), the CGSI allowed for opportunities for cross community knowledge sharing and engagement to take place in non-competitive and open spaces that also sought to strengthen community skills. This helped to reduce participant feelings of isolation and alienation in visioning and community planning.

3.2. Identification of Community Strengths

In an effort to build on the positive momentum gained from inter and cross community knowledge sharing and strength based dialogue, facilitated discussions around community strengths created and directed opportunities for community members in each of the pilot communities to submit (in writing or in picture form) their own thoughts and ideas about what it is that they value about their community. This method acknowledged and validated the strengths inherent in community connection. As Inuit continue to evolve and adapt to a changing world that impacts their environment, they are well positioned to identify the strengths that are integral to the continuation of their societies. NunatuKavut Inuit are deeply connected to the lands, waters, ice and kinship ties that make up their society and communities. Yet, they are often excluded from aspects of planning and decision-making on their lands. The identification of strengths by Inuit themselves has ensured that all sectors of society that are regarded as significant, have been included in the planning process and was an important part of ensuring a decolonized approach to community planning-one that acknowledges the various sources and sites of knowledge common to Inuit.

Submissions varied in length and individual participants described their connection to place and homeland. These submissions were compiled and integrated into three booklets. They are as follows: Why I love Black Tickle, Why I love Norman Bay, and Why I love St. Lewis. These stories were integral to deepening our understanding of community values in NunatuKavut. Below are two examples from the submissions that were compiled.

"The peacefulness. The beauty of the land. I love all what BT is. The way the bog smells in the spring when everything is starting to thaw, sitting out on the point and watching flock after flock of birds flying by. The smell of wetness in the air as you go in over the land berrypicking. The beautiful colours of bright green grass as you climb the hills in July, the sound of seagulls going crazy for a feed of fish when the fishermen come in with their catch. The way the lights dance on the water on a beautiful calm summers night. The way the town looks after its first snowfall. Seeing the kiddies going from

pond to pond to check the depth of the ice for skating time and the memories come racing in of when you were a child and the amount of hours you spent on them same ponds growing up".

"Norman Bay gave my husband and I a quiet, peaceful, and safe place to raise our children. Everybody's children played together. If you knew where one child was, you knew where the whole bunch was. I can honestly say I was never bored. The isolation from other communities never bothered me and still don't. I have always felt safe here. People would always be there to give help when it was needed, no matter what and it's still that way today. We don't have far to go for our wild foods and berries or wood for our heat".

The success of this strength-based exercise demonstrated the deep and enduring connection that individuals have to their homeland. In addition, by eliciting positive and strength-based versions of home and community, we strengthened and situated our collective understanding about what is most important to community members as they prepare and plan for the future. Community members became re-focused around what is most important to them during this process as well. Simultaneously, community sustainability coordinators were building on asset mapping skills they had learned during the workshops in HVGB and they each worked in their respective communities to identify assets in diverse areas like culture, social, human, financial, to name a few. Asset mapping, focused on community strengths, and served to reinforce that knowledge and expertise already exists within the communities. Participants began to see themselves reflected in this way and this furthered their ability to think about what they could achieve in their respective communities. This method further ensured the active inclusion of Inuit in the planning process and that Inuit values were reflected in the planning process. For example, we learned from participants that maintaining traditional skills, local knowledge of the land, including the use of knowledge passed down through generations, are key strengths and important considerations in sustainability planning work.

3.3. Strengthened Community Capacity

Through decolonized community engagement that used a strength-based approach, participant awareness, skillsets and capacity were strengthened in areas of interest and relevance to community members in pursuit of community planning. This further enabled the active participation and engagement of community sustainability coordinators in leading sustainability planning in their hometowns. Capacity strengthening exercises were conducted with the sustainability coordinators in the following areas: (a) community engagement, (b) community strengths and, (c) sustainability goals and visioning. This method has had positive implications for community, and it ensured that capacity strengthening efforts directly benefited the communities themselves. These measures were taken to avoid the pitfalls common to Western scientific research whereby external researchers enter a community, conduct the research, and then leave with the knowledge (gained through dialogue with Indigenous participants), and then analyze and use this knowledge outside of the community itself. By ensuring that capacity strengthening efforts focused directly on furthering the leadership of community members, we sought to avoid such colonial research practices.

Conversations and capacity strengthening opportunities took place with community sustainability coordinators and other participants from the three pilot communities. We talked about why participants were engaged in community sustainability work, why it was important for them, and for other community members, to be a part of change for the future in their respective communities. These conversations allowed us to better understand collectively why people remain connected to their community, and the values surrounding this connection. Together, we were better able to think of relevant and meaningful ways to engage communities in important conversations about the future, and in community planning projects. In reflecting on one of the workshops a participant stated: "What a strong group of community leaders. I'm so impressed by the ideas and the hard work that's going to propel these communities forward". As a result of these dialogue and working group efforts, community engagement ideas were compiled by sustainability coordinators to assist NCC and others who may seek to engage and work with communities in NunatuKavut.

The community sustainability coordinators furthered community asset mapping (a new skill learned during workshops in HVGB) within their respective communities. This allowed them to capture broad and insightful responses while expanding community vision through the identification of community strengths and opportunities. Working from a place of strength was integral to this study and facilitated discussions around strength-based approaches to community planning were successful.

During the workshops (in group and as a whole) sustainability goals were identified and then further verified and expanded upon within each community through visioning exercises. During the workshops in HVGB, visioning exercises were employed where representative community members in attendance worked in community groups to map out an ideal vision for their respective communities. In doing so, community members articulated (through drawings) their hopes for the future. Early discussions about strength-based planning aided participants in creating visions that were positive, realistic and hopeful. Overall, these early visions were well thought out and discussed in detail. They created opportunities for in-depth participant discussion about what worked well in the community in the past and present, and participants identified the skills, knowledge and expertise the community already has and that they deem relevant to pursuing sustainable community development. Participants identified practical goals like infrastructure and water security projects, to name a few (See Table A3 for detailed community goals). These goals are fundamental to economic development opportunities. In addition, participants identified economic development opportunities like bakeapple harvesting and processing, the fishery, sealing, and tourism in resource and culture rich areas (see result five). The practicality of these goals was further supported by the participants 'ability to locate existing assets in the community that could assist with achieving the goals. For example, abandoned structures, buildings, empty homes, and materials and skills that already exist in the community were identified as spaces and opportunities to further the economic development ideas. Visions for sustainable economic development like berry and seal harvesting and tourism development in Black Tickle, the construction of a multi-purpose building in St. Lewis that could accommodate a cultural Centre and growing tourism opportunities, and tourism growth potential in Norman Bay, all point to sustainability planning that seeks to incorporate aspects of community and cultural life that are relevant and meaningful to Inuit themselves.

3.4. Re-Connection to Community and Culture during the Planning Process

Strength based exercises that encouraged positive thinking and reflection also aided in the re-connection to and validation of home and culture. Strength based dialogue facilitated opportunities for participants to re-connect to those aspects of home and community life that are most valuable to them. Borrowing from Corntassel's [24] work related to the interconnections between Indigenous peoples connection to land and resurgence, these re-connections described by participants are also interpreted as acts of resurgence by Inuit. For example, one community member wrote:

> *"I love St. Lewis because it's a place I call home. I can teach our children traditional ways of living like hunting, fishing and trapping. Things I learned growing up as a kid and stuff I can pass on to them ... don't think they would learn these things if we lived in a city".*

There were ample stories (written and shared in discussions) that pointed to a high degree of pride in home across all three communities. It was obvious that by validating community and culture, people re-connected and became more engaged and responsive to thinking about the future from a place of strength and saw themselves as having a role in creating this vision for the future. Participants discussed some of the challenges and barriers that they continue to face in their communities, in a way that was solution oriented, as opposed to from a place of defeat and hopelessness, (a way of thinking apparent early on). For example, some community participants spoke about how policy and programming opportunities, or funding calls from provincial and federal governments, are often done without regard for the interests and goals of the communities. Some expressed how they felt invalidated over the years in their communities by provincial or federal governments and marginalized from funding and other crucial opportunities to pursue planning efforts that were important to them. Others felt that some government officials simply did not care about them or their communities and felt as though it was the tactic of government to have people relocate from their homes to lessen financial burden and responsibility of government. Yet by re-connecting to community and culture, participants were able to think outside of a pre-scripted box where programs and services are outlined by external actors, and were able to come up with ideas and goals that were directly related to the interests of the communities. We learned that community interests are integral to planning as many participants talked about, for example, the importance of ensuring the survival of tradition and life ways learned from their ancestors.

The strength-based exercises in this study were successful in validating the potential, expertise, and knowledge that exists in the study communities. This form of validation proved crucial to strengthening capacity and awareness for those involved in planning, and in overcoming feelings of defeat and isolation. Furthermore, the importance of community and cultural validation is a feature of sustainable self-determination that seek to counter colonial wrongdoings that deny people and communities their very Indigeneity. It appears that by re-connecting to community and culture in the planning process, participants become more engaged and take on a greater sense of responsibility for the future.

3.5. Sustainability Goals Identified, and Implementation Begun

The three pilot communities identified a range of community sustainability goals and priorities and they began to work towards design and implementation during the course of this study. (See Table A3 for more detail). The community goals and priorities identified illustrate that community members are aware of the need to provide for basic necessities in addition to priorities that impact holistic health and well-being. While these goals represent the voice and participation of Inuit, it is important to be alert to the ever-evolving realities that impact Inuit communities and the need for Inuit to evolve and adapt to these realities. This means that goals may change and evolve as well, and planning actors must be cognizant of this and capable of attending to the varying nature of planning in these communities. Participant work on the CGSI demonstrates a commitment to community and to ensuring the survival of communities. The sustainability work of the CGSI offered a dedicated space for community members to focus on key areas of interests as they relate to community survival. As a result, a community craft group was formalized, proposals for infrastructure development identified and furthered, proposals related to water security, as well as community craft and feast events, took place. Other long-term goals were identified and discussed including the diversification of industry for economic growth. Economic development ideas reflected the resources available to community, and the skills and knowledge of community members. For example, seal processing, berry processing and a range of tourism opportunities, were identified.

These goals and priorities came out of and were furthered through the asset mapping, visioning and engagement exercises. Further priorities and sustainability goals specifically included improvements to roads and transportation, water and sewer infrastructure (two of three communities lack water and sewer infrastructure entirely and the third, partially), infrastructure to support community development and growth (i.e., multipurpose community centre/fire hall), economic security, food and heat security initiatives, and culturally relevant education. Additionally, access to clean drinking water was identified as a goal across all three communities and the degree of urgency of this goal varied across communities, with the most urgent and priority need in Black Tickle. Each of these priority areas were considered important for community sustainability now and into the future.

Communities also identified initiatives that they felt could be undertaken immediately such as community gatherings and feasts to celebrate community (St. Lewis), art and craft sessions for communities and activities for youth (Norman Bay and Black Tickle). Community members identified these as opportunities to assist in sustaining the momentum around sustainability discussions that had been ongoing in their communities throughout the research. Community centred initiatives like these were also thought to positively impact collective well-being and promote togetherness, in turn reinforcing and further validating Inuit values. In this context, it is clear that community planning and development opportunities must adhere to principles that ensure the survival of community and culture in ways that respect and ensure the survival of the natural environment and all who live with it.

3.6. Limitations

The study faced some limitations and challenges such as geography. NunatuKavut spans a vast territory and the three pilot communities are not easily accessible to each other, nor for the research team. As a result, time in individual communities was limited due to costs associated with travel to remote coastal Labrador and in order to ensure that quality time was had in each community. Inadequate funding to support community sustainability coordinators beyond the life of this study due to the external funding opportunity being short term and project based was also a challenge for the longevity of continuing this work in communities.

4. Discussion and Conclusions

Topics of governance and sustainability, including community sustainability planning, are receiving increasing attention in Canada and across the globe. Yet, conflicts and tensions related to land and resources between Indigenous peoples and the state continue and often undermine Indigenous political autonomy [36]. When Indigenous political autonomy is undermined, so too are the sustainability of cultural landscapes and the social-ecological systems that Inuit are a part of. Booth and Muir [31] recognize that Indigenous planning is necessary in order for Indigenous peoples to effectively navigate their own terrain and to navigate federal and provincial forces on their land. Yet, these authors observe that little attention has been paid (in the literature, policy or practice) to this area. An Indigenous planning perspective is new and to some extent unrealized, though it remains necessary in overcoming some of the barriers and obstacles that face Indigenous peoples in planning for the future [31] and sustaining their communities and cultures.

This study illustrates decolonized and community led sustainability planning in action. Collaborative work with NunatuKavut Inuit has given rise to 'grounded decolonization' which refers to an approach that seeks to respect and honour the values, history and culture of those who belong to their homeland, in their place and time. It refers to decolonization that must take place in the context of people who live and are connected through generations. Simply put, it means that decolonizing efforts must be acutely aware, and cognizant of, the history and present of the people in their context-and on their own terms. From this vantage point, decolonization or decolonizing efforts must be designed, shaped and implemented in locally and context specific ways. Thus, grounding decolonization refers to the act of designing and implementing decolonizing efforts that have gained consensus and agreement from communities leading their own efforts. In the context of sustainability planning, decolonization can manifest as Indigenous consent and recognition of Indigenous priorities and expertise which are integral to the creation of sustainable communities.

Corntassel's concept of place further enlightens this study [24]. The community sustainability planning and capacity strengthening efforts of Inuit in NunatuKavut throughout the CGSI reflect the capacity and strength of Inuit to make decisions that impact them on their lands and informed by their own values and perspectives. The autonomy to make decisions that impact the future of Inuit communities in NunatuKavut, in a way that is indicative of Inuit values, world views and perspectives, is integral to decolonizing and self-determination efforts that are sustainable into the future. By building on the work of Corntassel in this area and applying key concepts and ideas to the work in NunatuKavut, we were able to assist communities in identifying short and long-term sustainability goals that positively impact community. Expertise and knowledge of generations past, of tradition, moving and living with changing seasons, all point to a reality in which people live in relation with the natural environment, not against it [4].

Study participants were active in achieving a number of the goals and objectives set out in their communities through the CGSI and it was clear that the health of people and communities, of lands and waters, was and is a stated priority. The priorities and goals set out by the communities in this study are meaningful, relevant and urgent. While they are not necessarily elaborate, it is important to understand these goals in context. In many ways, they reflect a desire for the basic and fundamental rights and privileges that most Canadians' already enjoy freely, including basic necessities necessary to support the planning and development of goals driven by the global economy (e.g., access to clean drinking water). Sustainability goals and priorities in this study point to inequalities and inequities that plague NunatuKavut Inuit in these areas, but these issues are not unique to them as Indigenous peoples. Water and food insecurity disproportionately impact Indigenous communities in Canada, and in particular, Northern Indigenous communities [37]. Thus, Indigenous led self-determination efforts that are locally driven and context specific are necessary for the planning of sustainable futures that promote equality and equity for Inuit.

Community asset mapping, engagement strategies, visioning exercises, and capacity strengthening initiatives provided spaces and environments for participants and communities to envision, for themselves, a future for their community. The idea behind capacity strengthening and thought-provoking exercises such as these was not to transport knowledge from one authoritative body onto community, but rather to open safe and meaningful spaces for communities to connect with, think about, and reflect upon what is possible in a way that positions community members as experts and knowledge holders in their own right. Following from the work of Eisenberg et al. [29], this research and the processes described in this study demonstrate that Indigenous peoples and communities are experts on their lands and their knowledge of place position them to make decisions to inform a future that is compatible with their own goals, ways of knowing and of being.

Overall, the work of the sustainability committees in communities set the stage for discussions whereby community people began to talk about governance and community planning from a community centered and value-based perspective. Several participants spoke to the way in which the sustainability committee in their community had allowed them to think about and move initiatives forward in a way that had not been possible before. Participants from all of the pilot communities spoke to the necessity of community involvement and leadership in decisions that impact them directly, emphasizing the importance of grounded, decolonizing approaches to community planning and visions for the future informed by Inuit goals and values, and shaped by their connection to people, place and history, rooted in their environment and culture.

Community knowledge, values and traditions, enlightened by communities themselves, has set an important expectation in motion-that in order to plan for a sustainable future, we must think about and reconnect with what it is that we value most about our communities. This approach allows community members to reflect and to think about positive aspects of a community (i.e., culture, values etc.), and to ensure that those facets of community are protected and considered in planning for the future. What is valued within and about community became the prominent factor in considering and determining community sustainability goals in these three pilot communities. This work situates grounded decolonization as that which creates, supports and fosters environments that allow communities and people to connect and re-connect to their communities in ways that are most meaningful to them. Decolonizing paths that seek to respond to the interests, priorities and values of people in their place and time, and not those ideals or values that come from outside the community, are particularly relevant. Grounded decolonization implies that these values about community should lead the community planning approach for the future.

Decolonized planning efforts are a necessary step to sustainable self-determination in NunatuKavut so as to ensure that community sustainability planning efforts come from a rights-based perspective. As a concept and point of discussion in modern day discourse and building on the work of Smith [20], decolonization can assist us in unpacking sites of colonial control (and even colonial relationships that have endured and continue to marginalize Indigenous governance systems). While Indigenous governance systems have much to contribute to the development of sustainable communities and societies, Indigenous communities are often faced with barriers due to a lack of interest in collaboration from dominant systems of control within society [22]. The implications of this work are that community sustainability for Indigenous communities under Indigenous led decolonization, as it is for the NunatuKavut Inuit, means that capacity is being strengthened, knowledge and awareness of Indigenous rights are becoming more prevalent, the desire and will to reclaim traditional aspects of culture and political society are more paramount, and the willingness to own, author and share one's story is becoming commonplace. This research study has been a witness to the power of culture, tradition and connection to community that has come as a result of decolonizing work, all of which are integral to beginning and maintaining decolonized community sustainability.

Author Contributions: A.H.: Supported by the NCC, I designed and led the CGSI, with the NCC and three NunatuKavut communities. I held a dual role as NCC employee, working with and for the communities, observing and reflecting on this process as a PhD student. The multiple roles of community member, researcher and employee of NCC held me accountable but also connected me with the communities and people who participated in this research. K.V.: My contributions to this paper are as a scholar in community sustainability, governance and development. As supervisor I provided guidance throughout the research, including the writing of this article and the related PhD dissertation, as well as specific input on the construction and content of this paper, as outlined below. Conceptualization, A.H.; methodology, A.H.; investigation, A.H.; resources, K.V.; writing—original draft preparation, A.H.; writing—review and editing, A.H., and K.V.; supervision, K.V. All authors have read and agreed to the published version of the manuscript.

Funding: The authors are grateful to the support of this work and the CGSI from NCC and participating communities, who provided essential input and in-kind support, and to the Atlantic Indigenous Mentorship Network: Kausattumi Grants Program scholarship.

Acknowledgments: The authors thank the communities of Black Tickle, Norman Bay and St. Lewis for their time, contributions, expertise and knowledge throughout this research. Special thanks to the NunatuKavut Community Council (NCC) for seeing the value in this research and for your support of this initiative throughout. The authors would also like to thank all of those who supported this research as partners, colleagues and friends, whose advice and guidance were invaluable to this research.

Conflicts of Interest: The authors declare no conflict of interest.

Abbreviations

NCC   NunatuKavut Community Council
CGSI   Community Governance and Sustainability Initiative
LIA   Labrador Inuit Association
ACOA Atlantic Canada Opportunities Agency
RIRSD Recognition of Rights and Self-Determination
HVGB Happy Valley-Goose bay

Appendix A. Community Characteristics

Black Tickle, Norman Bay, and St. Lewis were selected as pilot communities to pursue community sustainability planning with a vision towards identifying collective community goals, building on what is and has already been working well in the communities, in order to envision a future from a place of strength, Inuit values and perspectives. This process demonstrated that residents in the three communities are proud and eager to reclaim and strengthen a future that is bright and sustainable for their families for the years to come. The communities were selected based on remote geography in NunatuKavut, their vulnerability around economic development, food and water security concerns (although to varying degrees in each community), and rate of population decline, all of which affect community and cultural preservation. These communities are also rich in Inuit culture and their remoteness and lack of basic amenities give rise to continued subsistence living in a way that persistently demonstrates Inuit adaptation in the face of globalization. In sum, this research is driven by an approach to equity. Table A1 provides an overview of the remoteness of all three communities, highlighting the lack accessibility in and out of each community and a lack of primary industry that was once the economic driver in the communities.

Table A1. Community Characteristics.

	Black Tickle	Norman Bay	St. Lewis
Population [1]	110	20	185
Transportation	Fly-in/out, seasonal ferry (limited), small boat	Seasonal fly-in/out (helicopter), small boat	Road (TransLabrador Highway, TLH), fly-in/out
Major Industry	Fishery (local plant closed)	Fishery (travel to neighboring plant by boat for employment, no local plant)	Fishery (local plant closed)

[1] Population source: Community Town Council, Recreation Committee and Local Service District respectively. Other information in Table 1 reflects knowledge from study participants.

Table A2. Data Collection Activities (All Communities).

Activity Type	Participants (n)	Rationale	Impact
Focus group	Black Tickle: 7 Norman Bay: 2 St. Lewis: 6	Participant knowledge sharing and storytelling	Participant voices privileged. Increased understanding around community vision, goals and limitations.
Interviews	Black Tickle: 1 Norman Bay: 1 St. Lewis: 2 Other: 2	Standard data collection method	Less effective in accessing rich data. Not conducive to storytelling.
Survey	Black Tickle: n/a Norman Bay: 6 St. Lewis: 20	Baseline data collection	No surveys conducted in Black Tickle given the nature of researcher and community relationship (see methods). For others, increased researcher understanding of participant belonging to community (age, years in community, etc).
Community gathering	Black Tickle: 25 Norman Bay: 6 St. Lewis: 43	Appropriate Indigenous research method	Designed to enable researcher learning from participants.
Written submissions	Black Tickle: 12 Norman Bay: 12 St. Lewis: 26	Create space for positive and strength-based thinking around community	Re-connected community to positive attributes of community and culture. Increased understanding of participant values in relation to community and culture.
HVGB workshop 1 (strength-based planning, visioning, asset mapping, community engagement, proposal writing	Black Tickle, Norman Bay and St.Lewis: 10	Engage participants in positive and strength-based planning and visioning, identify range of community assets and engagement strategies, and highlight tips and best practices in proposal writing	Participants increasingly saw themselves as active agents and better identified positive attributes of communities integral to successful planning, identified planning opportunities and goals that were realistic and integral to core values around community life and culture, and identified and reflected on the many assets that already exist in communities. Strengthened community capacity and researcher learned best practices in engagement from communities.
HVGB workshop 2 (pilot community and NCC presentations, Q&A, networking and focus group)	Black Tickle, Norman Bay and St.Lewis: 10 Other: Approx 5	Privilege community participants as leaders, experts and knowledge holders expressing vison for their community, strengthen participant capacity and presentation skills, identify opportunities to advance goals, connect community participants with stakeholders, knowledge sharing and storytelling	Conversations revolved around stated community interests and needs, participants supported in efforts to pursue planning activities, centred feedback and opportunities around community planning interests and goals, provided opportunities to connect with potential funders, researchers, etc., increased researcher understanding around community planning goals and associated community values.

Table A3. A3.1 Black Tickle Community Goals and Progress; A3.2 Norman Bay Community Goals and Progress; A3.3 St. Lewis Community Goals and Progress.

A3.1 Black Tickle Community Goals and Progress		
Goal	**Rationale and Benefits**	**Progress**
Short-term: Local garden integrated with healthy eating program for children. Medium to long term: Enhanced food security and child development	- Will provide fresh source of local food. - Address local grocery store issues regarding fresh produce by providing local source of vegetables for purchase and sale. - Benefits for youth education and health	- School aged children/youth have begun participation in small scale gardening at school. - Community members continue to express interest in this goal.
Short-term: Community social events Medium to long-term: Intergenerational community engagement, holistic health, pride in culture and tradition.	- Events like winter carnivals and come home year celebrations provide opportunities to connect families to community and culture with lasting positive impacts for morale and health of community members. - Develops community planning skills.	- Local craft group formalized with the assistance of the CGSI, applying for funds to host social events regularly (e.g., Christmas and Easter events).
Short-term: Education programs related to traditional knowledge and life skills Medium to long term: youth and elder engagement, preservation of culture	- Educate children and youth in areas of traditional knowledge and life skills (e.g., traditional food preparation). - Ensure valued skills and knowledge are passed on will be important to community survival.	- Local craft group has begun partnering with NCC to deliver programs though NCC's Inuit Education Program and Community Grants Funding.
Short-term: Further investigate alternatives for water and sewer Infrastructure Medium to long term: Ensure reliable access to clean drinking water to community residents	- Benefits to overall health (mentally, physically, emotionally, etc.). - Access to clean drinking water is a right.	- Local Service District (LSD), with help from the CGSI, has developed and submitted a proposal and accessed funding to do feasibility work around water security options.
A3.2 Norman Bay Community Goals and Progress		
Goal	**Rationale and Benefits**	**Progress**
Short-term: Identify opportunities to upgrade and build needed infrastructure Medium to long-term: Infrastructure opportunities and upgrades to community centre, helicopter pad, winter snowmobile trail, garbage disposal site,	- Expand contact list and connections for partnerships. - Enhance community centre to meet community needs; - Enhance transportation means, enhance safety for travel and transportation of goods.	Volunteer labour has sustained the centre to date. Community looks forward to additional developments. - Discussions around funding opportunities have taken place.
Short-term: Community garden and Greenhouse development Medium to long-term: Communal access to local source of fresh foods	- Promote community connectedness, self-sufficiency and access to nutritious food. - Access to healthy food in light of need to travel for store bought goods. Increase self-sufficiency.	Small community garden infrastructure purchased through successful funding proposal.
Short-term: Potable Water Drinking Unit (PWDU) Medium to long-term: Reliable source of clean drinking water	- Access to clean drinking water is a right. - Increase access to clean water and particularly for aging population who otherwise rely on retrieving water with buckets from a brook.	No known progress to date.
Short-term: Equipment for Fire Fighting Medium to long-term: Increased capacity to respond to community crisis.	- Health and safety concern. - Increased self-sufficiency and response efforts during crisis.	No known progress to date.

Table A3. Cont.

A3.3 St. Lewis Community Goals and Progress

Goal	Rationale and Benefits	Progress
Short-term: Crafting Workshops and social events Medium to long-term: Increase community participation in culturally relevant activities	- Enhance community activity and skills building - Increase community cohesion and improve social and mental wellness across generations	Ongoing.
Short-term: Host community Feasts Medium to long-term: Provide opportunities to come together and share traditional foods	- Respond to community interests in like events. - Bring community together and support most vulnerable.	Ongoing.
Short-term Work towards necessary Infrastructure Upgrades Medium to long-term: Upgrades to museum and new build (fire hall)	- Enhance basic and necessary infrastructure for community planning and development - To address health and safety concerns of community members.	Ongoing discussions and identification of opportunities.
Short-term: Identify solutions to address gaps in water security Medium to long-term: Water and Sewer Infrastructure expanded	To address outstanding water insecurity in some parts of the community. - Provide access to clean drinking water to all community members.	Discussions ongoing.

References

1. Nilsson, A.E.; Larsen, J.N. Making Regional Sense of Global Sustainable Development Indicators for the Arctic. *Sustainability* **2020**, *12*, 1027. [CrossRef]
2. Whyte, K.P.; Brewer, J.P.; Johnson, J.T. Weaving indigenous science, protocols and sustainability science. *Sustain. Sci.* **2016**, *11*, 25–32. [CrossRef]
3. Johnson, J.T.; Howitt, R.; Cajete, G.; Berkes, F.; Louis, R.P.; Kliskey, A. Weaving Indigenous and sustainability sciences to diversify our methods. *Sustain. Sci.* **2016**, *11*, 1–11. [CrossRef]
4. Hibbard, M.; Lane, M.B.; Rasmussen, K. The Split Personality of Planning: Indigenous Peoples and Planning for Land and Resource Management. *J. Plan. Lit.* **2008**, *23*, 136–151. [CrossRef]
5. Ugarte, M. Ethics, Discourse, or Rights? A Discussion about a Decolonizing Project in Planning. *J. Plan. Lit.* **2014**, *29*, 403–414. [CrossRef]
6. McGregor, D. Representing and Mapping Traditional Knowledge in Ontario's Forest Management Planning. In *Reclaiming Indigenous Planning*; Walker, D., Natcher, D., Jojola, T., Eds.; McGill-Queen's University Press: Montréal, QC, Canada, 2013; pp. 414–435.
7. Porter, L. Indigenous Planning: From Principles to Practice. *Plan. Theory Pract.* **2017**, *18*, 639–640. [CrossRef]
8. Matunga, H. A Revolutionary Pedagogy of/for Indigenous Planning. *Plan. Theory Pract.* **2017**, *18*, 640–644. [CrossRef]
9. Mercer, N.; Parker, P.; Hudson, A.; Martin, D. Off-grid energy sustainability in Nunatukavut, Labrador: Centering Inuit voices on heat insecurity in diesel-powered communities. *Energy Res. Soc. Sci.* **2020**, *62*, 101382. [CrossRef]
10. Procter, A.; Chaulk, K. Our Beautiful Land: The Challenge of Nunatsiavut Land Use Planning. In *Reclaiming Indigenous Planning*; Walker, D., Natcher, D., Jojola, T., Eds.; McGill-Queen's University Press: Montréal, QC, Canada, 2013; pp. 436–456.
11. Gibson, R. Sustainability assessment and conflict resolution: Reaching agreement to proceed with the Voisey's Bay nickel mine. *J. Clean. Prod.* **2005**, *14*, 334–348. [CrossRef]
12. O'Faircheallaigh, C. *Negotiations in the Indigenous World. Aboriginal Peoples and the Extractive Industry in Australia and Canada*; Routledge: London, UK, 2016.

13. Archibald, L.; Crnkovich, M. *If Gender Mattered: A Case Study of Inuit Women, Land Claims and the Voisey's Bay Nickel Project*; Status of Women Canada: Ottawa, ON, Canada, 1999. Available online: http://publications.gc.ca/collections/Collection/SW21-39-1999E.pdf (accessed on 13 May 2020).

14. Diggon, S.; Butler, C.; Heidt, A.; Bones, J.; Jones, R.; Outhet, C. The Marine Plan Partnership: Indigenous community-based marine spatial planning. *Mar. Policy* **2019**, *5*, 103501. [CrossRef]

15. Hudson, A. Re-claiming Inuit Governance: Revitalizing Autonomy and Sense of Place in Self-Determined Decision Making in NunatuKavut. Ph.D. Thesis, Memorial University of Newfoundland, St. John's, NL, Canada, 2020.

16. Baxter, K.H.; Purcell, M. Community Sustainability Planning. *Munic. World* **2007**, *117*, 35–38.

17. Vodden, K.; Lane, R.; Pollett, C. Seeking Sustainability through Self-Assessment and Regional Cooperation in Newfoundland and Labrador. In *Sustainability Planning and Collaboration in Rural Canada*; Hallstrom, L., Beckie, M., Hvenegaard, G., Mündel, K., Eds.; University of Alberta Press: Edmonton, AB, Canada, 2016; pp. 321–346.

18. Corntassel, J. Toward Sustainable Self-Determination: Rethinking the Contemporary Indigenous Rights Discourse. *Alternatives* **2008**, *33*, 105–132. [CrossRef]

19. Palmater, P. *Indigenous Nationhood: Empowering Grassroots Citizens*; Fernwood Publishing: Halifax, NS, Canada, 2015.

20. Smith, L.T. *Decolonizing Methodologies: Research and Indigenous Peoples*, 2nd ed.; Zed Books: New York, NY, USA, 2012.

21. Wilson, S. *Research is Ceremony: Indigenous Research Methods*; Fernwood Publishing: Halifax, NS, Canada, 2008.

22. Jokhu, P.D.; Kutay, C. Observations on Appropriate Technology Application in Indigenous Community Using System Dynamics Modelling. *Sustainability* **2020**, *12*, 2245. [CrossRef]

23. Patrick, R.; Grant, K.; Bharadwaj, L. Reclaiming Indigenous Planning as a Pathway to Local Water Security. *Water* **2019**, *11*, 936. [CrossRef]

24. Corntassel, J. Re-envisioning resurgence: Indigenous pathways to decolonization and sustainable self-determination. *Decolon. Ind. Educ. Soc.* **2012**, *1*, 86–101.

25. Kovach, M. *Indigenous Methodologies: Characteristics, Conversations and Contexts*; University of Toronto Press: Toronto, ON, Canada, 2009.

26. Fanon, F. *The Wretched of the Earth*; Grove Press Inc.: New York, NY, USA, 1963.

27. Borrows, J.; Rotman, L.I. *Aboriginal Legal Issues: Cases, Materials, & Commentary*; Butterworths: Toronto, ON, Canada, 1998.

28. Napoleon, V.; Friedland, H. An Inside Job: Engaging with Indigenous Legal Traditions Through Stories. *McGill Law J. Rev. Droit McGill* **2016**, *61*, 725–754. [CrossRef]

29. Eisenberg, A.; Webber, J.; Coulthard, G.; Boisselle, A. *Recognition Versus Self Determination: Dilemmas of Emancipatory Politics*; UBC Press: Vancouver, BC, Canada, 2014.

30. Lambert, L. *Research for Indigenous Survival: Indigenous Research Methodologies in the Behavioral Sciences*; University of Nebraska Press: Lincoln, MI, USA, 2014.

31. Booth, A.; Muir, B.R. Environmental and Land Use Planning Approaches of Indigenous Groups in Canada: An Overview. *J. Environ. Policy Plan.* **2011**, *13*, 421–442. [CrossRef]

32. Walker, W.; Haasnoon, M.; Kwakkel, J.H. Adapt or Perish: A Review of Planning Approaches for Adaptation under Deep Uncertainty. *Sustainability* **2013**, *5*, 955–979. [CrossRef]

33. Hart, M.A. Indigenous Worldviews, Knowledge, and Research: The Development of an Indigenous Research Paradigm. *J. Indig. Soc. Dev.* **2010**, *1*, 1–16.

34. Cooper, E.J.; Driedger, S.M. Creative, strengths-based approaches to knowledge translation within indigenous health research. *Public Health* **2018**, *163*, 61–66. [CrossRef]

35. Fawcett, B.R.; Walker, R.; Greene, J. Indigenizing City Planning Processes in Saskatoon, Canada. *Can. J. Urb. Res.* **2015**, *24*, 158–175.

36. Lane, M.B.; Hibbard, M. Doing it for themselves: Transformative Planning by Indigenous peoples. *J. Plan. Educ. Res.* **2005**, *25*, 172–184. [CrossRef]
37. Hanrahan, M.; Sarkar, A.; Hudson, A. Exploring Water Insecurity in a Northern Indigenous Community in Canada: The "Never-Ending Job" of the Southern Inuit of Black Tickle, Labrador. *Arct. Anthropol.* **2014**, *51*, 9–22. [CrossRef]

Article

Placetelling® as a Strategic Tool for Promoting Niche Tourism to Islands: The Case of Cape Verde

Fabio Pollice, Antonella Rinella, Federica Epifani * and Patrizia Miggiano

Department of History, Society and Human Studies, University of Salento, 7-73100 Lecce, Italy;
fabio.pollice@unisalento.it (F.P.); antonella.rinella@unisalento.it (A.R.); patrizia.miggiano@unisalento.it (P.M.)
* Correspondence: federica.epifani@unisalento.it

Received: 2 May 2020; Accepted: 20 May 2020; Published: 25 May 2020

Abstract: This paper reports on the experience of the first Placetelling® training course in Santo Antão and Santiago, Cape Verde, promoted by Società Geografica Italiana and Fondazione Lelio e Lisli Basso. Placetelling® is a particular type of storytelling of places that promotes local development and helps to develop a sense of identity and belonging among the members of the community. Indeed, Placetelling® supports local communities to become directly engaged in the preservation of their common legacy in order to transmit it to coming generations. Tourism is the field where Placetelling® can best express its potential. This is particularly true for what concerns tourism to islands. The paper shows the first results of what we can define as a "maieutic reworking of local heritage" in Cape Verde, through the sharing of narrative and symbolic artifacts. Special attention will be dedicated to some crucial issues: The involvement of stakeholders through the lenses of empowerment, the discrepancies between how sense of identity is perceived by the locals and how it is communicated to tourists, and how and to what extent Placetelling® can change stakeholders' awareness of their own cultural heritage.

Keywords: Placetelling®; local heritage; islands; sustainable tourism; Cape Verde

---

1. Introduction

This research was conceived within the theoretical and methodological debate about Placetelling® and, more specifically, that it is inherent in the strategic role that Placetelling® can play in defining tourism-driven trajectories for local development. Reflections in this essay were inspired by two thematic pillars: Firstly, we considered which typologies of identity narratives could be identified depending on the purposes to be pursued and, secondly, we analyzed islandness as a complexifying variable in outlining new effective strategies for identity narratives.

Placetelling® is a method to create place narratives, a strategic asset to support communication and promotional processes. It was launched in 2016 thanks to the cooperation between the Centro Universitario Europeo per i Beni Culturali (European University Center for Cultural Heritage, depositary of the trademark) and the University of Salento. In 2017, the first School of Placetelling® took place at the University of Salento. Since then, two other editions have been organized, as well as a number of scientific and popular events. The main aim of the School of Placetelling® is to train a specialized professional, namely, the placeteller, who is an expert in the field of place-oriented storytelling, able to enhance places for their peculiar identities.

In this essay, we report the experience and the early results of the Placetelling® training course held in Santiago and Santo Antão, Cape Verde.

Cape Verde is an archipelagic state in the North Atlantic Ocean made up of 10 volcanic islands with a total population of more than 500,000.

Because of its strategic location (500 km from Senegal, close to the major north-south routes), it was colonized by Portugal in the 15th century, and became one of the most important slave trade centers. Since its independence, gained in 1975, Cape Verde has been one of the most stable African states, both economically and politically, yet it faces a number of geographic, economic, and social disadvantages because of which it is categorized by UN (United Nations) as SIDS (Small Islands Developing Countries) [1]. Hence, because of the severe lack of natural resources, the Cape Verdean economy is mainly based on development aid, foreign investments, remittances, and tourism [2]. Indeed, it is correct to say that the Cape Verdean economy is tourism-driven, largely depending on Eurozone countries. According to the Instituto Nacional de Estatistica Cabo Verde, tourism flows to the archipelago have increased significantly in the last decade; in 2017, more than half of the tourists came from four European countries (United Kingdom, France, Germany, and Holland) [3].

Although Cape Verde offers a wide variety of environmental and cultural resources, it is considered as a mere seaside destination, with severe impacts on resource shortage and landscape deterioration processes [4]. The current economic reforms, intended to diversify the economy in order to attract foreign investments and to boost employment, also include strategies to diversify tourism supply according to the principle of sustainable tourism, oriented to the crucial aim to preserve territorial carrying capacity [5,6].

Besides being the first occasion for testing Placetelling® abroad, this experience triggered both epistemological reflections and provided pragmatic evidence with regard to the relevance of the territorial element (in this case, islandness) and of identity in outlining shared, bottom-up narratives of the place and, consequently, how these elements could converge towards an effective strategic tool for local development. Place narratives for sustainable tourism are a pertinent example.

The essay starts with a theoretical overview regarding the intrinsic relationship between Placetelling® and sustainability, especially for what concerns cultural sustainability and how it can be interpreted considering the specific case of island tourism with its peculiar criticalities. Therefore, the taxonomy of place narratives and the mutual link between place narratives and nissology represent, in this specific study, a crucial issue. The link between Placetelling® and sustainability is self-evident: It arises from the need to build a narrative mechanism that makes the local community aware of the mutual relationship that binds it to the place. Such a reciprocity makes the local community a distinctive social organism. The adjective "local", in fact, means that the community is such when it recognizes itself in a territory and represents itself in symbiosis with it.

Secondly, we introduce the Placetelling® methodological framework, underlining its peculiar multidisciplinarity: Placetelling® is inspired by a prescriptive geographical approach, but it shows a robust attitude to hybridization concerning both theoretical aspects (e.g., semiotics, media studies, sociology) and practical applications (e.g., video making, photography, creative writing). Finally, we present the experience of the first Placetelling® training course in Cape Verde, which took place in April 2019 within the project "Cabo Verde: Historia, Cultura e Ambiente para um Turismo Sustentàvel", promoted by Fondazione Lelio e Lisli Basso and Fundacão Amilcar Cabral and funded by the European Union. The training course was designed by geographers from University of Salento (Lecce, Italy) and University of Tor Vergata (Rome).

Though the project has only just finished, some early results are already available. These results, which also represent the direct output of the training course, are intended to be the first step towards the development of a shared promotional strategy to be implemented by all the local operators according to the principles of sustainable tourism.

2. Theoretical Framework

The scientific reflection about Placetelling® is quite recent and, due to its potential multidisciplinary applications, it encompasses a wide range of suggestions. An essential trait is the relationship between sustainable tourism and Placetelling® and, in particular, how the enhancement of local bottom-up narratives could favor the development of new forms of sustainable tourism within a SIDS context.

UNWTO (United Nations World Tourism Organisation) pays close attention to the relation between tourism and SIDS. Starting from numbering the "three key characteristics: Small size, with implications for pressure on resources and limited economic diversity; remoteness and isolation, leading to challenges for trading but also to a unique biodiversity and cultural richness; and a maritime environment, leading to strong tourism assets but vulnerability to climate change", sustainable tourism, intended as a kind of tourism that takes full account of its current and future economic, social, and environmental impacts, addressing the needs of the visitors, the industry, the environment, and host communities is presented as a supporting tool for local development [7]. The scientific debate mirrors the prescriptive framework outlined within international organizations and offers theoretical robustness to the complexity of the issue. Only to cite the most recent studies, attention was paid to the analysis of the determinants of tourism toward SIDS, to the links between tourism and social issues, and to the way tourism affects environmental systems [8–11].

In addition, many social scientists have underlined the persistence of a sort of colonialist relationship between many SIDS with a colonial history and their former colonizing countries [12–14]. This calls into question a fourth dimension of sustainability, namely, cultural sustainability, defined as the effort "to respect and enhance the historic heritage, authentic culture, traditions, and distinctiveness of host communities". [15]. Such a statement appears particularly urgent if we consider tourism in relation to globalization, and we grasp not only its dimension as an economic phenomenon, but also as a social habitus fed by the desire to experience otherness in its many manifestations [16]. The resulting cultural interaction, however, can be corrupted by asymmetries that often have geo-economic genesis, and in the context of which we observe the action of dominant cultures capable not only of modifying the behaviors and lifestyles of the host community but also of contributing directly or indirectly to an exogenous and instrumental re-territorialization of the places of tourism. In the specific case of SIDS, as will be explained below, there is traditionally a prevalence of globalitarian forms of tourism, in which the need to attract capital induces a complete adaptation of tourist supply to global demand; the consequence is 'topophagy', a de-territorializing action whereby places become non-places [17,18].

On these bases, we argue that place narratives are deeply entwined with local sustainable development. Such an observation lies on a post-structuralist interpretation of development, and the attention paid by post-structuralist scholars to the quest for discourses, explanations, theories, and descriptions which are contextual and, consequently, plural [19]. These narratives are expected to be self-centered and representative of the local communities, so as to be functional to contrast what defines "imaginary geographies", often disseminated by dominant cultures.

3. Methodology for Identity Narratives

3.1. Orientative Narratives, Attractive Narratives, Hyperconnective Narratives

Hence, place narratives have a territorializing power, as they act as determinants within local development processes which are, in turn, immersed in complex systems of local relations. Questioning the implications that the narratives of a place, according to its hetero-directed or self-centered nature, have on the processes of territorial development, requires us to go beyond a purely semiotic investigation to try to understand how these implications materialize within a specific space, how they "take place". In this, the perceptive dimension referring to complex identity systems, within which the processes of construction, negotiation, and renegotiation of meanings take place is relevant. This is especially true when we focus not only on the fruition, but also on the modes of active perception of the dominant narrative, as well as on the forms of production of alternative and place-based narratives stemming from everyday life.

There are at least three typologies of place narratives, which can converge and/or diverge: Orientative, attractive, and hyperconnective.

Orientative narratives [20] stem from the territory for the territory, in order to build or rebuild its own identity dimension and hand it on to forthcoming generations, so as to make them aware of their heritage as well as make them responsible for its protection and enhancement [21]. Orientative narratives are based on two fundamental skills: "To make local community" and "nurture amor loci" [22]; these two skills are conditio sine qua non for setting off a sustainable, shared, "contextualized patrimonialization" [23]. This form of narrative is useful to guide or re-orient both individual and collective attitudes towards the local, in order to make them coherent and functional to a full comprehension of change processes occurring at local and global scale. In other words, it supports the resilience of local systems and, therefore, their capacity to adapt to changes in the global scenario, improving competitive performance and wealth levels among local community members.

Orientative narratives are focused on the quality of social networks, landscapes, heritage safeguarding, culture, and history, and consider local community as the core and the crucial element in designing strategies for the enhancement of the characteristic features of a place. The expressive forms generally used aim to give an overview made up of anecdotes and everyday stories considered as unique and essential, which are often hidden from and unknown to outsiders; such stories are chosen and recovered by local actors themselves (citizens, tour operators, local authorities), who show a proactive attitude towards a shared comprehension of symbols and values, as well as strengthening cohesion starting from the sense of belonging.

This kind of narrative, which underlies the rebuilding of a sense of belonging among the members of a local community as well as heritage regeneration, can in turn generate attractive narratives [1]. Attractive narratives are intended to transmit identity features to those who belong to other cultures and other contexts, so as to increase territorial attractiveness and stimulate a mutual empathic relationship [1]. As a consequence, functional cultural mediations are established, in order to respond to a touristic demand which is: (1) relational—the sharing of the same relational space favors interaction between tourist and territory and allows the former to develop a direct, not mediated, awareness of place cultural values; (2) experiential—a large part of intangible cultural heritage is accessible only through its local community. Therefore, interaction with the local community allows the tourist to experience the place in its intangible aspects; and (3) sustainable—it is not the territory that adapts to the needs of tourist flows; on the contrary, the tourist lives an immersive experience, a community experience, respectful of place values and environment. It is a community-involved approach for three reasons:

(1) Local community itself, as keeper of the intangible cultural heritage and cultural medium between tourist and territory, becomes the main territorial attractivity;

(2) Through awareness raising and empowerment, local community becomes the real protagonist of tourism supply, by managing, individually or collectively, the overall tourism services. For instance, we refer to the community hotel model as a strategy to reuse residents' houses as accommodation [20,21]. In general, the local community takes the lead in maintaining a territorial heritage that would otherwise be lost, and boosts active enhancement processes in order to contain both the decline of places—especially those facing abandonment and severe depopulation trends—and a museum-like approach to the preservation of whatever is considered as "authentic" [24].

Attractive narratives see the tourist as a "temporary" citizen who is "encouraged/invited" to: (1) Taste local foods and drinks originating in an invaluable, anciently rooted tradition [25]; (2)explore the history of the place through the in-depth knowledge of tangible and intangible assets layering within the territory; and (3) live the habits, handicrafts, celebrations, all the possible itineraries and workshops, sharing emotions and inclusive and original experiences with the local community.

Different from the first two narratives, the hyperconnective narrative gives absolute priority to the projection of a local system's core business within the global scale, aiming to enhance some specific tangible/intangible assets (a building, a museum, a festival, a site of pilgrimage, etc.) that are deemed appropriate to respond to specific supralocal needs. The fact that these elements, which often represent the main object of local investments, are able to satisfy the taste of a wide international audience may open up new creative and original scenarios and generate substantial multiplier effects and promising cultural bridges with other cultures. Nevertheless, many case studies have demonstrated that, in the long term, the precious "passe-partout" of cross-cultural influence can easily become a dangerous means to undermine tradition, here considered as a mere tool for territorial marketing; in this way, tradition is subjected to logics and models completely different to those characterizing sustainable tourism.

This approach underlies dangerous trends inspired by the current fashion, with the realistic risk of the damage or destruction of existential, sentimental, and emotional values which concur within a milieu and, as a consequence, to enhance the use value of that milieu to the detriment of its exchange value. Instead of being the pivotal protagonist of individual and collective growth of the local system, local community threatens to become nothing more than a bit player on a stage that can rapidly change from being the hot focus of external stakeholders to the most complete abandonment.

3.2. A Preliminary Methodological Note: Island Narratives between Mainstream Narratives and Placetelling®

As already stated, the foundational aim of Placetelling® is to achieve an identity narrative as a result of self-representational processes carried out by the local community. This appears to be particularly relevant with regard to the case of Cape Verde, given its connotation as an island state. Hence, islandness concurs in further problematizing the debate on an already complex issue, namely, the self-representation of local communities through self-determined re-appropriation and use of their own cultural and symbolic heritage. A further level of complexity comes from considering this process as the basis of local development strategies, as in the case of tourism.

The relationship between islandness and identity has been largely developed within the scientific debate, also with regard to tourism. Indeed, such a thematic triangulation (islandness-identity-tourism) identifies the focus of the most debated issues animating island studies. In particular, the emergence of a decolonial approach in island studies [26] and, more generally, the analysis of the relevance of colonial thinking in approaching islandness [27] have determined the theoretical set on which to base further reflections on the topic.

Tourism dynamics still reflect colonial links, mediated through the persistence of economic and political benefits, among former colonies and their former colonizing country, as well as language similarity. It is not the case that a huge number of small islands' economies are tourism driven; moreover, as already underlined in the literature [28,29], the effectiveness of tourism-led development strategies in island economies depends on their affiliation with more powerful countries, according to a center-periphery model [30,31] that affects both the impacts of tourism on island territories—whose carrying capacity is usually limited—and the touristic demand and supply system, often directed from the outside.

What has been said above finds further validation through the analysis of marketing narratives related to islands. As observed by dell'Agnese [32], islands are subjected to an aggressive branding operation. If we consider the place as a unique combination of objective elements and subjective perceptions, able to create what Tuan [33] defined as topophilia, islands must face a de-territorialized stereotyped imagery according to which an island—no matter where, no matter which island—represents nothing more than a pleasant, exotic refuge from the stress of urban and industrial civilization. In this imagery, historical, political, and cultural characteristics underlying island orientative narratives have no place; the hyperconnective narrative of the island as "warm water" [28] prevails on the attractive narratives which depict islands as univocally identifiable territorial entities.

In order to find evidence of the above assertions, we thought it would be interesting to see what would be the result if we entered the key word "Cape Verde tourism" on any browser, and then clicked on the most indexed links (Figure 1).

(a)	(b)

Figure 1. Example of conventional attractive narrative about Cape Verde. The two word clouds were generated using Wordclouds (www.wordclouds.com), a free online tool. Though it was not possible to customize the n-grams' dimension and word clouds in general may not be very accurate (for instance, semantic networks are not considered), what emerged was worthy of an in-depth analysis. The word clouds in Figure 1 refer to two of the most indexed links: Lonely Planet (**a**) https://www.lonelyplanet.com/cape-verde [34] and World Travel Guide [35] (**b**) https://www. worldtravelguide.net/guides/africa/cape-verde/. (Source: personal elaboration).

Word clouds are visual representations of text data, which show the frequency of a word within a text. The bigger the word, the higher the frequency within the text (corpus) analyzed.

In both cases, it was possible to observe that, besides local toponyms and, of course, *island/islands*, the most frequent words were related to leisure activities (*music, hikes, bars, tour*) and landscapes (*mountains, beaches, dunes, seaside, valleys*). Moreover, there was a clearly identifiable reference to what we could define as a general sense of remoteness (*abroad, foreign, travel*), as well as a wide range of lemmas describing beautiful, uncontaminated, peaceful landscapes (*peaceful, unspoilt, framed*), also by recalling colors (*white, green, indigo-blue*). Finally, a very few references to local culture (*moraleza, creole*) were recorded.

The same remarks are applicable to visual contents (Figure 2). By typing the search key "Cape Verde tourism" on Google Chrome and selecting the filter "images", the results were pictures of the same kind of landscape: Sandy beaches with sparkling blue waters, palm trees, tiny wooden boats on the shoreline, a rocky skyline. Green-blue, light blue, and white were the prevalent colors, respectively referable to the sea, the sky, and the sand. In the majority of the pictures, the absence of people within the framed space evoked the idea of a peaceful, desert island. The only human traces were straw parasols and deckchairs. If present, tourists were depicted swimming or sunbathing, while the presence of locals was quite rare, mainly reduced to a picturesque element embedded within the landscape.

The result was a stereotyped description of an earthly paradise whose attractive narrative lies in its power to instill in the reader a certain sense of peace and relaxation. However, such evocative potential derives from a mere aesthetic, and a static idea of the place, rather than a full comprehension of its relational and symbolic fabric resulting from unique processes of territorialization.

Landscape representations themselves, delivering the idea of stunning, captivating, striking, surreal places and views, were characterized by recurrent elements, none of which was uniquely attributable to Cape Verde. In other words, the representation delivered by the mainstream narration on Cape Verde was more like a setting, rather than a landscape or a place; the perfect setting for holidays and honeymoons, rather than a relational experience [36].

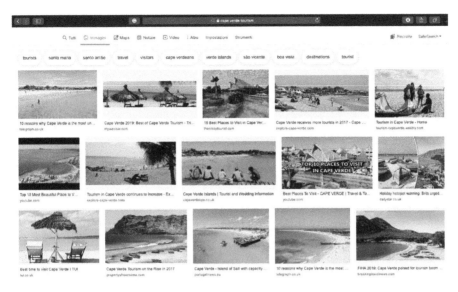

Figure 2. The first page of results after typing "Cape Verde tourism" on Google images. (Source: Google Chrome, 11 September 2019).

The risk [28] is to incur a neocolonialist-like nissology, written by outsiders—including both foreigners and islanders living outside the island—mainly for outsiders. For instance, in tourism narratives, elements like the language used (vehicular languages prevail over vernacular), the reference to a set of characteristic features as representative of a locality (those with a wider evocative potential on the large scale), even the kind of narrator (heterodiegetic narratives prevail over autodiegetic [37]), concur in preserving such tendencies.

Testing Placetelling® within an island *milieu* shows a clear epistemological value because islandness underlines the importance of the territorial variable in characterizing Placetelling® compared to other kinds of storytelling, whereas it inspires not only the goals, but also the design and implementation of narrative processes. With regard to our case study, it was a crucial Placetelling® aim to achieve an orientative and attractive narrative able to limit the globalist effects of hyperconnective narratives.

4. Case Study

4.1. Introduction

Placetelling® is a method for narrating places, which is:

- Identity-driven: It tells about the places, understanding their essence and their meaning, intended as a complex emotional attachment [38].
- Endogenous: Starting from the so-called genetic traits of a territory, which are born from the territory for the territory [1].
- Self-centered: It is a process that increases the capacity of a territory to elaborate narratives that operate on the territorialization and capitalization processes in compliance with the principles of sustainability [1,39].

A placeteller is able to develop new immersive narrative techniques, to act in supporting place interpretation. He or she is likely to be an expert in the field of information, communication, and education, in the subfield of infotainment and edutainment, who makes storytelling a powerful tool for the interpretation and enhancement of local heritage starting from the lived space shared by the dwellers.

From a methodological point of view (how can places be told?), Placetelling® gleans its theories and techniques from literature, narratology, aesthetics, and media studies, thus highlighting its interdisciplinary character.

To be effective, storytelling has to possess some specific characteristics.

Semiotics tell us much about these aspects. Indeed, in Placetelling® the relationship that is established in the narrative representation of reality between interpretation and hermeneutic processes is crucial.

With regard, in particular, to the functions of narrating, we will use a summary scheme that illustrates the factors at stake in the construction of a functional narrative, using an R to indicate the reference to reality, S the concepts and symbolic values (also of fundamental importance because they act on an emotional level), and F the formal values (good narrative form and quality of exposition/illustration).

Since no good narrative text is exclusively formal or symbolic, nor does it narrate reality without an attribution of meanings (otherwise it would be pure chronicle or documentary narrative), there are no cases of good narrative that properly relate to the vertices of the illustrated triangle. Instead, there are an infinite number of combinations that are placed within it and that describe, with different objectives and measures, the three fundamental components.

What is described here can, therefore, be summarized, as in Figure 3.

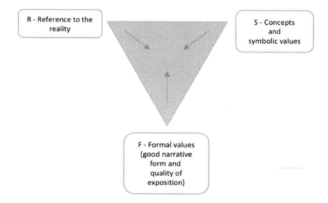

Figure 3. The scheme presents the functions of the narrative, *R*, indicating the reference to reality, *S* concepts and symbolic values (also of fundamental importance because they act on an emotional level), and *F*, the formal values (good narrative form and quality of the exhibition/illustration). (Source: Personal elaboration).

From the point of view of a narrative structured and finalized in this way, it is necessary to listen to the places before they are represented, experiencing them, letting the places speak, creating a universe of meanings in a sort of autopoietic process.

Placetelling® is, therefore, a precious instrument of transmission and conservation of the narrative heritage, nourished by written and oral stories of humanity.

At this point it is appropriate to make some clarifications to dispel any doubts about a possible overlap of meaning between Placetelling® and other methodologies that could present affinities, such as place-making [40]. We argue that Placetelling® can be functional to place-making, in the sense that the former can accompany the latter and make it a participatory and conscious process, capable of truly restoring the territorial identity and demands of the local community in planning terms. At the same time, Placetelling® is also an instrument capable of reducing the risk that place-making could become the expression of an intellectual elite, in an undemocratic planning process. Empowering a community's narrative attitude means making it aware of its own distinctive identity and of the material and immaterial manifestations that it assumes: An unavoidable prerequisite for the community to be involved in place-making and to become aware of its role and the purposes of planning.

In the same way, we think that Placetelling® cannot be assimilated to a technique of involvement of the local community because involvement is not the final aim of the narrative, but its presupposition. Placetelling® is a technique aimed at strengthening the identity dimension and projecting it outside. The story is a way of recognizing and representing oneself and these are actions that can contribute to strengthening the collective dimension and giving feedback on the level of participation in the political dimension.

4.2. Description

The first Placetelling® training course in Cape Verde was promoted within the project "Cabo Verde: Historia, Cultura e Ambiente para um Turismo Sustentàvel", implemented by Fondazione Lelio e Lisli Basso and Fundacão Amilcar Cabral and funded by the European Union. The project, launched in 2016, intended to trigger the diversification of Cape Verdean tourism according to the principles of sustainable tourism, so as to preserve and enhance environmental, social, and cultural conditions. Specific objects were identified in enhancing Cape Verdean cultural and environmental diversity by a high-quality tourism supply, able to create employment opportunities; improving awareness among local communities and local authorities with regard to the importance of preserving territorial capital as an asset for local development, also through tourism; and foster new community-oriented and rural-oriented tourism dynamics, in order to include peripheral areas with an attractive high potential in tourism chains [41].

The project's output included the creation of two itineraries: Circuito do Aguardente e derivados-O Grogue como produto cultura, in Santo Antao, and Circuito da Baixa da Praia-Das Águas Doces e Salgadas in Santiago.

Consequently, the training course was conceived as a maieutic process during which attendees were invited to identify the most significant elements of local culture and, on that basis, discuss how to develop a narrative product able to catch the attention of a specific target tourist interested in living a realistic experience of the two localities, starting from the narrative interpretation of the two itineraries.

In order to involve a larger number of participants and to develop a more place-oriented process, two three-day sessions were held in Sant'Antão and Santiago. Each session was attended by about 20 participants including local tourist operators, cultural operators, local administration officers, artists, women, and young people. The two training courses were conducted by Fabio Pollice (University of Salento), Simone Bozzato, and Marco Prosperi (University of Tor Vergata).

Placetelling® principles and methods, implemented in the creative process, were illustrated during the first part of the training, through short intensive classes. In particular, participants were invited to reflect on:

Experiential tourism and identity-driven constructs: How to favor the discovery of real local culture by promoting the place as relational space and, as a consequence, the tourist's immersive experience (embeddedness) and how to recover narrative local capital and identify the intangible and tangible characteristic elements which define place identities, so as to make them accessible both by insiders and outsiders? It has to be clarified that place identities are to be considered as a symbolic and relational complex, which could be perceived through the double dimension of sense of belonging (for the local community) and mutual recognition (among local community members). As a consequence, Placetelling® works as a "passe-partout code" for entering the place's relational networks [42].

Identity narratives and immersive narratives. This session calls into question the debate about a place's authenticity, which becomes relevant in outlining a truly place-based narrative. In this case, taking our cue from Massey's observations about the risk posed by the spread of a "regressive sense of place" [43], we recall Cohen and Cohen's definition of "hot" authentication as "an imminent, reiterative, informal performative process of creating, preserving and reinforcing an object's, site's or event's authenticity. It is typically an anonymous course of action, lacking a well-recognized authenticating agent. The process of "hot" authentication is emotionally loaded, based on belief, rather than proof, and is therefore largely immune to external criticism" [24]. In this way, we intend identity community narratives to be an inclusive, plural, and dynamic direct expression of daily life as lived by the locals, who represent, in turn, a plurality of subjective experiences and narratives, rather than a static, museum-like experience of the place.

Narrative forms, techniques and strategies. The definition of effective narrative place-based products and their implementation must consider four aspects.

(1)   Languages. Placetelling® can use different languages, each with its own code, characterized by a set of techniques, styles, and indications. The language is profoundly linked to the concept of form, according to which each language (textual, pictorial, photographic, cinematographic) corresponds to a precise communicative mode: A story or a film can tell a story or a plot because it develops along a temporal sequence (storytelling power); a photograph, a pictorial work or an image immediately suggests, instead, an idea, without developing a plot (evocative power). The choice of a language and of a form depends on a series of factors, such as the reference target, the objective, and the functions, and the medium or the media through which the narration will be conveyed.

(2)   Channels. This aspect appears particularly crucial with reference to tourism, especially if we consider Placetelling® as a way for generating shared narratives to share, in turn, with outsiders. More specifically, it is possible to identify direct (or face-to-face) channels (dealers, hoteliers, restaurant owners, taxi drivers, tour guides, residents, other tourists) and indirect (or mediated) channels (websites, books, films and documentaries, brochures). In fact, the principle of specificity remains the fundamental paradigm of Placetelling® for the public as well as for the language, channels, and integrated media for communication. Not all audiences, in fact, have the same characteristics: They can be distinguished by target group (e.g., age groups) or, better still, by composition of the public, that is, by homogeneous groups as regards socio-demographic or psychological characteristics. This allows the placeteller to identify the correct communication strategy to convey geo-historical content related to the enhancement of cultural heritage. A preliminary investigation allows the placeteller to trace the profile of the users/recipients provided, for example, to choose the most inclusive and captivating language (e.g., performative, literary, or audiovisual), the channel, and the media that will promote wider participation (e.g., social network, TV, cinema, etc.) and with them also the techniques of narration.

(3)   Multi-sensorial narratives. The narrative and communication techniques used in Placetelling®—with particular reference to Immersive Placetelling®—aim to outline a narrative path able to catch the evocative potential of the place, in order to stimulate the perceptive sphere of the tourist and involve him or her in a totally immersive experience. In other words, a place can be experienced, for example, through the sense of smell (typical smells, local scents), hearing (soundscape, environmental sounds, typical local instruments), taste (traditional food and drinks), sight (landscapes, skylines, colors, typical architectures), and touch (materials, artifacts, and local fabrics). This does not mean mystifying the place, but rather favoring the understanding of its symbolic layer through the use of the narrative strategy deemed most appropriate for that specific case and for that particular audience.

(4)   Narrative strategies. The placeteller must be able to grasp the very essence of the place and outline a specific story for a variety of objectives. In order to achieve this, it could be useful to recall the principle of edutainment as it combines the two essential aspects of the experience connected to a cultural asset: Entertainment, on the one hand, and learning (education), on the other. Such a combination guarantees the coexistence between the didactic, communicative, and emotional components. If we think that enhancement consists in any activity aimed at improving the conditions of knowledge and conservation of cultural heritage and increasing public enjoyment, we understand how evident the need for integrated action is to reconcile heritage enhancement and local development by assigning a priority role to local communities.

The second part of the course, subsequent to the theoretical investigation of Placetelling® principles and methods, consisted of site visits to the two itineraries created during the project. On the basis of theoretical classes and group discussions, participants were asked to elaborate a product able to guide the tourist through an inclusive experience of the two localities. As a result, the groups outlined two "10 things to do" lists for tourists, as analyzed in the next paragraph.

5. Results and Discussion

The difficulty of guides and tour operators of the Cape Verdean islands where the course was held was to tell the specificity of their culture and to focus on its attractiveness, while trying to direct tourism towards a conscious and experiential use of its territory.

In fact, it must be considered that, differently from Sal and Boa Vista, the islands that welcome 90% of the tourist flow directed to the Cape Verde archipelago, which owe their attraction to their beaches and to the large and well-equipped tourist villages that rise along the coast, Santiago has few beaches, totaling a couple of kilometers, while Santo Antão has no beaches at all.

The Placetelling® course, therefore, worked on identifying the elements of local culture that can take on an attractive value for the tourist and how they can be told so that their attractive value produces its effects on international demand.

The metaphor used was the "cachupa", a traditional food of these islands: A corn and bean stew with meat or fish, a smart combination of elements that brings together all that these islands can offer.

In order to promote experiential tourism, the decision was made to tell this culture starting from the experiences that the tourist must have to feel Cape Verdean: A "to do-list" capable of communicating the richness of Creole culture, what to eat and when, where to go to see the sunset or dawn, when and where to drink grog, how to prepare cachupa, etc.

The two "10 things to do" lists (one for Santo Antão, one for Santiago) (Table 1) were the result of group discussion about what are the most relevant elements that compose the place identities of Santo Antão and Santiago and what kinds of activities can help an outsider to better interpret the spirit of the place. The lists represent an example of bottom-up narrative, whose elaboration required a shared reworking of traditional heritage by the locals (orientative narrative).

The "10 things to do" lists can be delivered both through direct and indirect channels, and their evocative communication mode aims to arouse people's curiosity in investigating, through an immersive experience, the daily, hidden aspects of the two Cape Verdean localities (attractive narrative).

Table 1. The two "10 things to do" lists.

10 Things to Do in Santo Antão	10 Things to Do in Santiago
1 Keep smiling	1 Visit Cidade Velha, the birthplace of Cape Verdean nationality
2 Experience the everyday life of Santo Antão	2 Learn the typical dances of Santiago, Batuque and Funaná
3 Learn to cook Cachupa	3 Live the everyday experience of the rural world
4 Go to the Grogue rum factory and have a go at using the trapiche press	4 Take a sea bath or rain showe
5 Learn to dance contradança and mazurca	5 Join a pilgrimage party
6 See the artisanal manufacture of violins and play violin in Fontainhas village	6 Join the Tabanka party procession
7 See the sunset in Sinagoga	7 Play traditional music in the community
trapiche press8 Make tracks along the mountain trails	8 Listen to traditional storytelling in Santiago
9 Visit the waterfalls of Paul and Caibros	9 Discover the village of Rabelados
10 Go fishing with the local artisan fishermen	10 Try Cachupa, the traditional dish

Although an early Placetelling® experiment, the "things to do" show important differences compared to the mainstream narrative. Every item suggests an activity to achieve, so we can record a significant frequency of verbs like *"learn"*, *"make"*, *"experience"*, *"participate"*, and *"know"* that sound like an evocative imperative, implying a direct, experiential involvement of the tourist. Besides stimulating the emotional sphere, most of the identified activities aim to preserve what we can call a "local skill" (e.g., *"learn to cook Cachupa"*): Once learned, the tourist can replicate it after his/her experiential trip.

There are several specific references to the places: toponyms (besides the two localities, *Fontainhas village, Paul and Caibros, Cidade Velha, Rabelados*), food traditions (*Grogue, Cachupa*), celebrations (*pilgrimage parties, Tabanka party procession*), and leisure activities (*dance Batuque and Funanà, playing traditional music, learning to dance contradança and mazurka*) suggest *where to go* and *why*, allowing the tourist to outline an experiential mapping. Moreover, references to local toponyms and cultural features play an evocative function; in contrast, this function in mainstream narrative is played by the large use of adjectives like *stunning, striking* (cf., the paragraph "Identity narratives and islandness"), which are totally absent in Placetelling® narratives. We can conclude that Placetelling® enhances the evocative potential of the place, and evocativeness becomes site specific (Figure 4).

Figure 4. Word cloud generated using the two "10 things to do" lists.

Finally, differences between the mainstream narrative and Placetelling® in Santo Antão and Santiago can also be identified with regard to principles and methods, as summarized in the Table 2.

Table 2. Comparison between mainstream narratives and Placetelling®.

Aspects	Mainstream Narrative	Placetelling®
Narrative	Attractive, hyperconnective	Orientative/attractive
Attractiveness	Remoteness; sense of exotic	Exploration
Reference to the context	Weak. Many of the items refer to elements which are not unique to Cape Verdean identity. Place mystification/disneyfication.	Strong. Large presence of toponyms, traditional activities and dishes.
Involvement within the hosting context	Limited	Embeddedness

Table 2. *Cont.*

Aspects	Mainstream Narrative	Placetelling®
Narrative forms, techniques and channels	Communication form is mainly evocative, excites desire. Channels are mainly indirect. No diversification of target strategies.	Communication form is mainly evocative, so as to arouse curiosity. Channels can be both direct and indirect. Target diversification.
Governance	Outsiders. Tourist holdings.	Insiders: local community, local stakeholders.
Perspectives	Subjected to the dynamics of mainstream tourism markets.	Spread of Placetelling® experiences in other contexts. Development of a shared bottom-up narrative on which to base new branding strategies.

6. Conclusions

As already stated, the debate on Placetelling® is very recent, and still needs to be further developed. This means that practical applications like the one reported in this essay are useful to spot a number of weaknesses and criticalities, both in the processes and in the results that definitely deserve an in-depth reflection.

It is worth mentioning that besides the limited amount of time, a huge difficulty can be identified in the use of a non-native language, which could lead to misunderstandings in the dissemination and interpretation of symbolic meanings at the base of intangible cultural heritage.

Nevertheless, these experiences are also fundamental to show the potential of Placetelling® as a preparatory action for the enhancement of the cultural heritage of a place, a way to stimulate and direct it.

In other words, it is precisely through narration that the value of a place is created and communicated, because the community is put in a position to recognize it and reclaim it, building its own development project around it [1].

Therefore, it is necessary to know how to increase the appeal, or the charisma of places, through actions capable of stimulating the tourist's emotional imagination, especially given the fact that the tourist is increasingly motivated by the desire to achieve an intense cultural experience [31].

In fact, before travelling, the tourist imagines: So he/she wishes to satisfy an expectation built through the recourse to narratives of various kinds, mostly mediated.

Then it is a true promesse de bonheur composed of material elements, but also and above all of the intangible elements (emotions, lived experiences) that play a fundamental role in choosing the destination since the tourist phenomenon sinks its roots in the complex territory of desires [36].

The unique and unrepeatable characteristics of a place that give rise to the desire in the tourist towards that place are, therefore. fundamental elements to start an identity process and a sustainable economic and social development.

On the other hand, the tourist is not the only protagonist of Placetelling® processes: The main actor is the place itself, here considered as the result of symbolic negotiations whose tangible effects are a multiplicity of layers (material infrastructures, stories, celebrations, customs, and habits) through which the place has historically built itself. When dealing with place enhancement, the worst threat to incur is what we can define, to recall the relational approach, a regressive sense of place, namely inspired to an idea of place like something given, fixed, rather than articulated moments in networks [44,45]. This means that every Placetelling® strategy should be implemented avoiding every form of museification or Disneyfication of the place and, consequently, of local community. This also suggests a reflection about the local community's level of empowerment and, more specifically, the narratives about the local people's ability to meet their needs and set up self-determined development trajectories [46,47].

Another final aspect to be investigated is the launch of new projects through the use of narration.

Placetelling®, in fact, using a geographical approach of a prescriptive type, directs the development and growth direction of a place, becomes a harbinger of new projects and, therefore, somehow reconstructs the place we are describing through the stories that originate from it (maieutic action). In this sense, the debate on cultural sustainability ties in with territorial narratives and also affects the debate on place-making [40], as Placetelling® is here seen as a supporting tool to the implementation of community-based forms of tourism, so that it is able to enhance the symbolic narrative component. More specifically, in the case of Cape Verde, Placetelling® can act as the first step of a shared process for a new territorial branding, different from the one disseminated by the mainstream narrative. Here, territorial branding is seen as a "collective reflection on territorial identity and its representation, contributing to the strengthening of the sense of belonging and creating the basis for a strategic convergence among local actors" [46], and an appropriate territorial narrative can assume the role of "cornerstone" of the subsequent phases (regulation, circulation, exchange, consumption), which, in turn, will strengthen its evocative value and propulsive force [37].

In this way, a new direction for individual and collective behavior is imprinted, through interpretation and work on perceptions—which plays a role of meaning of social behavior.

Placetelling® tells us how to see the future of a place, intercepts potential, and gives a direction to development.

Author Contributions: Conceptualization, F.P.; methodology, F.P., A.R., F.E., and P.M.; formal analysis, F.E.; investigation, F.E., A.R., and P.M.; writing—original draft preparation, F.E., A.R., and P.M.; writing—review and editing, F.E. and P.M.; visualization, F.P.; supervision, F.P. and A.R.; project administration, F.P. Sections 1, 2 and 3.2. are to be attributed to F.E.; Section 3.1. to A.R.; Section 4 to P.M.; Sections 5 and 6 to F.P. All authors have read and agreed to the published version of the manuscript.

Funding: The experience was developed within a wider cooperation project between the Fondazione Basso foundations and the Fondaçao Amil Cabral.

Conflicts of Interest: The authors declare no conflict of interest. The funders had no role in the design of the study; in the collection, analyses, or interpretation of data; in the writing of the manuscript, or in the decision to publish the results.

References

1. Briguglio, L. Small island developing states and their economic vulnerabilities. *World Dev.* **1995**, *23*, 1615–1632. [CrossRef]
2. Central Intellingence Agency. Available online: https://www.cia.gov/library/publications/the-world-factbook/ (accessed on 13 November 2019).
3. Instituto Nacional de Estatística. Available online: http://ine.cv (accessed on 18 May 2020).
4. Sánchez Cañizares, S.M.; Castillo Canalejo, A.M.; Núñez Tabales, J.M. Stakeholders' perceptions of tourism development in Cape Verde, Africa. *Curr. Issues Tour.* **2016**, *19*, 966–980. [CrossRef]
5. Nadarajah, Y.; Grydehøj, A. Island studies as a decolonial project (Guest Editorial Introduction). *Isl. Stud. J.* **2016**, 437–446.
6. UNWTO; UNEP. World Tourism Organization (2012), Challenges and Opportunities for Tourism Development. In *Small Island Developing States*; UNWTO: Madrid, Spain, 2005; p. 17.
7. Vítová, P.; Harmáček, J.; Opršal, Z. Determinants of tourism flows in Small Island Developing States (SIDS). *Isl. Stud. J.* **2019**, *14*, 3–22. [CrossRef]
8. Scheyvens, R.; Momsen, J.H. Tourism and poverty reduction: Issues for small island states. *Tour. Geogr.* **2008**, *10*, 22–41. [CrossRef]
9. Holden, A. *Tourism, Poverty and Development*; Routledge: London, UK, 2013.
10. Teelucksingh, S.S.; Watson, P.K. Linking tourism flows and biological biodiversity. *Small Isl. Dev. States (SIDS)* **2013**, *18*, 392–404.
11. Ricker, B.A.; Johnson, P.A.; Sieber, R.E. Tourism and environmental change in Barbados: Gathering citizen perspectives with volunteered geographic information (VGI). *J. Sustain. Tour.* **2013**, *21*, 212–228. [CrossRef]
12. Persaud, I. Unpacking Paradise: Geography Education Narratives from the Seychelles. Ph.D. Thesis, University College London, London, UK, January 2017.
13. Nielsen, H.P.; Rodríguez-Coss, N. Island studies through love and affection to power and politics. In *Gender and Island Communities*; Routledge: London, UK, 2020; pp. 158–173.

14. Parsons, M.; Nalau, J. Adaptation policy and planning in Pacific small island developing states. In *Research Handbook on Climate Change Adaptation Policy*; Edward Elgar Publishing: Cheltenham, UK, 2019.

15. World Tourism Organization, Challenges and Opportunities for Tourism Development. In *Small Island Developing States*; UNWTO: Madrid, Spain, 2012.

16. Turco, A. *Turismo e Territorialità. Modelli di Analisi, Strategie Comunicative, Politiche Pubbliche*; Unicopli: Milano, Italy, 2012.

17. Pollice, F.; Urso, G. Turismo vs. globalitarismo. In *Filiere etiche del turismo. Territori della vacanza tra valori, politiche e mercati*; Unicopli: Milano, Italy, 2014.

18. Magnaghi, A. *Il Progetto Locale*; Bollati Boringhieri: Torino, Italy, 2000.

19. Vanolo, A. *Geografia Economica del Sistema-Mondo; Territori e reti nello scenario globale*; Utet: Torino, Italy, 2010.

20. Pollice, F. Alberghi di comunità: Un modello di empowerment territoriale. In *Territori della Cultura*; CUEBC: Ravello, Italy, 2016; Volume 25, pp. 82–95.

21. Pollice, F. Placetelling® per lo sviluppo di una coscienza dei luoghi e dei loro patrimoni. In *Territori della Cultura*; CUEBC: Ravello, Italy, 2017; Volume 30, pp. 106–111.

22. Pileri, P.; Granata, E. *AmoAmor Loci. Suolo, Ambiente, Cultura Civile*; Libreria Cortina: Milano, Italy, 2012.

23. Emanuel, C. Patrimoni paesistici, riforme amministrative e governo del territorio: Svolte e percorsi dissolutivi di rapporti problematici. In *Bollettino della Società Geografica Italiana*; Società Geografica Italiana: Rome, Italy, 1999; Volume 4, pp. 295–318.

24. Cohen, E.; Cohen, S.A. Authentication: Hot and Cool. *Ann. Tour. Res.* **2012**, *3*, 1295–1314. [CrossRef]

25. Finocchi, F. *Geografie del Gusto*; Aracne: Roma, Italy, 2010.

26. McElroy, J.L. Tourism development in small islands across the world. *Geogr. Ann.* **2003**, *85*, 231–242. [CrossRef]

27. McElroy, J.L.; Pearce, K.B. The advantages of political affiliation: Dependent and independent small island profiles. In The round table. *Commonw. J. Int. Aff.* **2006**, *95*, 529–539. [CrossRef]

28. Baldacchino, G. Studying Islands: On Whose Terms? Some Epistemological and Methodological Challenges to the Pursuit of Island Studies. *Isl. Stud. J.* **2008**, *3*, 37–56.

29. McElroy, J.L.; De Albuquerque, K. Problems for managing sustainable tourism in small islands. *Isl. Tour. Sustain. Dev.* **2002**, *85*, 15–31.

30. Scheyvens, R.; Momsen, J. Tourism in small island states: From vulnerability to strengths. *J. Sustain. Tour.* **2008**, *16*, 491–510. [CrossRef]

31. Dallen, T.J. Tourism and the Personal Heritage Experience. *Ann. Tour. Res.* **1997**, 751–754.

32. Dell'Agnese, E. *Bon Voyage. Per una Geografia Critica del Turismo*; Franco Angeli: Milano, Italy, 2018.

33. Tuan, Y.F. *Topophilia*; Columbia University Press: New York, NY, USA, 1974.

34. Cabo Verde Travel Africa—Lonely Planet. Available online: https://www.lonelyplanet.com/cape-verde (accessed on 6 November 2019).

35. Welcome to Cape Verde—World Travel Guide. Available online: https://www.worldtravelguide.net/guides/africa/cape-verde/ (accessed on 6 November 2019).

36. Giordana, F. *La Comunicazione del Turismo tra Immagine, Immaginario e Immaginazione*; Franco Angeli: Milano, Italy, 2004.

37. Pollice, F.; Rinella, A.; Rinella, F.; Epifani, F. C'era una volta e c'è ancora: La narrazione dell'autenticità nel Progetto "Comunità Ospitali" dell'Associazione Borghi Autentici d'Italia (BAI). *Geotema* **2019**, in press.

38. Greiner, A.L.; Dematteis, G.; Lanza, C. *Geografia umana. Un Approccio Visual*; Utet Università: Turin, Italy, 2014.

39. Pollice, F. Valorizzazione dei centri storici e turismo sostenibile nel bacino del Mediterraneo. *Bollettino della Società Geografica Italiana* **2018**, *1*, 41–56.

40. Dupre, K. Trends and gaps in place-making in the context of urban development and tourism. *J. Place Manag. Dev.* **2019**. [CrossRef]

41. Bozzato, S.; Prosperi, M.; Pollice, F. *Projecto "Cabo Verde: Historia, Cultura e Ambiente para um Turismo Sustentàvel"*; Unpublished Report; 2019.

42. Rinella, F. Dal rito locale della "Taranta" alla "Pizzica globale". *(S)radicamenti Società di studi geografici Memorie Geografiche* **2017**, *15*, 335–340.

43. Massey, D. Places and their pasts. History. *Workshop J.* **1995**, *39*, 182–192. [CrossRef]

44. Nunkoo, R.; Gursoy, D.; Devi Juwaheer, T. Island residents' identities and their support for tourism an integration of two theories. *J. Sustain. Tour.* **2010**, *5*, 675–693. [CrossRef]

Sustainability **2020**, *12*, 4333

45. Massey, D. *Space, Place and Gender*; John Wiley & Sons: Hoboken, NJ, USA, 2013.

46. Pollice, F. Spagnuolo, F. Branding, identità e competitività. *Geotema* **2009**, *37*, 49–56.

47. Pocock, D.; Relph, E.; Tuan, Y.-F. *Classics in Human Geography Revisited*: Tuan, Y.-F. 1974: Topophilia. Englewood Cliffs, NJ: Prentice-Hall. *Prog. Hum. Geogr.* **1994**, *3*, 355–359. [CrossRef]

 *sustainability*

Article

Different Worldviews as Impediments to Integrated Nature and Cultural Heritage Conservation Management: Experiences from Protected Areas in Northern Sweden

Carl Österlin [1,*] , Peter Schlyter [2] and Ingrid Stjernquist [1]

[1] Department of Physical Geography, Stockholm University, 106 91 Stockholm, Sweden; ingrid.stjernquist@natgeo.su.se

[2] Department of Spatial Planning, Blekinge Institute of Technology, 371 79 Karlskrona, Sweden; peter.schlyter@bth.se

* Correspondence: carl.osterlin@natgeo.su.se

Received: 26 March 2020; Accepted: 24 April 2020; Published: 26 April 2020

Abstract: In the management of protected nature areas, arguments are being raised for increasingly integrated approaches. Despite an explicit ambition from the responsible managing governmental agencies, Swedish Environmental Protection Agency and Swedish National Heritage Board, attempts to initiate and increase the degree of integrated nature and cultural heritage conservation management in the Swedish mountains are failing. The delivery of environmental policy through the Swedish National Environmental Objective called Magnificent Mountains is dependent on increased collaboration between the state and local stakeholders. This study, using a group model building approach, maps out the system's dynamic interactions between nature perceptions, values and the objectives of managing agencies and local stakeholders. It is identified that the dominance of a wilderness discourse influences both the objectives and management of the protected areas. This wilderness discourse functions as a barrier against including cultural heritage conservation aspects and local stakeholders in management, as wilderness-influenced objectives are defining protected areas as environments "untouched" by humans. A wilderness objective reduces the need for local knowledge and participation in environmental management. In reality, protected areas depend, to varying degrees, on the continuation of traditional land-use practices.

Keywords: integrated environmental management; cultural landscapes; stakeholder participation; landscape planning; systems thinking; group modeling; participatory modeling; conservation; wilderness; wilderness discourse

1. Introduction

The division of society and nature into separate realms is a frequent Western perspective for describing and understanding nature [1,2]. Some scholars (e.g., Cronon (1996) [3]) have pointed out how the idea of wilderness is, to a certain extent, a construction and description of an alien but supposedly pristine environment, which, as a response to guilt over environmental degradation, should be protected from human actions such as industrialization. Whether or not these areas constitute a wildernesses in the sense that they are essentially unaffected by the influence of human actions has been a long-standing and polarized academic [4] and popular debate [5], both internationally and in Scandinavia [6,7].

This view of nature is also linked to a romantic, and to some extent nationalistic, view of nature [8], exemplified by the creation of national parks for the preservation of wilderness. Or, in the words of the Committee Report, which formed the basis for subsequent Swedish national park legislation, (translation from Swedish) " ... certain areas are to be set aside where life in nature may develop perfectly undisturbed by the influence of culture" [9].

Although there have been ample efforts to define what wilderness is, there is no universal definition. However, North American perspectives originating from the US Wilderness Act of 1964 have been influential in the parameters that are often included in the various definitions of wilderness. Human perceptions of nature and spatial scale are typically the foundations of the term wilderness rather than ecological parameters. For example, the experience of solitude and large and remote areas, in combination with a landscape typically perceived as unaffected by human activities, are common components in the term wilderness (see, e.g., Lupp et al. 2011 [10] for a more comprehensive review of wilderness descriptions). In a Scandinavian context, an early analogue was the identification of "wilderness core areas" in a government inquiry (SOU 1971:75) [11], defined as contiguous areas larger than 1000 km^2, more than 15 km from any road or railway and without designated hiking trails and facilities for overnight stays. Protections for these areas have subsequently been developed in the Swedish Planning and Building Act (Ch. 4, §5 Planning and Building Act) by delimiting areas of "unbroken mountain areas", i.e., large contiguous areas unbroken by roads and with severe restrictions on building. Similarly, in Norway, areas may be designated as wilderness-influenced and given special status and management, but ecological uniqueness or biodiversity are less relevant components in the ideas of wilderness compared to the absence of human interference [12]. Thus, as the examples above illustrate, it is not ecological qualities per se that are meant to be preserved in the idea of wilderness, but rather the absence of human use and artifacts.

Despite a strong prevalence of perceptions that certain areas constitute a wilderness, the concept is often based on a misconception of the social-ecological processes that shape natural environments [13]. In Scandinavia, this has certainly been the case too. Emanuelson (1987) [14] concluded in 1987 that the Swedish mountain region has long been incorrectly described as a wilderness and that, as a region, it cannot be seen as "untouched" by humans. Instead, the human impact on the landscape, through various forms of traditional and indigenous land uses, for example reindeer husbandry, could be observed already from the 17th century onwards.

Nevertheless, wilderness perspectives on the Swedish mountains prevail. For example, Wall Reinius (2009) [15] exemplifies how this wilderness perspective has also been present in the Swedish context, where protected areas in the northern mountain region have been described, for example, as "Europe's last great wilderness". Further, Wall Reinius (2009) [15] highlights how even in the mixed natural and cultural World Heritage Site (WHS) of Laponia, which is co-managed with local Sámi communities, this wilderness description of an area that is effectively a Sámi cultural landscape still continues to be held, unreflectedly, by responsible managing agencies like the Swedish Environment Protection Agency (SEPA) and the Swedish National Heritage Board (SNHB) [15].

Cultural heritage conservation has a long tradition in Sweden. The managing agency SNHB was founded as early as 1630, and by 1666 the scope of the board had widened from documentation to the beginning of one of Europe's first legislations to actively protect cultural heritage features in the landscape from exploitation and destruction. The legislative framework (previously the Ancient Monuments Act (1942:350), replaced by the Historic Environment Act 1988:950 and the Historic Environment Ordinance (1988:1188)) is strong and has protected, in principle and rather radically, all objects considered ancient, whether or not they have been identified or designated as such. By tradition, the preserved objects have typically been point features (e.g., a rune stone, a burial site etc.) and have only recently and rarely come to include larger areas or landscapes.

Nature conservation is, on the other hand, a much more recent pursuit—in many respects a reaction to the industrial transformation of traditional pre-industrial landscapes, and typically concerned with protecting ecosystems and landscapes, i.e., a wider area rather than point features (though early protection also had a component of point feature protection, e.g., of large erratic boulders or very old oak trees). In 1909 Sweden created the first national parks in Europe. The early national park system had, area-wise, a clear focus on the mountainous areas of northern Sweden and was established on state-owned (then Crown-owned) land. This northern bias may be explained by a perceived need to protect what was seen as an unspoilt original pristine nature against industrial resource exploitation [16,17]. This "northern approach" was also financially expedient, as the land was state-owned and no economic compensation to landowners or users was thus required. Later legislation enabled other forms of area protection and, from the early 1960s onwards, the establishment of an extensive network of smaller nature reserves, often located on private land [17]. A weaker supporting legislation and the need to negotiate with, and economically compensate, private owners for the establishment of reserves has resulted in a more deliberative, activist and negotiating culture within nature conservation in comparison with the cultural heritage field, where deliberation and negotiations were not really needed and, as a default, protection was at hand. While the establishment of nature reserves addressed to some extent the issue of a northern bias in conservation, the larger contiguous parks and reserves are still, by and large, a northern feature of protected areas in Sweden, as apparent in Figure 1.

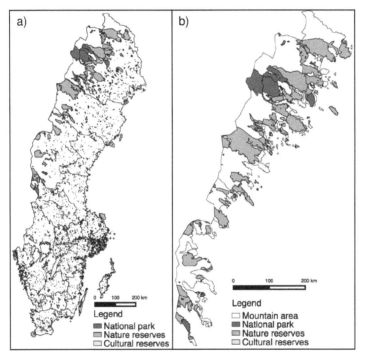

Figure 1. (**a**) National parks, Nature reserves and Cultural reserves in Sweden. The largest contiguous areas of protected nature are found in the northwestern mountain regions. Also clearly visible is that Nature reserves cover a much larger area compared to National parks, and the fact that Cultural reserves are barely identifiable at this scale. (**b**) Protected areas, within the mountain area (defined by the *Productive Forest Line*). Protected areas constitute a large proportion of the mountains region. In northern Sweden, Nature reserves cover a notably larger area than National Parks. There are three Cultural reserves located within the mountain area but are hardly visible.

From the 1970s onwards, as a consequence of increasingly mechanized and industrial-style forestry, there have been increasing public demands on forestry to include environmental concerns in its operations. The current Forestry Act, from 1992, equates the environmental and economic objectives [18]. Similar ambitions to integrate environmental aspects in all societal activities are expressed in the broad National Environmental Objectives (NEO) unanimously adopted by the Swedish parliament in 1999. The NEO approach is supposed to allow a transition from reactive regulation-based environmental governance to a more proactive approach based on objectives [19]. The NEO approach is further based on the assumption that all stakeholders (government agencies, municipalities, private and public corporations, individual owners and the public) should contribute to the attainment of the 16 national objectives, and the objective fulfillment of the objectives is monitored through national and regional indicators [20].

In character, the sixteen national environmental objectives vary significantly. Some of them may be described as science-driven, while others are more value-driven [21]. Emmelin (2005) [22] also points out how value-driven objectives are more contested by the various actors that need to relate to these objectives as compared to science-driven objectives. The need for a discussion on how to operationalize and concretize the rather fuzzy objectives is, according to Emmelin (2005) [22], underpinned by a situation where actors may agree on the objectives on a rhetorical level while at the same time disagreeing on the operationalization of the objectives, the chosen policy measures or the legitimacy and the balance between other environmental objectives or other societal goals on a regional or local scale. One objective characterized as a vague value-driven objective is the Magnificent Mountains objective, focused on the mountainous landscape of northern Sweden.

The Magnificent Mountains objective is one of the few, as opposed to most of the other Environmental Objectives, that can be achieved, in principle, by decisions and actions taken within Sweden—the other NEOs are all to various degrees dependent on collective European, or global, action. Nonetheless, progress reports about the state of the Swedish mountain environment have consistently emphasized that the objective will not be achieved within the given timeframe. Either the conditions in the mountain area are deteriorating rather than improving and maybe even worsening, or there is not even enough knowledge to know whether the situation is deteriorating or not [23–26].

A group modeling-based systems analytic study by [27] showed, however, that practically all stakeholders of relevance to the Magnificent Mountain objective had doubts about the relevance and utility of designated sub-objectives to the overarching objective, as well as with regard to the relevance of the chosen official indicators. Furthermore, one of the main reasons for poor objective attainment was the effect of focusing too specifically on the environmental aspects of the sustainability trinity, to the detriment of the economic and social aspects. The study highlighted the importance of a more integrated approach in the NEO work, i.e., a call for a better coordination between concerned authorities as well as the need to involve local stakeholders in the planning and local resource management of the mountain environment. Finally, the study included a stakeholder defined transdisciplinary research agenda. One concrete outcome of the latter was a joint call on "Integrated nature and cultural heritage conservation" from the SNHB and the Swedish Environmental Protection Agency (SEPA).

Sayer et al. (2013) [28] and Reed et al. (2017) [29], for example, illustrate the importance of including multiple stakeholder perspectives in a landscape with competing land-use interests and where broader environmental concerns are balanced against other societal objectives in order to develop appropriate management strategies. In addition, to achieve a better integration between natural and cultural value power balances between managing governmental agencies is also a key issue, as such power balances determine who can define the landscape to be managed [30]. Dawson et al. (2017) [31] have demonstrated how systems analysis and causal loop diagrams may be efficient methods to include such multiple stakeholder perspectives in a landscape-based setting.

In a northern Swedish context, there have been arguments for increased community-based management or co-management, made by scholars as well as by stakeholders involved in land-use management in and around the mountain area (see for example [27,32,33]). Despite some examples of the adoption of such approaches (Laponia WHS, Vilhelmina Model forest), there is an evident disconnect and discontent at a local level with the possibilities for local actors to be involved in land-use management [34,35]. The Magnificent Mountains objective setting is a clear example of when the perspectives of multiple stakeholders are desirable, as it is comprised of a goal-oriented proactive environmental policy, whose delivery is dependent on multiple stakeholders with their own agendas—acting within the boundaries of a defined region, in which environmental qualities should be preserved, but also weighed against other societal objectives. After almost 20 years since its adoption the environmental objective Magnificent Mountains is far from being achieved [36]. Despite an explicit will and a common idea from both of the responsible governmental agencies, SEPA and SNHB, about increasing the integration of nature and cultural heritage conservation management, this practice does not yet appear to be taking place.

The aim of this study is to identify and understand current challenges and barriers to an integrated nature and cultural heritage conservation management in protected areas, using the protected areas in the northern Swedish mountains as reference, and in particular the influence of different discourses on management policy and the possibility to identify potential leverage points for actors to overcome these challenges and barriers.

2. Background

The Swedish mountain area is a large region. A number of definitions of what constitutes the Swedish mountain regions have been used in various settings (see, e.g., Naturvårdsverket (2019) [37] for a summary of commonly applied definitions). One commonly occurring definition, also used in this study, is the administrative boundary for productive forests, called the *Productive Forest Line*. The area above this boundary is a vast region just above 100,000 km^2 and accounts for approximately a quarter of the country. It is also a sparsely populated region in general, and few people live within the protected areas. Despite a sparse population pressure, it is a region of "contested landscapes" (exemplified by, e.g., Horstkotte (2013) [38]). These contestations come from shifting views on how the land should be used—for example reindeer husbandry, nature protection, cultural heritage management, tourism and industrial exploitation [35]—which are not necessarily compatible. The fact that the northern Swedish mountain landscape is also strongly shaped by long and continuous traditional land-use, both through reindeer grazing by indigenous Sámi communities as well as mountain and summer farming is well established in the literature (e.g., in [14,39–42]). Reindeer husbandry as a traditional indigenous Sámi land-use is regulated by the Reindeer Husbandry Act (1971:437) and it can, during certain periods of the year, be practiced by reindeer herding communities within the large reindeer husbandry area—an area that covers approximately half the size of Sweden. The Swedish mountain region is also part of the "year-round grazing area", which means that reindeer husbandry is allowed there during the whole year. In all protected areas in the mountain region, reindeer husbandry is thus allowed, with the exception of the southernmost national park of the mountain range—Fulufjället national park. Mountain farming, despite being significantly smaller in spatial extent than reindeer grazing, has created considerable biological values, of high interest for conservation. Both reindeer husbandry and mountain farming are thus cultural legacies in the landscape as well as prerequisites to maintaining conservation values.

In Sweden, generally speaking, nature conservation in national parks and nature reserves is the responsibility of the SEPA and the County Administrative Boards' Environmental Units, sometimes through various sui generis co-management governance structures, e.g., Laponia WHS, involving local stakeholders. Similarly, cultural heritage preservation is the responsibility of the SNHB and the County Administrative Boards' Cultural Heritage Units. As previously noted, the two types of conservation practices have different traditions and approaches, one with a focus on the preservation of remains largely of a point character, the other on the acquisition of land for conservation, management and sometimes restoration. A more integrated nature and cultural heritage conservation management through better or more efficient and locally connected collaboration has been an explicit governmental desire. Accordingly, the government has instructed SEPA [43] and SHB [44] to initiate this development, possibly due to the fact that integrated management in the region has been proven to be a challenge [35]. However, the challenge of integrating conservation management efforts based on natural and cultural values is not unique to the northern mountain area. Wu et al. [30] show that this has been a challenge in landscape management in southern Sweden as well.

The majority of protected nature in Sweden is currently in the form of nature reserves, accounting for 9.3% of the land area [45]. National parks are the second most common form of protected nature, with 1.5% of the country protected [45]. Most of Sweden's protected nature is within the mountain area, with nature reserves accounting for a much larger area (31,181 km^2) than national parks (6457 km^2). As a comparison, the total area of cultural reserves in the mountains is only 0.1% (approx. 38 km^2) of the area protected as nature reserves. This effectively means that if integrated environmental management is going to take place, in practice, in the protected areas of the mountain region, it is within the nature reserves this will have to develop.

3. Methods

In order to understand the dynamics behind natural and cultural environmental management in the mountains, and to be able to provide decision support for policymakers based on a systems understanding [46,47], this study was conducted using group modeling sessions [48–51] combined with follow-up interviews and modeling sessions with selected key stakeholders.

3.1. Pre-Modeling Session Preparations

In a stakeholder-based analysis by Sverdrup et al. 2010 [35] of the Magnificent Mountain objective set for the Swedish mountains, significant efforts were made to identify all the key actors involved in the dynamic processes shaping the mountain region. This study draws on that rigorous stakeholder identification process.

The study was divided into four steps. First, stakeholders directly related to nature and culture heritage conservation were chosen, as already identified in Sverdrup et al. 2010 [35]. These included various public agencies on different levels: national agencies like SEPA, SNHB and the National Property Board, and on the regional level the county administrative boards of the four mountain counties, two municipalities from the southern and northern parts of the mountain area and representatives of museums and cultural foundations for landscape heritage conservation. Two NGOs were also chosen: the Swedish Reindeer Herding Association and the Swedish Hamlet Users Association.

Secondly, the 17 identified stakeholders (see Table 1) were invited to participate in a one-day group modeling session; thirdly, four of the key actors were interviewed for validation of the model; and fourthly, management plans for nature reserves were analyzed using Geographical Information Systems (GIS) as an additional validation step.

Table 1. Stakeholders that participated in the group-model building.

Stakeholder	Participant
Ájtte—Swedish mountain and Sámi museum	Head of museum
Association of Swedish Mountain Farmers and Hamlet users	Representative
County administrative board—Dalarna	Unit for Nature Protection
County administrative board—Västerbotten	Unit for Cultural heritage
County administrative board Jämtland	Head of unit for Nature Protection
County administrative board Norrbotten	Curator
Gaaltije—Centre for Southern Sámi culture	Head of operations
Laponia World Heritage Site	Representative
Malung-Sälen Municipality	Head of Spatial Planning
Nätverket Norden—Association for settlers in the mountain region	Representative
Särna-Idre och Transtrands sockenförening—Mountain farming association	Representative
Storuman municipality	Representative
Swedish Environmental Protection Agency	Representative
Swedish National Heritage Board	Curator
Swedish National Heritage Board	Representative
Swedish National Property Board	Cultural environments specialist
Swedish Reindeer Herding Associations (SSR)	Chairman
Swedish Society for Nature Conservation—Västerbotten	Representative

3.2. Group Modeling

The participants were introduced to systems thinking, the group modeling process and Causal Loop Diagram (CLD) notation before the modeling session began. Several of the participants were already familiar with the process owing to their participation in the Sverdrup et al. (2010) [35] study. At the beginning of the workshop the actors were asked to separately list which kind of natural and cultural environments needed more protection, as well as what they perceived as the biggest threats to these areas. The result was used as a basis for the modeling.

As the aim of the workshop was to understand the dynamics of drivers contributing or impeding integrated management, the following questions were addressed:

- What is the objective for integrated nature and cultural heritage conservation management?
- What modes of cooperation are available to meet this objective?
- To what extent is it possible for local actors to participate in these modes of cooperation?
- Is there an arena for local cooperation?

The results of the workshop generated a first draft of a CLD describing the dynamics behind natural and cultural environmental management in the mountains.

3.3. Follow-up Interviews for Validation and Model Revision

After the group modeling session, the resulting CLD presented a logic that was closely connected to the activities of four key actors. These actors were the SNHB, The units for Natural environments at the County Administrative Board of Västerbotten, The units for Cultural environments at the County Administrative Board of Västerbotten and The Swedish Hamlet Users Association. Follow-up interviews and individual modeling sessions were therefore conducted with each actor. The draft CLD generated during the workshop was used as a basis for the follow-up sessions, where the model was validated through a combination of the revision of the CLD and interview-style discussions on logics expressed in the CLD. These sessions were concluded with a general discussion on the stakeholders' experiences of integrating nature and cultural heritage conservation. In addition, a qualitative validation test of the final CLD was employed using a review of official management plans for protected nature.

3.4. Management Plan Analysis

Based on a GIS analysis, all the nature reserves within the area defined as mountain area were selected, resulting in 104 protected areas. From the mandatory management plans of these nature reserves, the 25 which were also identified in the Hayfield and Meadow Survey [52] were chosen for a management analysis.

Management plans include information on the history of the nature reserve, a discussion of values worth preserving/managing as well as the objectives and the intended management to reach them. If cultural heritage is taken into consideration in the management of nature reserves in the mountain area, one would expect (1) that all management plans for reserves where nature and cultural heritage values were identified by the Hayfield and Meadow Survey should at least mention the existence of such values, discuss them or include them as objectives, and (2) that a significant number of reserves, irrespective of character, should mention nature and cultural heritage values associated with traditional reindeer husbandry and remaining physical features associated with reindeer herding and the Sámi culture.

4. Results

The group modeling identified the presence of two alternative views on nature. The dominant public discourse for the mountain areas views nature and human land-use as separate, indeed incompatible, and the mountain areas as exponents of pristine nature. This view on ecosystems and landscapes was termed "the wilderness perspective" by the participants in the group modeling. The concept is hereafter referred to as the wilderness discourse (cf. Foucault 1969) [53], as the discourse concept reflects in many respects the powerful and pervasive influence of the wilderness perspective. These different, indeed competing discourses on the development of nature constitute the two major drivers in the conservation system affecting objectives, means and outcomes. The model (Figure 2) illustrates the dynamics of what may be described as a discourse system where the balance of power affects the perceived legitimacy and need for local participation—and thus, in the end, for the availability of an arena for local interaction and participation in the management of nature and cultural landscapes/heritage.

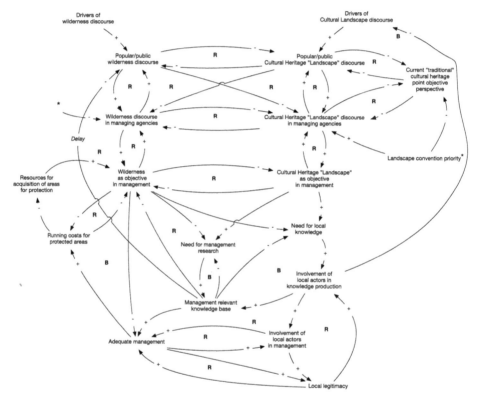

Figure 2. CLD synthesizing the stakeholders model over the management of nature and cultural heritage conservation. A + symbolizes a step in the same direction, while a − symbolizes a step in the opposite direction of the preceding driver. **R** indicates a reinforcing behaviour of a loop, and **B** a balancing behaviour of a loop. A bold line symbolizes a stronger influence in that connection, while a dashed line symbolizes a weaker influence.

4.1. Analysis of Final CLD—The Nature Conservation Discourse System

When the wilderness discourse becomes dominant, it reinforces both the public arena, through conservation NGOs, and the managing agencies (SEPA and County administrations). It also influences the degree to which the wilderness perspective influences official management objectives for protected areas, nature reserves and national parks in the mountains.

Additionally, since nature is perceived as pristine and therefore best managed by being left alone—i.e., unused and unmanaged, thereby remaining in a state of pristinity, i.e., wilderness—there is, as a consequence, little need for any local knowledge about land-use history and practices, or for research on previous and current land-use and resource management. Indeed, such information would pose a threat to the wilderness discourse. Furthermore, if human land-use and activities are by definition detrimental to the wilderness objective, there is clearly little need for developing local knowledge or involving local stakeholders and actors, as that would pose a threat to the objective. This, in turn, leads to less involvement of local actors in the management of protected areas and, as a consequence, to a reduction in legitimacy among local actors of the wilderness management approach. When local actors are less involved in knowledge production, it also means that less local knowledge is supplied to the knowledge base relevant for management, which further reinforces the wilderness objective, since relevant knowledge about local and historic land-use is not supplied.

The wilderness discourse within managing agencies is also affected by the influence of the opposing *Cultural heritage "Landscape" perspective discourse*. Conceptually, the two-discourse system represents a zero sum game. The dominance of the wilderness discourse in managing agencies has been at the expense of a *Cultural heritage "Landscape" discourse*. This is also reinforced by what is best described as a "traditional" point feature perspective in cultural heritage conservation, which refers to what stakeholders describe as a tendency among managing agencies for cultural heritage to focus on specific point features of cultural heritage value rather than on cultural landscapes as a whole, with or without these point features (expressed as *Cultural heritage "Landscape" discourse* in the CLD). Current cultural heritage practices can rather be described as isolated islands of cultural heritage features amid a wilderness ocean.

An illustration of the dominance of the wilderness discourse is the reoccurrence, in modeling sessions and interviews, of what is termed the "free development" paradigm for managing protected nature in the mountains, a management concept where nature is managed by being left without any management. Several stakeholders pointed out that this paradigm has had a significant impact on management objectives and practices. Another illustration of the dominance of the wilderness discourse is the fact that within the mountain area there are currently 104 nature reserves, covering approximately 31,000 km^2 of land, and only three cultural reserves, covering just 38 km^2.

4.2. Validation: Objectives vs. Outcomes

In order to validate, to some extent, the interpreted "strong influence" of the wilderness discourse and free development paradigm on the objectives and outcomes, a test was devised using official surveys and management plans for nature reserves in the mountain area (Table 1). Official "Hayfield and Meadow Surveys" (Ängs och hagmarksinventeringar), initiated by SEPA, have been performed at the county level in order to identify and map the extent of former and existing grazing and meadow areas with high nature conservation values. Very high biodiversity *and* cultural heritage values are associated with these grazing and hay meadow ecosystems, both within and outside the mountain areas. Such areas are the outcome of, and require, traditional management, i.e., they are the product of a clear cultural influence through previous and currently active land-use and require management in order to retain their character or to restore it.

Of 104 nature reserves within the mountain area, at least 25 have some overlap with areas identified in the Hayfield and Meadow Survey. A reasonable assumption is that the management plans for these 25 reserves ought to mention, in some way or other, cultural values related to grazing or hay production activities in the background descriptions or objectives for the protected area. However, only 7 of these 25 nature reserves mentions human influence in the background descriptions or the objective of protecting or recreating the cultural environments. On the other hand, 17 of these 25 nature reserves explicitly mention that part of the protection is aimed at preserving the area with intact/untouched/original nature, i.e., with wilderness as the stated objective. When it comes to the means for managing the 25 areas, only 8 mention that active grazing should be part of the management, as seen in Table 2.

Table 2. Validation of the dominance of the wilderness discourse based on a review of 104 management plans for nature reserves in the mountain region.

Management Plans Mentioning Reindeer Husbandry	Nature Reserves with Areas Included in Hayfield and Meadow Survey	Management Plans Mentioning Protecting, or Recreating the Cultural Environments as an Objective	Management Plans Mentioning that Active Grazing Should be Part of the Management
30	25	7	8

The 25 reserves should constitute a best-case scenario in that the objects had been officially identified, mapped and designated as valuable in a published survey of nature conservation values that required management, i.e., cultural influence. The fact that so few management plans even note managed/cultural values and that the majority defines wilderness as the only management objective, in spite of identified cultural values, constitutes, in our opinion, a robust confirmation that the wilderness discourse is the dominant discourse of the official nature conservation in the mountain area. Additionally, it is important to underline the fact that reindeer grazing is a land-use that takes place throughout practically the entire area of nature reserves. Active reindeer husbandry does not only produce a grazed landscape, but it also contributes to the cultural values of the mountain landscape. Reindeer husbandry is taking place in practically all the 104 nature reserves, and it is noteworthy that the practice was only mentioned in 33 of the 104 management plans. Also worth noting is that reindeer husbandry is mentioned in several instances as being allowed, or not considered as a hindrance to the conservation objectives and methods. In view of the fact that reindeer husbandry is the traditional land-use, the generator of the Sámi cultural landscape, the fact that reindeer grazing is only mentioned in 6 of 104 management plans is nothing short of remarkable. Whereas wilderness was explicitly mentioned as part of the objectives in 36 of the 104 plans, and the wording "untouched" (translation from Swedish, where "untouched" could also mean *unimpacted by human actions*) was mentioned in 71 management plans. This lends further support to our conclusions about the dominance of the wilderness discourse in operational conservation practices and about how it excludes human influences on the landscape when defining the objectives for environmental management.

5. Discussion

There is an often-stated desire for an increased integration of nature and cultural heritage conservation management in protected areas (see, e.g., Drew & Henne (2006) [54] and Linnell et al. (2015) [55]). The governmental decisions for increased integration between natural and cultural heritage in conservation [43,44] and the vision document of the SNHB [56] are clear examples of this, as was the research call funding this study. Nevertheless, what this desired integration should more specifically entail remains unclear both at a theoretical and practical level. It is not clearly expressed or communicated by the responsible governmental agencies. One interpretation could be that this reflects ambitions of administrative efficiency and savings more than it does the integration in question.

In practice, increasing integration has proven to be difficult, and not only for the mountain region. For example, when worldviews differ between agencies, ambitions of integration have been abandoned in favour of traditional opinions. As Wu et al. (2017) [30] show, this has been the challenge in landscape management in southern Sweden (where human influence is obvious) as well. This has forced SNHB into disputes and trade-off with SEPA regarding priorities of what to protect, revealing a lack of integration as well as the power imbalance between the agencies. Further, Wu et al. (2017) [30] argue, based on southern Swedish cases, that although there is a willingness among both SNHB and SEPA to achieve an integrated natural and cultural perspective in landscape management, they are also strongly expert-driven organizations where interactions and partnerships with public and local actors are treated half-heartedly. Wu et al. (2017) [30] demonstrate that neither SEPA nor SNHB, when forced into trade-offs, had been willing to compromise and develop an integrated approach, as they were "quickly locked into their conventional positions due to their divergent interests and understandings of the landscape".

The dominance of the wilderness discourse has a significant impact on the kinds of management objectives that are formulated. With regard to management, the means are in principle limited to a hands-off, "free development", approach. In practice, the means and the objective become one.

Further, the prevalence of objectives influenced by a wilderness perspective on nature and protected areas precludes or reduces the possibilities of creating an arena open to the nature conservation and cultural heritage administrations, as well as to the participation of local actors. Consequently, when no arena for participation exists there are few possibilities to generate common objectives between managing agencies or to involve local actors in conservation efforts, local/indigenous knowledge production or co-management, which clearly is a prerequisite for initiating a more integrated environmental management [57–59] on a general level. As shown in Figure 2, the absence of an arena for participation impacts the local perception of how legitimate management practices and objectives are. This is problematic because, as pointed out in Sandström et al. (2008) [19], in order to sustainably achieve environmental objectives, the state is dependent on the involvement of local actors in the environmental management and operationalization of objectives. The strong influence of a wilderness perspective has been noted, in a Swedish context, by, e.g., Wall Reinius [15], even in the Laponia WHS, which is arguably the most integrated environmental management attempt there is in the Swedish mountain region. If the dimensions of historical use and the cultural values that traditional land-use in the mountains have produced are not emphasized, the praxis and management risk marginalizing the groups practicing traditional land-uses [15], like reindeer husbandry and mountain and summer farming. There is the obvious risk with this strong and prevalent wilderness discourse, in the protected areas, that artefacts of traditional land-use will in practice be excluded in areas that cannot under any compelling argument be considered as wilderness. Further, one important difference worth mentioning is that for the Laponia WHS site there is a co-management arrangement in place with local Sámi communities, which might help to counterbalance the risk of excluding the historical and cultural perspectives. The nature reserves in the mountain area, while representing the largest protected area, tend to go under the radar when compared to the national parks and the WHS site, in that they lack co-management arrangements. Here, the risk is larger that the wilderness perspective will remain unchallenged. It is clear that current conditions for participating in consultations or co-management with stakeholders involved with cultural features or landscapes in the mountain area are far from satisfactory (see, e.g., Reed (2008) [60] and Sverdrup et al. (2010) [35]).

It has long been evident that climate change modeling points to (in, e.g., SOU 2007:60 [61]) major habitat changes in decades to come in the mountain area—where alpine/sub alpine heaths, in particular, are likely to develop into tree or shrub dominated habitats. The region is currently experiencing rapid on-going climate change which is affecting land-use and ecosystem composition and distribution [62]. If open habitats are to be retained, land-use interventions may be needed, and it is worth noting in this context that reindeer grazing mitigates the effects of the warming climate in the Swedish mountains by, for example, reducing tree-line advancement [63] and maintaining biodiversity [64].

Finally, the wilderness discourse, as expressed in management objectives, is also an impediment to developing and funding management-relevant research that includes human and local agency in the landscape. As is evident in the model, the more prevalent the wilderness ideals are in the objectives, the less need for local knowledge. This creates a self-reinforcing situation where new knowledge challenging the dominant discourse is less likely to emerge.

6. Conclusions

The dynamics of nature and cultural heritage conservation management, as modeled by the stakeholders in the group modeling session and subsequent interviews (Figure 2), in combination with the literature, allows the following conclusions:

- Generally speaking, there are two main discourses on how to understand the mountain environment. One, clearly evidence-based, argues that the mountain region has been affected for centuries by human land-use, albeit with limited modernistic impacts. The other, with a more ideological character, claims that the mountains constitute a "wilderness" untouched by human activities, which should be left unmanaged, i.e., without human interference or use.
- The dominant "wilderness" discourse is both marginalizing the "cultural landscape" discourse and defining management objectives for the vast majority of the protected areas, i.e., Nature reserves and National Parks.
- With "wilderness" as the dominant objective, there is little or no legitimacy and scope for local knowledge, participation or co-management of nature and cultural heritage values in the protected mountain areas.
- The dominant discourse and attendant objectives are a clear hindrance to a more integrated nature and cultural heritage conservation, as it excludes, in practice, the latter and provides no shared arena for cooperation.

There is little expressed disagreement between managing agencies when it comes to a stated willingness to increase the integration of nature and cultural heritage conservation management. However, what this call for integration should entail with regard to the involved agencies is unclear. What is clear after the stakeholder-based group modeling and interviews is that any substantive and successful integration between nature and cultural heritage conservation is currently unlikely. The dominance of the wilderness discourse is an impediment to an integrated nature and cultural heritage conservation, as it precludes the development of a joint understanding of landscapes and their conservation values, objectives and management approaches. Furthermore, the dominant discourse is an impediment to the development of arenas for local stakeholders to engage with the conservation authorities.

Given the current state and dynamics of the conservation discourse system in the Swedish mountains, any form of deeper integration between nature and cultural heritage conservation is unlikely to develop until the dominance of the wilderness discourse is reduced.

Author Contributions: Conceptualization, C.Ö., P.S. and I.S.; methodology, C.Ö., P.S. and I.S.; formal analysis, C.Ö., P.S. validation, C.Ö., P.S.; writing—original draft preparation, C.Ö., P.S.; writing—review and editing, C.Ö., P.S. and I.S. All authors have read and agreed to the published version of the manuscript.

Funding: This research (Dnr 3.2.-3093-2013) was funded by the Swedish National Heritage Board (SNHB) and the Swedish Environmental Protection Agency (SEPA) as a part of the call "Integrated nature and cultural conservation management in the Swedish mountains".

Acknowledgments: The authors would like to express their gratitude to the stakeholders that participated in this study and contributed with their valuable insights.

Conflicts of Interest: The authors declare no conflict of interest. The funders had no role in the design of the study; in the collection, analyses, or interpretation of data; in the writing of the manuscript, or in the decision to publish the results.

References

1. Latour, B. *We Have Never Been Modern*; Harvard university press: Cambridge, MA, USA, 1993; Volume 12, ISBN 0674948386.
2. Sluyter, A. Material-conceptual landscape transformation and the emergence of the pristine myth in early colonial Mexico. In *Political Ecology*; Zimmerer, K.S., Bassett, T.J., Eds.; The Guilford Press: New York, NY, USA, 2003; pp. 221–239.
3. Cronon, W. The Trouble with Wilderness: Or, Getting Back to the Wrong Nature. *Environ. Hist.* **1996**, *1*, 7–28. [CrossRef]
4. Nelson, M.P.; Callicott, J.B. *The Wilderness Debate Rages on: Continuing the Great New Wilderness Debate*; University of Georgia Press: Athens, GA, USA, 2008; ISBN 9780820331713.

5. Budiansky, S. *Nature's Keepers: The New Science of Nature Management*; Free Press: Winnipeg, ME, USA, 1995; ISBN 0029049156.

6. Riseth, J.Å. An indigenous perspective on national parks and Sámi reindeer management in Norway. *Geogr. Res.* **2007**, *45*, 177–185. [CrossRef]

7. Rybråten, S. "This is not a wilderness. This is where we live.". In *Enacting Nature in Unjárga-Nesseby, Northern Norway*; University of Oslo: Oslo, Norway, 2013.

8. Brockington, D.; Duffy, R.; Igoe, J. *Nature Unbound: Conservation, Capitalism and the Future of Protected Areas*; Earthscan: London, UK, 2008; ISBN 1844074404.

9. Parliament, S. *Betänkande Rörande Åtgärder Till Skydd för Vårt Lands Natur Och Naturminnesmärken*; Nordiska bokh.: Stockholm, Sweden, 1907.

10. Lupp, G.; Höchtl, F.; Wende, W. "Wilderness"—A designation for Central European landscapes? *Land Use Policy* **2011**, *28*, 594–603. [CrossRef]

11. Civildepartementet. *Hushållning med Mark och Vatten. Inventeringar. Planöverväganden om vissa Naturresurser. Former för Fortlöpande Fysisk Riksplanering. Lagstiftning*; Civildepartementet: Stockholm, Sweden, 1971.

12. Skjeggedal, T. Orört och konfliktfyllt i norsk naturförvaltning. In *Omstridd Natur*; Sandström, C., Hovik, S., Falleth, I., Eds.; Borea: Umeå, Sverige, 2008; pp. 63–82.

13. Krech, S. *The Ecological Indian: Myth and History*; W.W. Norton & Co.: New York, NY, USA, 1999; ISBN 9780393321005.

14. Emanuelsson, U. Human influence on vegetation in the Tornetrask area during the last three centuries. *Ecol. Bull.* **1987**, *38*, 95–111.

15. Wall Reinius, S. Protected Attractions—Tourism and Wilderness. In *the Swedish Mountain Region*; Stockholm University: Stockholm, Sweden, 2009.

16. Ödmann, E.; Bucht, E.; Nordström, M. *Vildmarken Och Välfärden: Om Naturskyddslagstiftningens Tillkomst*; 1. uppl.; LiberFörlag: Stockholm, Sweden, 1982; ISBN 9138600056.

17. Bernes, C.; Lundgren, L.J.; Naylor, M.; Naturvårdsverket. *Use and Misuse of Nature's Resources: An Environmental History of Sweden*; Swedish Environmental Protection Agency: Stockholm, Sweden, 2009; ISBN 9789162012755.

18. Schlyter, P.; Stjernquist, I. Regulatory challenges and forest governance in Sweden. In *Environmental Politics and Deliberative Democracy—Examining the Promise of New Modes of Governance*; Bäckstrand, K., Kahn, J., Kronsell, A., Lövbrand, E., Eds.; Edward Elgar: Cheltenham, UK, 2010; pp. 180–196.

19. Sandström, C.; Falleth, I.; Hovik, S. *Omstridd Natur: Trender & Utmaningar i Nordisk Naturförvaltning*; Sandström, C., Hovik, S., Falleth, I., Eds.; Borea: Umeå, Sverige, 2008; ISBN 978-91-89140-60-8.

20. Edvardsson, K. Using goals in environmental management: The Swedish system of environmental objectives. *Environ. Manag.* **2004**, *34*, 170–180. [CrossRef] [PubMed]

21. Emmelin, L.; Cherp, A. National environmental objectives in Sweden: A critical reflection. *J. Clean. Prod.* **2016**, *123*, 194–199. [CrossRef]

22. Emmelin, L. Att synas utan att verka—Miljömålen som symbolpolitik. In *Konflikter, Samarbete, Resultat—Perspektiv på Svensk Miljöpolitik. Festskrift till Valfrid Paulsson*; Lundgren, L.J., Edman, J., Eds.; Kassandra: Brottby, Sweden, 2005; pp. 19–43.

23. Naturvårdsverket. *Miljömålen – når vi dem? de Facto 2004– når vi dem?* Naturvårdsverket: Stockholm, Sweden, 2004; ISBN 91-620-1237-1.

24. Naturvårdsverket. *Miljömålen i halvtid, de Facto 2009*; Naturvårdsverket: Stockholm, Sweden, 2009; ISBN 978-91-620-1272-4. ISSN 1654-4641.

25. Naturvårdsverket. *Förslag till en Strategi för Miljökvalitetsmålet Storslagen Fjällmiljö*; Naturvårdsverket: Stockholm, Sweden, 2014; pp. 3–220.

26. Naturvårdsverket. *Miljömålen—Årlig Uppföljning av Sveriges Nationella Miljömål 2017*; Naturvårdsverket: Stockholm, Sweden, 2017.

27. Schlyter, P.; Stjernquist, I.; Sverdrup, H. Handling Complex Environmental Issues—Formal Group Modelling as a Deliberative Platform at the Science-Policy- Democracy Interface. In Proceedings of the 30th International Conference of the System Dynamics Society, Gallen, Switzerland, 22–26 July 2012.

28. Sayer, J.; Sunderland, T.; Ghazoul, J.; Pfund, J.-L.; Sheil, D.; Meijaard, E.; Venter, M.; Boedhihartono, A.K.; Day, M.; Garcia, C.; et al. Ten principles for a landscape approach to reconciling agriculture, conservation, and other competing land uses. *Proc. Natl. Acad. Sci. USA* **2013**, *110*, 8349–8356. [CrossRef]

29. Reed, J.; van Vianen, J.; Barlow, J.; Sunderland, T. Have integrated landscape approaches reconciled societal and environmental issues in the tropics? *Land Use Policy* **2017**, *63*, 481–492. [CrossRef]

30. Wu, C.J.; Isaksson, K.; Antonson, H. The struggle to achieve holistic landscape planning: Lessons from planning the E6 road route through Tanum World Heritage Site, Sweden. *Land Use Policy* **2017**, *67*, 167–177. [CrossRef]

31. Dawson, L.; Elbakidze, M.; Angelstam, P.; Gordon, J. Governance and management dynamics of landscape restoration at multiple scales: Learning from successful environmental managers in Sweden. *J. Environ. Manag.* **2017**, *197*, 24–40. [CrossRef]

32. Larsen, R.K.; Raitio, K.; Stinnerbom, M.; Wik-Karlsson, J. Sami-state collaboration in the governance of cumulative effects assessment: A critical action research approach. *Environ. Impact Assess. Rev.* **2017**, *64*, 67–76. [CrossRef]

33. Lawrence, R.; Larsen, R.K. The politics of planning: Assessing the impacts of mining on Sami lands. *Third World Q.* **2017**, *38*, 1164–1180. [CrossRef]

34. Sandström, C.; Widmark, C. Stakeholders' perceptions of consultations as tools for co-management—A case study of the forestry and reindeer herding sectors in northern Sweden. *For. Policy Econ.* **2007**, *10*, 25–35. [CrossRef]

35. Sverdrup, H.; Belyazid, S.; Koca, D.; Jönsson-Belyazid, U.; Schlyter, P.; Stjernquist, I. *Miljömål i Fjällandskapet*; Naturvårdsverket: Stockholm, Sweden, 2010.

36. Naturvårdsverket. *Årlig Uppföljning av Sveriges Nationella Miljömål 2019—Med fokus på Statliga Insatser Reviderad Version*; RAPPORT 6890; Naturvårdsverket: Stockholm, Sweden, 2019.

37. Naturvårdsverket. *Storslagen Fjällmiljö—Underlag till den Fördjupade Utvärderingen av Miljömålen 2019*; Naturvårdsverket: Stockholm, Sweden, 2019.

38. Horstkotte, T. *Contested Landscapes Social-Ecological Interactions between Forestry and Reindeer Husbandry*; Department of Ecology and Environmental Science, Umeå University: Umeå, Sweden, 2013; ISBN 9789174595352.

39. Aronsson, K.-Å. Fjällen som kulturlandskap. In *Hållbar Utveckling och Biologisk Mångfald i Fjällregionen Rapport från 1997 års Fjällforskningskonferens*; Olsson, O., Rolén, M., Torp, E., Eds.; Forskningsrådsnämnden: Stockholm, Sweden, 1998; pp. 115–122.

40. Bryn, A.; Daugstad, K. Summer farming in the subalpine birch forest. In *Nordic Mountain Birch Ecosystems*; Wielgolaski, F.E., Ed.; Parthenon Publishing group: New York, NY, USA, 2001; pp. 307–316.

41. Herder, M.D.; Niemelä, P. Effects of reindeer on the re-establishment of Betula pubescens subsp. czerepanovii and Salix phylicifolia in a subarctic meadow. *Rangifer* **2003**, *23*, 3–12. [CrossRef]

42. Tunón, H.; Sjaggo, B.S. *Ájddo-reflektioner Kring Biologisk Mångfald i Renarnas Spår*; Centrum för biologisk mångfald, Sveriges lantbruksuniversitet: Uppsala, Sweden, 2012.

43. ESV. Regeringsbeslut M2015/2531/Nm. Available online: https://www.esv.se/statsliggaren/regleringsbrev/?RBID=16739 (accessed on 25 March 2020).

44. ESV. Regeringsbeslut Ku2014/2121/RFS. Available online: https://www.esv.se/statsliggaren/regleringsbrev/?RBID=16181 (accessed on 25 March 2020).

45. SCB. Skyddad Natur 2016. Available online: http://www.scb.se/hitta-statistik/statistik-efter-amne/miljo/markanvandning/skyddad-natur/pong/statistiknyhet/skyddad-natur-2016/ (accessed on 19 March 2018).

46. Vennix, J.A.M. Group model-building: Tackling messy problems. *Syst. Dyn. Rev.* **1999**, *15*, 379–401. [CrossRef]

47. Rouwette, E.A.J.A.; Korzilius, H.; Vennix, J.A.M.; Jacobs, E. Modeling as persuasion: The impact of group model building on attitudes and behavior. *Syst. Dyn. Rev.* **2011**, *27*, 1–21. [CrossRef]

48. Maani, K.E.; Cavana, R.Y. *Systems Thinking and Modelling: Understanding Change and Complexity*; Prentice Hall: Auckland, NZ, USA, 2000.

49. Vennix, J.A.M. *Group Model Building*; Wiley: New York, NY, USA, 1996.

50. Haraldsson, H.V.; Sverdrup, H.U.; Belyazid, S.; Holmqvist, J.; Gramstad, R.C.J. The tyranny of small steps: A reoccurring behaviour in management. *Syst. Res. Behav. Sci.* **2008**, *25*, 25–43. [CrossRef]

51. Sterman, J.D. *Business Dynamics: Systems Thinking and Modeling for a Complex World*; McGraw-Hill Education: New York, NY, USA, 2000.

52. Jordbruksverket. *Ängs- och Betesmarksinventeringen 2002–2004*; Jordbruksverket: Jönköping, Sweden, 2005.

53. Focault, M. *The Archaeology of Knowledge*, 1989th ed.; Routledge: London, UK, 1969.

54. Drew, J.A.; Henne, A.P. Conservation biology and traditional ecological knowledge: Integrating academic disciplines for better conservation practice. *Ecol. Soc.* **2006**, *11*, 34. [CrossRef]

55. Linnell, J.D.C.; Kaczensky, P.; Wotschikowsky, U.; Lescureux, N.; Boitani, L. Framing the relationship between people and nature in the context of European conservation. *Conserv. Biol.* **2015**, *29*, 978–985. [CrossRef]

56. Swedish National Heritage Board. *Swedish National Heritage Board Proposals for Imple- Mentation of the European Landscape Convention in Sweden*; Swedish National Heritage Board: Stockholm, Sweden, 2008.

57. Margerum, R.D. PROFILE: Integrated Environmental Management: The Foundations for Successful Practice. *Environ. Manag.* **1999**, *24*, 151–166. [CrossRef]

58. Margerum, R.D. Integrated Environmental Management: Moving from Theory to Practice. *J. Environ. Plan. Manag.* **1995**, *38*, 371–392. [CrossRef]

59. Born, S.M.; Sonzogni, W.C. Integrated environmental management: Strengthening the conceptualization. *Environ. Manag.* **1995**, *19*, 167–181. [CrossRef]

60. Reed, M.S. Stakeholder participation for environmental management: A literature review. *Biol. Conserv.* **2008**, *141*, 2417–2431. [CrossRef]

61. *Sweden Facing Climate Change—Threats and Opportunities*; SOU 2007:60; Ministry of the Environment: Stockholm, Sweden, 2007.

62. Moen, J. Climate change: Effects on the ecological basis for reindeer husbandry in Sweden. *Ambio* **2008**, *37*, 304–311. [CrossRef]

63. Olofsson, J.; Oksanen, L.; Callaghan, T.V.; Hulme, P.E.; Oksanen, T.; Suominen, O. Herbivores inhibit climate-driven shrub expansion on the tundra. *Glob. Chang. Biol.* **2009**, *15*, 2681–2693. [CrossRef]

64. Kaarlejärvi, E.; Eskelinen, A.; Olofsson, J. Herbivores rescue diversity in warming tundra by modulating trait-dependent species losses and gains. *Nat. Commun.* **2017**, *8*, 1–8.

Article

Urban Heritage Conservation and Rapid Urbanization: Insights from Surat, India

Chika Udeaja [1], Claudia Trillo [1], Kwasi G.B. Awuah [1], Busisiwe C.N. Makore [1,*], D. A. Patel [2], Lukman E. Mansuri [2] and Kumar N. Jha [3]

1   School of Science, Engineering and Environment, Salford University, Salford M5 4WT, UK;
    c.e.udeaja@salford.ac.uk (C.U.); c.trillo2@salford.ac.uk (C.T.); k.a.b.gyau@salford.ac.uk (K.G.B.A.)
2   Department of Civil Engineering, Sardar Vallabhbhai National Institute of Technology, Ichchhanath,
    Surat 395007, India; dapscholar@gmail.com (D.A.P.); erlukman@gmail.com (L.E.M.)
3   Department of Civil Engineering, Indian Institute of Technology Delhi, Hauz Khas, New Delhi 110016, India;
    kumar.neeraj.jha@civil.iitd.ac.in
*   Correspondence: B.C.Makore@salford.ac.uk

Received: 17 February 2020; Accepted: 8 March 2020; Published: 11 March 2020

Abstract: Currently, heritage is challenged in the Indian city of Surat due to diverse pressures, including rapid urbanization, increasing housing demand, and socio-cultural and climate changes. Where rapid demographic growth of urban areas is happening, heritage is disappearing at an alarming rate. Despite some efforts from the local government, urban cultural heritage is being neglected and historic buildings keep being replaced by ordinary concrete buildings at a worryingly rapid pace. Discussions of challenges and issues of Surat's urban area is supported by a qualitative dataset, including in-depth semi-structured interviews and focus groups with local policy makers, planners, and heritage experts, triangulated by observation and a photo-survey of two historic areas. Findings from this study reveal a myriad of challenges such as: inadequacy of urban conservation management policies and processes focused on heritage, absence of skills, training, and resources amongst decision makers and persistent conflict and competition between heritage conservation needs and developers' interests. Furthermore, the values and significance of Surat's tangible and intangible heritage is not fully recognized by its citizens and heritage stakeholders. A crucial opportunity exists for Surat to maximize the potential of heritage and reinforce urban identity for its present and future generations. Surat's context is representative of general trends and conservation challenges and therefore recommendations developed in this study hold the potential to offer interesting insights to the wider planners and conservationists' international community. This paper recommends thoughtful integration of sustainable heritage urban conservation into local urban development frameworks and the establishment of approaches that recognize the plurality of heritage values.

Keywords: urban heritage conservation; historic urban landscapes; urban planning and management; cultural heritage; Surat's heritage; sustainable development

1. Introduction

The challenges faced by urban areas in South Asia today are steep and are at the forefront of the development of inclusive cities. Today, South Asian urban areas are among the largest and densest in the world, home to approximately 1.77 billion people, with the Indian urban population projected to double by 2050 from 410 million urban residents in 2014 to a staggering 857 million in 2050 [1]. Consequently, the urban fabric is experiencing issues such as growing informality, housing shortages and increasing rural to urban migration. India is arguably known as one of the most popular destinations for cultural tourism with rich and varied histories and traditions that allow for the exploitation of opportunities offered by cultural heritage [2]. The country has a considerable number of heritage assets, including 38 sites inscribed on the World Heritage list with 30 cultural properties, seven natural sites, and one mixed site, as well as over 3,600 centrally protected monuments under the Archaeological Survey of India (ASI) [3]. Additionally, there are 13 elements of intangible cultural practices and expressions on the UNESCO list. However, this rich heritage is facing major threats in urban areas and structures considered to be of national, state, or local importance in India, and remain under threat from urban pressures, neglect, vandalism, and demolition. Despite the intensification of urban growth in India's cities, restoration efforts to safeguard valuable heritage assets remain visible at only a few places of historic significance [4–6], and cultural heritage issues have not been mainstreamed into the overall urban planning and development framework

International consensus exists on the role played by heritage in achieving sustainable development. In 2015, the Sustainable Development Goals (SDGs) were unanimously adopted by United Nations (UN) member states resulting in a wide-ranging set of 17 goals and 169 targets aimed at poverty reduction, leaving no-one behind, and advancing the health and well-being for all by 2030 [7]. Out of the finalized SDGs, Goal 11 is the United Nation's strongest expression of the vital role cities and urban environments play in the global landscape. There are sporadic explicit references to cultural aspects in the 17 goals and these include: target 11.4, which focuses on the strengthening of efforts to protect and safeguard the world cultural and natural heritage; target 4.7, which gives emphasis to the promotion of knowledge and skills and the appreciation of cultural diversity; targets 8.9 and 12.b, which focus on sustainable tourism and local culture aligned with target 14.7, which gives attention to the sustainable use of aquaculture and tourism [8]. All of the targets have specific implications in the field of culture. These targets give light to the growing consensus that the future of our societies will be decided in urban areas of which culture plays a key role [1,9,10]. The 2016 United Nations New Urban Agenda recognizes both tangible and intangible heritage as a significant factor in developing vibrant, sustainable, and inclusive urban economies, and in sustaining and supporting urban economies to progressively transition towards higher productivity [1,7,11].

Furthermore, the global discourse has focused on this crucial role of heritage in the context of urban development and heritage conservation. In particular, the UNESCO recommendation on the Historic Urban Landscape (HUL) [12,13] has synthesized these elements by proposing a holistic understanding of urban historic areas through all-inclusive approaches [14]. On the 10 November, 2011, UNESCO's General Conference adopted the new recommendation on the HUL as an additional tool, a "soft-law" to be implemented by Member States on a voluntary basis. This document conceptualizes urban heritage as the multi-layering of cultural and natural values and attributes that go beyond the notion of "historic center" or "ensemble" to encompass a much broader urban geographical context [12,13,15]. This value is often constructed through processes of selection criteria appropriated internationally or nationally and objectified to become worthy of political, economic, and touristic attention and conservation. There is therefore a need to safeguard and respect the inherited values and significance of cultural heritage in cities.

In line with the main entry points for culture heritage in the achievement of sustainable development, this paper aims to explore the landscape of urban heritage conservation in the Indian city of Surat as an instrument to a better understanding of challenges and pressures that threaten the implementation of heritage conservation policies within rapidly growing urban contexts, beyond the international principles and criteria.

This paper is structured in seven sections. Following the introduction (Section 1), a focus on Surat allows for proper contextualization of this study (Section 2). The research methodology is presented in Section 3. The chosen research strategy stems from the epistemological paradigm of interpretivism based on the empirical qualitative dataset (Section 4), including policy and planning documents, semi-structured interviews, focus groups, and direct observation of two sub-cases, i.e., the two historic precincts of Surat (1) Gopi Surat Central Zone and (2) Rander Gamtal. Section 5 discusses the data by articulating the six thematic areas emerged from the findings (Sections 5.1–5.6):

- Surat's built heritage,
- Urban heritage management,
- Valuing heritage: intangible dimension,
- The emerging local community awareness of heritage conservation,
- Urban development and real estate pressures,
- Cultural heritage and climate change.

As a result, this paper draws recommendations for the development of a sustainable urban heritage framework that includes: (6.1) holistic urban heritage legislation, (6.2) identifying and mapping the city's heritage values and preserving local identity and sense of place, and (6.3) developing local urban heritage and planning expertise, skills, and knowledge. Inclusive community and stakeholder engagement are central to the successful integration of urban heritage conservation. Section 7 concludes the paper by summarizing the findings and detailing areas of future research.

2. Setting the Context: The City of Surat and Its Heritage, an Overview

The city of Surat (Figures 1 and 2) is currently urbanizing rapidly with demands of urban sprawl and development [3,16]. According to the census taken in 2011, Surat's urban district had a population of 4,849,213 people although the actual population may exceed these figures due to rapid development in Surat's metropolitan region [17,18]. Surat's urban context includes social cohesion challenges, increasing rural to urban migration, rising housing demands, and considerable stress on city management and resources [19]. Yet, in the context of these urban pressures, there is an evolution of approaches recognizing tangible and intangible heritage as strategic assets in creating cities that are more resilient, inclusive, and sustainable [10,20,21]. Surat has a diverse and vibrant heritage that has created and shaped the cultural identity of the city. Historic social practices and processes have remained interdependent and reciprocal with Surat's built fabric. However, urban heritage conservation is not perceived as a priority when considering other urban development objectives [16,22]. Cultural heritage continues to remain marginal in urban development agendas, often overlooked in the context of urban poverty, social inequalities, and a severe lack of basic infrastructure [11]. Although it is evident that effort is being made to improve sustainable planning and heritage conservation [17,19,22], there exist significant challenges that limit the impact and scope of these initiatives.

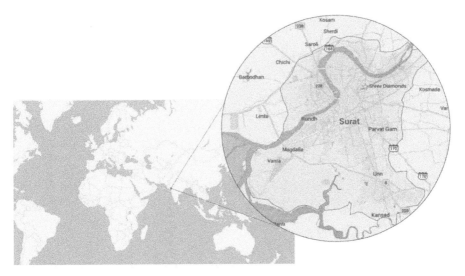

Figure 1. The city of Surat [Source: Author (modified arcGIS) map].

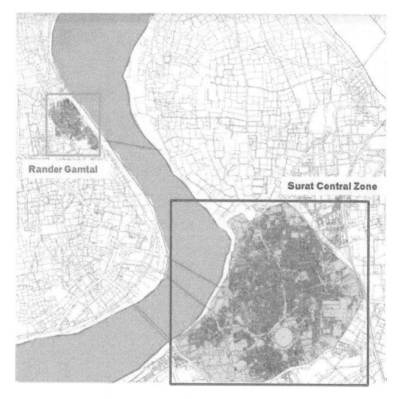

Figure 2. Surat Central Zone and Rander Gamtal. [Source: Author (modified arcGIS) map].

Having survived numerous historic invasions and power structures, Surat is presently in the top ten largest cities in India and recognized as one of the fastest growing cities (Figure 1) [19]. The strategic location of the city aided in forming historic overseas links with the rest of Asia, Europe, Africa, and the Middle East, which date back from 300 BC. These trading connections influenced the living patterns and built heritage in Surat, particularly in the historic precincts Gopi Surat Central Zone and Rander Gamtal. Historically, Surat's heritage conservation was mostly concerned with safeguarding the remains of architectural monuments. Key historic monuments include major development by Malek Gopi, a rich trader in 1496-1521 AD, the establishing of silk and cotton factories from the 1600s, the construction of the inner-city wall in 1664 AD, and the outer-city wall in 1715 AD [18]. The city of Surat grew in the 17th and 18th centuries to become an established and formidable export and import center of India. Settlement in Surat continued to develop with custom houses and gardens along the River Tapi and Surat's fort. By 1901 AD, the diamond cutting industry was established and began exporting diamonds to the United States of America from the 1970s. Currently, 80 percent of diamonds of the world are cut in Surat [19] and the jewelry and textile industry has allowed a steady flow of wealth into the city. The evolution of the concept of heritage preservation has developed in parallel with the evolution of Surat, becoming a practice that goes beyond tangible assets and possesses a human and socio-cultural element [15]. However, the practice of conservation in Surat still lags behind the actualization of this diverse concept. The city lacks an official holistic values-based approach that specifies the significance of Surat's historic areas whilst taking into account the existing built environment, intangible heritage, cultural diversity, socio-economic and environmental factors, and local community values [14,15].

The Surat Municipal Corporation is the main government body in Surat responsible for urban planning schemes, alongside the Surat Urban Development Authority (SUDA) (which includes the municipal corporation area) and the Hazira Development Authority, which governs the port and industrial hub located downriver from Surat city [23]. SUDA is responsible for preparing the area development plan and for controlling unauthorized developments. The South Gujarat Chamber of Commerce and Industry is influential in Surat's governance structure as it takes the lead on several critical regional and city development initiatives [24]. Achieving urban sustainability is of significance in Surat as it is particularly vulnerable to the effects of climate change. The city lies in a flood plain area and the southwest area of the city hosts a number of creeks. Natural disasters have been recurrent and devastating, such as a plague in 1994 and floods in 2006 and 2008. Surat's climate change predictions and risk profile all indicate an increase in rainfall, with monsoons dominated by heavy spells of rain combined with longer dry spells, leading to an increase of floods [19,23].

3. Research Methodology

The research strategy of this study is based on a single case study, i.e., the city of Surat. This city was purposely selected for two reasons. Firstly, this research is funded by the Arts and Humanities Research Council investigating Surat as a case study for urban heritage conservation. Secondly, the richness and diversity of Surat's heritage combined with the city's rapid urbanization trend reflects a need for investigation into the conservation of its urban heritage and the challenges being faced. This study area therefore presents an opportunity for the development of holistic and sustainable approaches towards the preservation of Surat's urban heritage.

In line with the research goals of articulating a discourse on challenges and issues related to the implementation of heritage conservation policies in growing cities, the research strategy stems from the epistemological paradigm of interpretivism and mainly rests on qualitative research methods. These latter are often seeking to understand processes and cultural and contextual meanings. Therefore, giving emphasis to the need for enquiry through an inductive approach that attempts to understand the experiences with a goal to present a credible representation of the interpretations of those experiences [25]. A variety of sources were used to obtain data for triangulation purposes. The main advantage of using triangulation is that it allows for the evaluation of different sources of information to investigate concepts on the basis that a consensus of the findings will yield more robust results [26].

As anticipated, the empirical dataset is based on qualitative data, including policy and planning documents, interviews, focus groups, and direct observation of two sub-cases, i.e., the two historic precincts of Gopi Surat Central Zone and Rander Gamtal (Figure 2). Policies and strategies have been chased at multiple scales (national and city wide), while direct observation of physical urban fabric has been conducted at a neighborhood scale. A visual illustration of the research methodology and instruments is offered with Figure 3. Qualitative primary data was collected in September 2018 by a team of 3 UK and 3 Indian researchers. A total of 34 stakeholders participated in focus groups and 10 stakeholders in the interviews prepared by UK and Indian researchers (Figure 3).

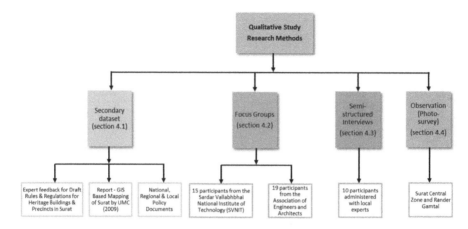

Figure 3. Dataset Outline.

4. Data Presentation

4.1. Secondary Dataset: Policy and Planning Documents, Strategies, Regulations

First, documents relevant to India's cultural heritage and the city of Surat were collected and analyzed (Figure 3). All relevant policies and regulations in force nationally, regionally, and locally were systematically gathered and considered, including national laws, policies, and governance of heritage conservation in India, Gujarat, and Surat. This also included key reports discussing the impact of climate change on the city of Surat including the vulnerability assessment (2010) on Surat undertaken by The Rockefeller Foundation's Asian Cities Climate Change Resilience Network (ACCCRN) [23] and Surat Resilience Strategy [19]. All relevant previous surveys and investigations on Surat heritage were systematically collected and analyzed. Incidentally, the team of researchers provided the City of Surat with expert feedback on the draft regulation "Rules & Regulations for Heritage Buildings & Precincts in Surat" [27]. This is a local regulation aimed at the conservation of all the listed heritage buildings and sites and identified precincts, as listed by Surat Municipal Corporation (SMC) in 2009. The project team conducted an in-depth study of the GIS-Based Mapping of Living Heritage of Surat For Improved Heritage Management in Surat prepared by the Urban Management Center in 2011 [17], which still forms the basis of the knowledge of the local heritage in Surat. The desk analysis of the documents was complemented with primary data collected in Surat in September 2018 and is discussed in the next sections. The document analysis revealed a failure to encompass a broader urban geographical context for urban heritage when considering the preservation of cultural heritage. The perspective demonstrated from the analysis suggests heritage conservation in Surat is side-lined when considering other urban development objectives such as housing and infrastructure. The documents were discussed with local practitioners and with city planners with the aim of checking the level of accuracy of the work, how the studies were generating impact on actual heritage conservation policies, and how far the current situation was with respect to the study.

4.2. Focus Groups

Two focus groups were organized with local academics, decision makers, and practitioners to capture different views and perspectives on heritage conservation in Surat (Figure 3). The goals of the two focus groups were more general and categorized into two sections of discussion. Firstly, discussing with local experts about heritage conservation to gauge their view on principles and criteria applied in Surat. The second goal was the exploration of how to raise awareness about the importance of heritage conservation for local identity. Gathering a total of 15 participants, focus group 1 was arranged at the Sardar Vallabhbhai National Institute of Technology (SVNIT). Focus group 2 was arranged by the Association of Engineers and Architects and gathered 19 participants. Engaging with the stakeholders in heritage was imperative for discussing key urban conservation issues in Surat. The diversity of participants in focus group 1 allowed for an exploration of the challenges in urban heritage conservation in Surat. The discussion in focus group 2 centered around national and local initiatives to develop a smart, sustainable, and resilient Surat. In both focus groups, the researchers understood that though conservation of heritage was considered important in principle, still different views on what should be included in heritage and how to conserve persisted. This revealed a disconnection between the national legislative framework for heritage conservation and local guidance provided by local authorities. The focus groups proved to be crucial for the facilitation of understanding meanings attached to issues in contexts that had not been interrogated in advance by the project team. The transcripts and informal notes taken were analyzed as a means of providing a coherent method for reading the interview material in relation to the questions. The aim of the analysis was to draw out salient dimensions related to urban heritage conservation in Surat.

4.3. Semi-Structured Interviews

As a final step, further qualitative empirical data was conducted with 10 semi-structured interviews administered with local experts on heritage conservation (Figure 3), sampled by selecting them across both the public and private sector (Table 1). Furthermore, the focus groups assisted in providing a diverse sample for the expert interviews. The interviews were used to undertake in-depth exploration of emerging issues from the focus groups, observation, and documentary evidence. Perspectives were sought for the interviews from respondents from the built environment as well as those concerned with intangible heritage. This included a local yoga teacher who drew on his experience and skills and enabled a kind of storytelling about Surat's heritage. These additional perspectives assist in highlighting the interconnection between tangible and intangible heritage. Other stakeholders who contribute significantly to strategic planning of heritage in Surat such as the local Government (Surat Municipal Corporation) officials, heritage architects, and consultants were consulted. The details of the interviewees are listed below in Table 1.

Table 1. Semi-structured local expert interviewees.

Interview Code	Local Expert Group	Affiliated Organization	Role
I1 & I2	Local Government	Surat Municipal Corporation	Heritage experts from Surat Museum
I3	Public University	Sardar Vallabhbhai National Institute of Technology (SVNIT), Town and Regional Planning	Heritage Consultant
I4	NGO	Indian National Trust for Art and Cultural Heritage (INTACH)	Heritage Architect
I5	Private & Local Government	Local Organisation & Surat Municipal Corporation	Art historian & Heritage Cell Officer
I6	Private	Local Organisation	Heritage Architect
I7	Local Government	Surat iLAB & Surat Smart City	Officer
I8	NGO & Local Government	Resilience Surat as part of the Rockefeller Foundation 100 Resilient cities project	City Resilience Officer
I9	Public University	National Institute of Technology (SVNIT), Town and Regional Planning	Heritage Proprietor & Industrialist
I10	NGO	Patanjali Yog Prashikshan Samiti, Surat	Yoga expert

The interview schedule consisted of three broad thematic sections. The first section contained points of discussion exploring the conceptualization of heritage in India and Surat and the heritage conservation landscape, including questions such as: "Can you describe how international frameworks (e.g., UNESCO World Heritage Convention) have shaped efforts towards conserving Surat's heritage?". The second section discussed the challenges in conserving Surat's heritage, e.g., "What are some of the challenges you face in integrating cultural heritage in your practice and how do you overcome these challenges?". The third section encouraged the respondents to provide recommendations on how to develop sustainable heritage conservation approaches in Surat. The semi-structured interviews provided an opportunity for the respondents to discuss these themes in greater depth with reference to their practice and experience.

4.4. Direct Observation and Photo-Survey of Two Chosen Historic Precincts

The observation of the city was undertaken as a visual tool to support the understanding of heritage conservation in Surat (Figure 3). Fieldwork was conducted in Surat's historic areas, Gopi Surat Central Zone and Rander Gamtal (Figure 2). The observation was based on the study conducted by the Urban Management Centre for the Surat Municipal Corporation [17]. GIS maps of the historic areas were used to identify sub-areas in the two historic areas showing highest concentration of historic buildings and further investigate the state of conservation and actual context situation. Fieldwork was conducted both by car/motorbike and by walking during working days, morning and afternoon. Photographs were taken both to document the state of conservation of the built environment and to capture people using it. The direct observation and photo-survey of the two areas allowed understanding of some of the main challenges to heritage conservation in Surat. Although there are some efforts to restore key monuments such as the fort and castle restoration, Surat's heritage remains neglected and increasingly in desperate need of urgent attention. This is further discussed in the following sections, covering findings from all the empirical data gathered by the team.

5. Data Discussion and Findings

This section discusses the findings from evidence gathered through the secondary dataset, interviews, focus groups, and photo-survey discussed above. Data analysis has been conducted through content coding of interviews, focus groups, and direct observation notes. Photos shown in this section are taken as part of the direct observation and photo-survey of the two chosen historic precincts. Furthermore, the representation of cultural heritage from the document analysis is included in this data discussion. Six thematic areas of discussion emerged from the findings. Two dominant paradigms of heritage conservation exist in Surat. The first is a traditional paradigm in which built heritage (Section 5.1) is a central focus with restoration efforts concerned with monumentalism and heritage experts largely responsible for maintaining and preserving heritage assets. This is reflected in the lack of integration of heritage conservation within local planning documents (Section 5.2). The second paradigm is underdeveloped in practice and exists largely in emerging discourse. It is concerned with values-based approaches to heritage and the holistic inclusion of intangible attributes (Section 5.3). The findings revealed that the local community lack the understanding of the values of heritage and how to care for Surat's heritage assets (Section 5.4). As a result, development projects for new infrastructure are usually insensitive to the authenticity and integrity of cultural heritage (Section 5.5). The final thematic area is concerned with the relationship between cultural heritage and climate change (Section 5.6).

5.1. Surat's Built Heritage

Surat's built fabric reflects the powers that have historically dominated and influenced the city, including the Hindus, Muslims, French, Dutch, Portuguese, and the British. As a port city located on the western part of India in the state of Gujarat (Figure 1), Surat has an established heritage with a diverse portfolio of tangible heritage assets. Although, the city does not have a UNESCO World Heritage site, six sites are listed by the Archaeological Survey of India (ASI) and acknowledged as "Monuments of national importance" in Surat. These include (1) Dargah known as Khawaja Dana Saheb's Rouza; (2) Old English Tombs; (3) Tomb of Khawaja Safar Sulemani; (4) Old Dutch & Armenian Tombs & Cemeteries; (5) Ancient site comprising S.Plot No.535 and (6) Fateh Burj [3]. This markedly adds to the promotion of Surat's urban heritage. This pride in Surat's heritage was demonstrated in the interviews, as illustrated in the quote below by the Heritage Consultant.

Surat is one of the oldest economic hubs and hence the impact of various cultural eras from all over the world. This has been the result of our old city houses and buildings (I3, Heritage Consultant, Expert Interviewee).

Despite this recognition, Surat does not have an official register of heritage assets of historical importance or protected monuments. Heritage sites across the city reflect elements and motifs that tell its own individual story through its design, material, woodwork, cornicing, paint, color, and landscaping of that era. Building materials evolved depending on the influence at that time. Local traditional houses used timber for the main house construction. Indeed, the use of other construction materials such as brick and concrete demonstrated external influence as shared by the Heritage Architect.

The construction techniques of the housing are quite similar … , but the decoration is different. The housing inside are very simple but the façades instead are very different, because they are an expression of social distinction and power (I6, Heritage Architect, Expert Interviewee).

Surat's built heritage also has a historic economic impact. It reflects cultures of the settlers as well as the economic growth and status of their owners. The house form has evolved over the centuries responding to modernization and contemporary living and the rise of industry. Indeed, some historic buildings no longer exist; however, in terms of boundaries identification of the two main historic areas in the city of Surat and in terms of heritage classification, including the articulation of the historic traditional houses into 4 typological influences (i.e., vernacular, colonial (Gothic and Renaissance), Art Deco, and Arabesque) are still current. Different architectural languages are visible in the house form such as the facades, the layout, plan form, and hierarchy of spaces. In particular, the front façade is a crucial reflection of the owners sociocultural, political, and economic status and beliefs. The vernacular architecture depicts houses built from local resources and with local traditions, often with wooden facades, large brackets, and overhanging eaves. The carvings in the wooden columns are highly decorated, reflecting animal, bird, and floral patterns. Surat's colonial influence resulted in forms of Gothic and Renaissance styles (Figure 4). The Arabesque style includes the use of repetitive geometric patterns on the facades and the buildings are made completely in brick and lime. Façade divisions using decorative art forms built with modern industrial material reflect the influence of the Art Deco style.

The design and ornamentation of certain structural elements are great examples of the cross-cultural influences in Surat and richness of its patrons. For example, columns and brackets can be found in Surat's heritage buildings, with detailed carving and embellishment often bearing floral, animal, and bird carvings and general geometric patterns with associated meanings. Figure 5 shows the beautification applied to carvings on the Chintamani temple column. Additional elements of focus central to Surat's heritage architecture are the windows and doors (Figure 5). These are often found to be symbolically decorated with meaningful motifs, dominating the façade in a predominantly symmetrical composition.

Restoration efforts for Surat's built heritage have focused predominantly on monuments as these were deemed to have historical and architectural importance [17]. As a result, heritage properties not fitting this criteria had a lack of maintenance and investment, thus amplifying their vulnerability. The findings from the direct observation as recorded by the photo-survey demonstrated that Surat's heritage is increasingly at risk, neglected, and in desperate need of urgent attention as shown in Figure 6. The SMC has made notable yet limited efforts to restore key monuments such as the fort and castle restoration (Figure 7). An example of the commitment to heritage restoration is the development of the city's first heritage precinct at Chowk Bazar [16]. Under this project, 11.5 hectares of land around Surat's fort are currently being redeveloped including Surat's castle and moat, Suryaputri Udyan up to the river edge, Frazer promenade, and Shanivari along the river bank. The field visits and discussions with a local conservation architect and 60 selected architecture students from across India established the core focus of efforts on monument restoration such as Surat's fort. Surat's fort was built in the year 1540–41 for protection against the Portuguese raids. The fort currently has twelve-meter-wide battlements and four-meter-thick walls.

Figure 4. House façades in Rander and Gopipura showing colonial style influences and elements with Art Deco influences (Source: authors' photos).

Figure 5. Interior temple pillar decoration in Gopipura (left) and door and window design and decoration of heritage houses in Gopipura (Source: authors' photos).

Figure 6. Dilapidated heritage buildings in need of restoration in Rander Gamtal (Source: authors' photos).

Figure 7. Redevelopment occurring alongside the old Surat fort walls.

5.2. Urban Heritage Management

While there is a superabundance in policies and practices on heritage at an international level, the context is different in India. In fact, India also differs from other countries in the Asian region. For example, countries such as Sri Lanka and Bhutan have clearly defined policies regarding urban heritage [1]. India in contrast has an institutional framework dedicated to heritage protection, but lacks a strategic focus on urban heritage. Heritage legislation has largely developed as a result of a fear that development changes and pressures will erase the history of places [28,29]. The urban development models followed since independence have irrevocably altered many historically important towns and cities [27]. The decentralization of power to local bodies is given in the 74th amendment to the Constitution. This therefore empowers local bodies to act proactively and develop processes and practices that suit their context. These local mechanisms feed into the state's acts and legislation.

The fragmentation and complexity of the current governance systems have not provided a favorable ground for culturally sensitive urban development strategies. The national system does not allow for the translation of fundamental steps in heritage conservation at a local level such as the identification of heritage and the provision of regulations that prevent demolition and regulate new developments [30].

At a state level, Gujarat's inclusive urban development policies lack consistent integration of heritage issues [30]. At a local level, Surat's policy instruments on heritage conservation are underdeveloped and there are no specific local policies or strategies on heritage conservation in place yet. An attempt to produce guidelines for the conservation of heritage based on a study conducted by the Urban Management Centre for the Surat Municipal Corporation (SMC) [17] has been made, but still the local authority struggles to implement it. The SMC considered the National Institute of Urban Affairs (2015) studies in preparing the draft for the "Rules & Regulations for Heritage Buildings & Precincts in Surat" [27]. The team of researchers provided the SMC with expert feedback on the draft regulation as part of the documentary analysis. The analysis revealed a significant focus on monuments in Surat, overlooking associated intangible attributes. Additionally, the regulation failed to encompass a broader urban geographical context of urban heritage that goes beyond monuments and integrates the multi-layering of cultural and natural values. This implies that heritage conservation is not perceived as a priority when considering other urban development objectives [16,22]. The existence of a top-down approach to governance in Surat leads to the exclusion of communities in the practice and processes of urban planning. Cultural heritage continues to remain marginal in discussions about urban development agendas, often overlooked in the context of urban poverty, social inequalities, and a severe lack of basic infrastructure [11]. Although it is evident that effort is being made to improve sustainable planning and heritage conservation, there exist challenges that limit the impact and scope of these initiatives.

5.3. Valuing Heritage: Intangible Dimension

The city of Surat has a diverse and vibrant economic and sociocultural fabric (Figure 8). Tangible and intangible cultural heritage is represented, developed, and protected in Surat, and is depicted as diverse and multidimensional [31,32]. Heritage is a concept that is difficult to define, what it means and how it has been presented, re-presented, developed, and protected, set against a back-drop of demands and motivations is multidimensional [33,34]. In the drive to define traditions and identities in a community [34], the notion of "heritage" is developed [21]. Living expressions and practices of heritage are also often misunderstood and treated as ambiguous due to its complexity and variation [35,36]. The interrelationship between history/the past [33] and heritage is recognized in literature-defining heritage as elements of the past for contemporary society to inherit, record, conserve, and pass on to future generations [5,37]. In this landscape, urban heritage plays a fundamental role in reinforcing cities' identities through the integration of heritage and historic urban area conservation, management, and planning strategies into local development processes and urban planning aids [20,38]. It allows for the broader urban context to be considered with the interrelationships of heritage and its physical form, spatial organization, connection, and values. Throsby [39] highlights the need for acknowledging the "interconnectedness of economic, social, cultural, and environmental systems". Thereby positioning cultural heritage as the "glue" among the multidimensions of sustainable development. This approach extends beyond the notion of monuments and historic centers and includes social and cultural practices and values, economic processes, and the intangible dimensions of heritage as related to diversity and identity [38]. It reinforces the integral role cultural heritage can play as a key resource in urban sustainable development.

Figure 8. Vibrant city of Surat (top left—Station road known as Rajmarg Surat); Street markets of Surat (bottom left—Chauta bazar); Daily life embedded in urban fabric (middle); Residents using urban traditional areas (Rander Gamtal) for small retail or everyday traditional activities (right) (Source: authors' photos).

The concept of Surat's heritage is associated closely with broader notions of local identity, memory, and nationalism [40–42]. Scholars [4,37,43–45] have argued that heritage is an essential element of national representation with the potential to perpetually remind citizens of the symbolic foundations upon which a sense of belonging is based. It is therefore presented or re-presented as something of special value or significance relating to the past. This dynamic history has created and shaped the cultural identity of the city of Surat. Historic social practices and processes have remained interdependent and reciprocal with Surat's built fabric. The built environment is a crucial space for expressing traditional and spiritual activities that are still actively imprinted on urban life as shown in Figures 8 and 9. As the heritage architect explains below, Surat's tangible and intangible heritage are interdependent.

The city of Surat has kept the heritage and survived invasion and calamities. The people's spirit is inclusive and festive … Surat is an amalgamation of many traditions and communities. It is a base for many crafts. The city has a lot of harmony, which has its footprints in a way of amalgamation in the built heritage and intangible heritage (I4, Heritage Architect, Expert Interviewee).

The photos (Figure 8) below show how lively Rander Gamtal historic area is and the role played by tangible and intangible heritage in shaping the place and in adding quality to the urban environment and in enabling the consolidation of the social bonds.

At present, there is no standard classification and valorization approach towards the cultural heritage in Surat. The paucity in recognizing the pluralistic values of Surat's tangible and intangible heritage leaves to question what type of heritage should be preserved, for what reason, and by whom [14,47]. Consequently, heritage assets that have significant attached values to citizens are left out of local government efforts to raise awareness and promote heritage tourism. The articulation of heritage values allows for the consideration of decisions for heritage assets to give a "heritage status and significance" and therefore the assessment of these values attributed to heritage is a very important activity for the achievement of sustainable urban conservation [48]. Despite the fact that values are widely understood to be critical to heritage conservation, there is still a paucity of knowledge about how plural heritage values can be used to assess tangible and intangible heritage [49]. Expert interviewees pointed, as described below, that local politicians are not concerned with the value assigned to Surat's cultural heritage. Without political buy-in and commitment, heritage is left at the margins of urban development.

> *Elected people, local leaders … They even do not bother about the value of these heritage buildings, they would rather demolish them and replace with new buildings* (I3, Heritage Consultant, Expert Interviewee).

> *So far, we have not been able to capitalize the value of the history and of the heritage, this city has been always well known for trade and commerce, not for its history* (I8, City Resilience Officer, Expert Interviewee).

Without the acknowledgment and appreciation of Surat's culture and values, opportunities for establishing social cohesion and connectivity are missed. Surat's urban fabric is under consistent pressure to "modernize", leading to the continuous disappearance of traditional skills and crafts that are part of the intangible cultural heritage [21,36]. Expert interviewees commented on the depreciation of a sense of place and belonging in Surat due to the various physical environmental challenges mentioned above and the increase in population.

> *Most of the heritage sites are present in the middle of the city, but due to blindly following the Western culture, people neglect their own heritage and culture* (I9, Heritage Proprietor & Industrialist, Expert Interviewee).

Social connectivity and cohesion are weak and therefore there is a lack of interest in engaging with Surat's heritage [19]. This challenge is exacerbated when considering migrant populations who have settled in Surat primarily for industrial activities and have no inherited sense of responsibility to conserve and value Surat's heritage.

Figure 9. Festivals and traditions are still very lively and fully embedded in the city's everyday life. (Source: [46] Uttarayan—The festival of kites left and middle); (Source: authors' photos top and bottom right).

5.4. The Emerging Local Community Awareness of Heritage Conservation

There was agreement in the findings that the local community lack education, language, and understanding about the values of tangible and intangible heritage and how to care for these heritage assets. Cultural heritage can promote contact, exchanges, and reciprocity, particularly when people engaging with heritage are not considered as passive consumers but as creators, distributors, and decision makers [50]. Expert interviewees highlighted the need for citizen participation in urban heritage conservation as illustrated by the quotes below.

> *Surat is experiencing constant dense growth of the CBD and acute migration. There is a need for an active dialogue with people and making them aware of our rich history. People's participation will bring awareness about the many layers of history. It will facilitate the connection of the footprints about history and the immediate past* (I6, Heritage Architect, Expert Interviewee).

> *The local community has a crucial role to play in promoting the pride of our heritage. There should be more involvement of various activities related to heritage* (I3, Heritage Consultant, Expert Interviewee).

The rise in modern practices leaves little room for recognition of traditional activities and processes [51]. Some efforts to build heritage awareness have already been created as discussed in the sections above. However, there is no existing formal strategy to engage with urban communities about Surat's diverse heritage and how to preserve it. Increased awareness about history, story, and the reality about heritage monuments and intangible heritage can instill a sense of pride in the local community [52,53]. Younger generations with digital access to global agendas on sustainability and heritage identity have a growing interest in visiting and taking steps to restore heritage sites in Surat [54]. However, Surat's underdeveloped heritage tourism industry reduces the interest and exploration of heritage. Thus, contributing to the paucity of understanding of the significance and value of heritage [22]. An expert interviewee representing the local municipality commented on the need to develop the tourism industry with a view to stimulating interest from the locals and to urge them to understand and appreciate the value of the heritage.

> *Now tourists are coming to the city for business and go away after the visit, so we are trying to offer something that might induce those people to go with the family and to spend time and money around the city. Surat should not only be for business, but also for tourism. If tourists were paying attention to the buildings, then the locals would understand and appreciate the value of the heritage* (I1 & I2, Surat Municipal Corporation Museum, Expert Interviewee).

The fieldwork revealed that the compartmental thinking and fragmentation in Surat's heritage landscape is largely attributed to the absence of skills and knowledge amongst decision makers in Surat's local government and heritage organizations. The current approaches to heritage conservation in Surat are described in the interviews as "artificial" and "copying the West". Without proper training that focuses on solutions and techniques catering to the uniqueness of Surat's urban context, heritage assets will continue to decay and vanish. There is a need to innovate and develop solutions through communication, cooperation, and collaboration with multiple disciplines. Few of the heritage experts and decision makers can use the digital technology [55] needed for restoration, and there is generally a lack of interest and awareness to learn these crucial skills [56,57]. Therefore, heritage conservation strategies lack any digital innovation and technique.

5.5. Urban Development and Real Estate Pressures

Surat faces the urgent task of providing new infrastructure to meet the needs of a growing population. People from rural areas and other less-developed towns and cities are migrating to Surat in search of employment opportunities in expanding and established sectors such as the textile trade and diamond business [19]. Consequently, Surat is experiencing real estate pressures for new infrastructure and commercial developments that can house more people and add increased value to the land (Figure 10). There is an existing conflict between the need to preserve heritage and its urban fabric and modernization projects to meet economic objectives. Providing urban infrastructure to meet the rise in population while protecting the integrity and authenticity of its heritage remains a distinct challenge [58]. Development projects for new infrastructure and commercial developments are often based on standardized solutions that are intended to generate immediate revenues [1]. However, they are usually insensitive to the authenticity and integrity of cultural heritage [28,59]. The interpretation given by local experts on the impact of such a rapid urbanization on local heritage was twofold as illustrated below. Through the analysis of both interviews and focus groups data, the researchers understood that (1) rapid urbanization boosts the property market to produce more housing, hence old buildings are replaced with new buildings with higher densities and (2) the replacement of newcomers weaken the affection that local communities still have for local heritage, since newcomers are often not aware about the heritage value and local identity.

Figure 10. New development and heritage building (Rajmarg) (Source: authors' photos).

Surat is experiencing an increasing population at a very fast rate and very rapid urbanization. This creates significant problems to create heritage awareness, identifying and awaiting opportunities (I7, Officer, Surat iLAB & Smart City, Expert Interviewee).

The problem is not just about land value, is also about money. They go up and up because they do want to rent to more and more people (I4, Heritage Architect, Expert Interviewee).

Urbanization is a threat because young generation left the historic city and new owners replaced traditional owners, and found old housing unsuitable to accommodate contemporary lifestyle (I1 & I2, Surat Municipal Corporation Museum experts, Expert Interviewee).

The rise in the real estate market has increased the land value in certain areas resulting in housing that are unaffordable for low-income groups and therefore remaining vacant. Developers are buying land in the historic areas, demolishing heritage buildings and replacing them with modern housing with higher density to increase the land value (Figure 10). As a result, heritage buildings and their surrounding areas are falling rapidly into decay. Furthermore, the attractiveness of contemporary ways of living are leading to many people leaving traditional houses and the historic parts of Surat because of unsuitability [17]. Some heritage houses, as designed according to the Indian tradition, lack adequate infrastructure such as toilets, sewage systems, and water pipes. Implementing contemporary infrastructure such as an air conditioning unit, bathroom, or flush toilets that is compatible with the old fabric in heritage buildings can be a challenge.

The photo sequence Figure 11 refers to different buildings captured in the same day (Figure 11a–d). However, it shows the typical trend happening in the two areas of Rander Gamtal and Gopi Surat Central Zone (Figure 2). Historic traditional buildings are often 2 or 3 storey buildings, built of traditional materials such as bricks (Figure 11a). In a leapfrogged but yet systematic way, they are replaced by individual landowners/builders with concrete buildings, allowing to push the density higher. Figure 11b shows a single traditional building demolished. This is happening in a leapfrogged way, ending up in chunks of the historic precincts being replaced with scattered interventions. On one hand, this makes the process of destruction of the traditional heritage slower, on the other hand, this process is happening silently but in a growingly pervasive manner and is spoiling the identity and the value of the historic urban fabric. Figure 11c shows the typical higher rise building replacing the previously existing traditional one. The last image (Figure 11d) clearly shows how the new building follows a kitsch aesthetic, replacing the sophisticated elegance of traditional architecture with bombastic, inconsistent, and ungrounded architectural features. Still, it also clearly shows how the owner considers such replacement aesthetically appealing since the façade looks quite willingly manicured. This corroborated the finding from the interviews and focus groups, regarding the necessity to raise awareness across the locals on the value of traditional heritage.

(a)   (b)

(c)   (d)

Figure 11. Photo sequence of new developments in historic area, Rander Gamtal. (**a**) Brick traditional buildings, (**b**) demolished single traditional building, (**c**) high rise development, (**d**) Replacement building with new features (Source: authors' photos).

5.6. Cultural Heritage and Climate Change

The impact of climate change on heritage has wide consequences ranging from structural damage, atmospheric moisture and temperature changes, and new interactions between natural and anthropogenic factors to more socioeconomic factors such as tourism demand and supply. Findings from the literature analysis suggest that the cultural aspects (social and spatial) are increasingly being considered for achieving environmental sustainability [23,60,61]. At a state and local level, there is a paucity of evidence of policies/measures that take into account traditional and local community knowledge in assessing the possible impact of climate adaptation on cultural heritage elements and practices. Indeed, at an urban scale, assessing the value of heritage resources is required for various reasons, such as, assessing vulnerabilities, adequately defining conservation priorities and directing funding [61]. Disaster management and risk mitigation policies with a heritage focus remain largely insufficient, particularly in view of Surat's vulnerability to repetitive flooding. Indeed, there lacks an established discourse in the area of sustainability and inclusive urban development concerning the relationship between cultural heritage and climate change. Efforts worth mentioning in addressing this gap include the 100 Resilient Cities (RC) Challenge which seeks to work with cities around the world to build resilience and tackle social, economic, and physical challenges that are faced by cities in an increasingly urbanized world. As a result, Surat introduced the Surat Resilience Strategy [19] as a platform to help address the critical question of what can be done to protect and improve the way of life of citizens of Surat in the present and in the future. The approach for developing and implementing this strategy is one of diverse collaboration, involving stakeholders such as the Surat Climate Change Trust (SCCT) and Surat Heritage Cell. However, there is an absence of specific measures on heritage sites to reduce the exposure and vulnerability of people and ecosystems to the risks and hazards of climate change. The strategy focuses heavily on social sustainability but fails to consider traditional and local community knowledge in assessing the possible impact of climate adaptation on heritage elements and practices. Similarly, the vulnerability assessment on Surat undertaken by The Rockefeller Foundation's Asian Cities Climate Change Resilience Network (ACCCRN) [23] highlights heritage as a strength in the profile description of Surat with no further explorations concerning cultural heritage and climate change.

6. Recommendations: Sustainable Urban Heritage Framework

Surat's heritage conservation efforts need to be located within the context of the city's socioeconomic and physical infrastructural urban pressures, needs, and demands. The diverse challenges discussed in the section above indicate the crucial necessity for a focus on sustainable urban heritage conservation in Surat. As highlighted by the UN Sustainable Goals (SDG 11), cultural assets represent an essential resource for sustainable and inclusive human development and to progress cities' social resilience [1]. At a local level, the recommendations discussed in this section demonstrate that Surat is a relevant qualitative case study for exploring the Historic Urban Landscape (HUL) approach in different urban contexts that experience similar elements of heritage conservation [12,13,62]. The findings have shown that Surat is a diverse urban settlement with multi-layers from the physical socio-cultural environment. Therefore, learning from the qualitative study of Surat and the HUL approach, more general recommendations can be drawn to address the inclusive local management of heritage resources as illustrated in the framework in Figure 12. These recommendations include developing holistic urban heritage legislation (6.1), identifying and mapping the city's heritage values, preserving local identity and sense of place (6.2), and developing local urban heritage and planning expertise, skills, and knowledge (6.3). Inclusive community and stakeholder engagement are central to the successful integration of urban heritage conservation (Figure 12). Using the HUL approach as a guiding framework [12], Surat Municipal Corporation and other local heritage stakeholders together with the inclusive participation of Surat's residents, can reinforce local identity, local distinctiveness, and local tangible and intangible values.

Figure 12. Integrated Urban Heritage Model.

6.1. Holistic Urban Heritage Legislation

On a national level, heritage policies need to be integrated with planning interfaces. The national Planning Act has good capacity for spatial control and regulation, but needs to broaden when dealing with cultural assets [63]. Surat has committed to becoming a resilient, smart, and sustainable city facilitated by international and national programs, and therefore, the protection of cultural heritage should be central to fulfilling these goals. To this extent, the recognition and appreciation of both tangible and intangible cultural heritage will enhance social cohesion and create a sense of place and belonging. These benefits can only truly be actualized through the development of urban heritage policies that integrate heritage protection into urban planning legislation and practice. Not only monuments but also traditional housing and local heritage should be targeted by local planning policies, by embedding heritage conservation principles within the local planning instruments such as plans and guidelines. Surat's local policies must go beyond monumentalism and instead address the heritage and its urban fabric as well as associated interdependent intangible heritage [10,21]. This can be financially viable by combining in an integrated strategy of the concepts of resiliency and heritage conservation. Intersections between heritage conservation, social cohesion, resilience, and local identity (Resilient City and heritage conservation) may support interventions leading to a better appreciation of the value of traditional housing and local heritage and elicit a more responsible approach from developers/local owners. Furthermore, disaster management and risk mitigation policies with a heritage focus will establish a discourse in the area of sustainability and inclusive urban development concerning the relationship between cultural heritage and climate change. Still, limitations and constraints to the demolition of traditional buildings must be included in the local planning policies in support of two areas, firstly, a better understanding of what must be valued by the community, and secondly, to make sure that conservation policies are endorsed consistently in the two historic areas of Rander Gamtal and Gopi Surat Central Zone (Figure 2). It should not be expected that the real estate market will acknowledge the value of heritage unless constraints and limits are imposed by local authorities when a gap in national conservation policies exists.

6.2. Identifying and Mapping Heritage Values and Preserving Local Identity and Sense of Place

A vital part of any sustainable approach is to recognize and understand the values linked to Surat's heritage. Thus, moving away from a material-based approach, also referred to as "authorized heritage discourse" [34,64,65] or an expert-driven approach that places the conservation of heritage solely in the hands of heritage authorities. Universal solutions that solely focus on monuments and do not embrace the intangible associations with heritage sites, nor their management systems and practices tend to oversimplify the complex reality of Surat's heritage landscape. A values-based approach places the people of Surat at the core of conservation. This approach is largely based on the Burra Charter (ICOMOS) and has been further developed to recognize the plurality of values, voices, and perspectives in the practice and interpretation of heritage conservation. The inclusion of the local community in decision making about Surat's heritage is prioritized in the discussions of solutions. This is with the view to democratize heritage and increase community participation. Initiatives such as U-Turn awareness programs reflect significant action from the local people of Surat to organize resistance to prevent the demolition of heritage buildings [19]. In this context, a values-based approach builds on the growing momentum and makes concerted effort to engage the whole range of stakeholder groups throughout the conservation process [66]. The youth have a crucial role to play in the success of community awareness. Intergenerational approaches encourage older people and the younger generations to share and learn about heritage together and in a meaningful and impactful way. Surat's educational institutes, schools, and colleges can facilitate this learning and allow for a high level of engagement with tangible and intangible heritage.

In rapid urbanization conditions, local communities are often replaced at a rapid pace too, by becoming less resilient to change and therefore not capable to advocate for their own identity preservation. Again, it should not be expected that disenfranchised local communities will be strong enough to advocate for local heritage conservation, it is a duty of local authorities to impose limits and constraints to the demolition of local heritage. Successfully integrating the historic environment into urban planning management includes identifying and recognizing the complex elements that make Surat distinctive and create a sense of place and identity.

6.3. Local Urban Heritage and Planning Expertise, Skills and Knowledge

Heritage buildings are perceived for the most part as a financial liability and non-priority topic in Surat's investment discourse. This is partly due to the costs, skills, and resources needed to restore the buildings and the surrounding urban fabric. Surat's heritage practitioners lack a strong evidence base for their decision-making in heritage improvements and the quantification of damage to historic materials [19]. The effective use of technology in the heritage sector in Surat has significant potential to contribute to an accurate and informed understanding of the heritage sites, buildings, and interiors. Therefore, heritage professionals and decision-makers need to gain skills and knowledge to identify innovative solutions as well as to seek synergy with other disciplines and fields of work. Resilience building can be combined and associated with heritage conservation, to empower local administrators in their role of endorsing heritage conservation. Organizations such as the ASI and INTACH need to develop formal systems that recognize and support the conservation of heritage as an interdisciplinary effort [67,68]. The Smart City program could support the implementation of new technologies facilitating knowledge sharing on local heritage.

7. Conclusions

This paper has examined the context of the challenges in Surat and the efforts made with the view to make heritage an integral of part of urban planning and management. A presentation of the conceptualization of urban heritage conservation within the city of Surat has been made. The discussion is situated in the context of a growing global discourse on the crucial role culture plays in sustainable urban development. The city of Surat is explored as an exemplar study through qualitative fieldwork. Although Surat has made deliberate steps in addressing its urban heritage, the existing challenges are considerable. The findings from this study highlight the need for decision-makers in the heritage sector to acknowledge Surat's multi-layered and diverse cultural heritage as a critical resource preserved through community engagement. Furthermore, the findings reflected the diversity in Surat's built architectural heritage that demonstrate the typological influences (i.e., vernacular, colonial (Gothic and Renaissance), Art Deco, and Arabesque). The absence of structured approaches can be presented as an opportunity for the design of locally defined participatory processes that promote the diverse transformation of cultural heritage.

Future research can focus on community-based negotiation of urban cultural representation. Surat is not an isolated case, the narrative on this case study reflects current trends and challenges on conservation of heritage assets in rapidly-growing urban areas. Thus, considerations and recommendations are indeed relevant to the larger heritage cities' planners and the conservationist international community. The inclusive development of urban heritage has the potential to foster a shared cultural identity experiencing both material (tangible) and socio-psychological (intangible) remnants of the nation's past and bringing pasts, peoples, places, and cultures into performative contestation and dialogue. Unifying these separate elements to present a coherent story and sustainable representation of urban heritage, however, remains a priority area for future research. Additionally, this paper recommends future research should be supported with comprehensive statistical and geo-spatial heritage data that can allow for the investigation of the role of urban heritage with broader urban issues.

Author Contributions: Conceptualization, C.U., C.T., K.G.B.A., B.C.N.M., D.A.P. and L.E.M.; Formal analysis, C.U., C.T., K.G.B.A., B.C.N.M., D.A.P. and L.E.M.; Methodology, C.U., C.T., K.G.B.A., B.C.N.M., D.A.P., L.E.M. and K.N.J.; Supervision, C.U.; Writing—original draft, B.C.N.M.; Writing—review & editing, C.U., C.T., K.G.B.A., B.C.N.M. and L.E.M. All authors have read and agreed to the published version of the manuscript.

Funding: This research is financed by the Arts and Humanities Research Council (AHRC), UK [project reference: AH/R014183/1] and the Indian Council of Historical Research (ICHR), New Delhi, India. This work is a part of an ongoing project "IT INDIAN HERITAGE PLATFORM: Enhancing cultural resilience in India by applying digital technologies to the Indian tangible and intangible heritage".

Acknowledgments: The authors express their sincere gratitude to the Municipal Commissioner of Surat, Deputy Municipal Commissioner and their team of Heritage Cell, Sardar Patel Museum and Science Center of Surat Municipal Corporation (SMC), Archeological Survey of India (ASI) Vadodara Circle, SURATi iLAB (Surat Smart City Development Ltd.), Gujarat Tourism, Surat-Gopipura Heritage Conservation Society, Surat Heritage Trust, Tapi Trust Surat, Indian Building Congress (IBC) Surat Chapter, Southern Gujarat Chamber of Commerce and Industry (SGCCI) Surat, Resilience Strata Research and Action Forum Surat, Institute of Civil Engineers and Architects (ICEA) Surat, Sarvajanik Education Society Surat, Patanjali Yog Prashikshan Samiti Surat, Raman Bhakta School of Architecture, Tarsadi, and CREDAI Surat for their generous interest and support for this research.

Conflicts of Interest: The authors declare no conflicts of interest. The funders had no role in the design of the study; in the collection, analyses, or interpretation of data; in the writing of the manuscript, or in the decision to publish the results.

References

1. UNESCO. *Global Report on Culture for Sustainable Urban Development*; United Nations Educational, Scientific and Cultural Organization (UNESCO): Paris, France, 2016.
2. Menon, A. Heritage Conservation in India: Challenges and New Paradigms. In Proceedings of the SAHC2014—9th International Conference on Structural Analysis of Historical Constructions, Mexico City, Mexico, 14–17 October 2014; Available online: www.hms.civil.uminho.pt (accessed on 2 February 2019).

3.   Archaeological Survey of India. World Heritage Sites. 2019. Available online: http://asi.nic.in/ (accessed on 2 February 2019).

4.   Sharma, A. Exploring Heritage of a Hill State—Himachal Pradesh, in India. *AlmaTourism* **2015**, *12*. [CrossRef]

5.   Sharma, A.; Sharma, S. Heritage tourism in India: A stakeholder's perspective. *Tour. Travelling* **2017**, *1*, 20–33. [CrossRef]

6.   Sharma, M. *3 Intangible Cultural Heritage Elements from India Inscribed on UNESCO's List till Date*; Government of India, Ministry of Culture: New Delhi, India, 2018.

7.   United Nations. *The New Urban Agenda*; United Nations: New York, NY, USA, 2017.

8.   United Nations. Sustainable Development Goals. 2016. Available online: http://www.un.org/sustainabledevelopment/sustainable-development-goals/ (accessed on 2 February 2019).

9.   Hosagrahar, J. *UNESCO Thematic Indicators for Culture in the 2030 Agenda for Sustainable Development, in Analytical Report of the Consultation with the Member States, 2019*; UNESCO World Heritage Centre: Paris, France, 2019.

10.   Bandarin, F.; Oers, R. *The Historic Urban Landscape: Managing Heritage in an Urban Century*; Wiley-Blackwell: Hoboken, NJ, USA, 2012.

11.   Hosagrahar, J. Cultural heritage, the UN Sustainable Development Goals, and the New Urban Agenda. *Bdc Boll. Del Cent. Calza Bini* **2016**, *16*, 37–54.

12.   UNESCO. The HUL Guidebook. In *15th World Conference of the League of Historical Cities*; UNESCO: Bad Ischl, Austria, 2016.

13.   UNESCO. *The UNESCO Recommendation on the Historic Urban Landscape*; UNESCO: Paris, France, 2011.

14.   Lorusso, S.; Cogo, G.M.; Natali, A. The Protection and Valorization of Cultural and Environmental Heritage in the Development Process of the Territory. *Conserv. Sci. Cult. Herit.* **2016**, *16*. [CrossRef]

15.   Dastgerdi, S.A.; Luca, G.D. Specifying the Significance of Historic Sites in Heritage Planning. *Conserv. Sci. Cult. Herit.* **2018**, *18*, 29–39.

16.   Rakeshkumar, G.; Padhya, H.; Naresh, R. Heritage—A Case Study of Surat. In *International Journal of Advanced Research in Engineering, Science & Management (IJARESM)*; pp. 1–9. ISSN 2394-1766. Available online: www.ijaresm.net (accessed on 1 March 2018).

17.   Baradi, M.; Malhotra, M. *At the Core: Understanding the built heritage of Surat and Rander*; UCD: New Ranip, India, 2011.

18.   Directorate of Census Operations. *District Census Handbook Surat*; Village and Town Wise Primary Census Abstract (PCA): New Delhi, India, 2011.

19.   TARU Leading Edge. *Surat Resilience Strategy*; TARU Leading Edge: Surat, India, 2017.

20.   Girad, L. Toward a Smart Sustainable Development of Port Cities/Areas: The Role of the "Historic Urban Landscape" Approach. *Sustainability* **2013**, *5*, 4329–4348. [CrossRef]

21.   Jigyasu, R. *The Intangible Dimension of Urban Heritage, in Reconnecting the City: The Historic Urban Landscape Approach and the Future of Urban Heritage*; Bandarin, F., Oers, R.V., Eds.; Wiley-Blackwell: Hoboken, NJ, USA, 2014.

22.   Rameshkumar, P.M. Heritage Route Optimization for Walled City Surat Using GIS. *Sarvajanik Education Societyl* **2017**. [CrossRef]

23.   ACCCRN. *Phase 2: City Vulnerability Analysis Report Indore & Surat*; Edge, T.L., Ed.; Asian City Climate Change Resilience Network (ACCCRN): Surat, India, 2010.

24.   Bhat, G.K. Addressing flooding in the city of Surat beyond its boundaries. *Environ. Urban.* **2013**, *25*, 429–441. [CrossRef]

25.   McGregor, S.L.T.; Murnane, J.A. Paradigm, methodology and method: Intellectual integrity in consumer scholarship. *Int. J. Consum. Stud.* **2010**, *34*, 419–427. [CrossRef]

26.   Proverbs, D.; Gameson, R. Case study research. In *Advanced Research Methods in the Built Environment*; Knight, A., Ruddock, L., Eds.; Blackwell Publishing Ltd: Chichester, West Sussex, UK, 2008.

27.   National Institute of Urban Affairs. *Compendium of Good Practices, Urban Heritage in Indian Cities*; National Institute of Urban Affairs: New Delhi, India, 2015.

28.   Madgin, R. *Cultural Heritage and Rapid Urbanisation in India*; Dorset Press: Dorchester, UK, 2015.

29.   Kant, A. Indian Heritage's Economic Potential. 2017. Available online: www.livemint.com (accessed on 2 February 2019).

30. Anthony, A. *City HRIDAY Plan Heritage City Development and Augmentation Yojana (HRIDAY) Dwarka, Gujarat*; Urban Management Center: Dwarka, Gujarat, India, 2016.

31. Nocca, F. The Role of Cultural Heritage in Sustainable Development: Multidimensional Indicators as Decision-Making Tool. *Sustainability* **2017**, *9*, 1882. [CrossRef]

32. Giraud-Labalte, C. *Cultural Heritage Counts for Europe Report*; CHCfE Consortium: Krakow, Poland, 2015.

33. Harvey, D.C. Heritage Pasts and Heritage Presents: Temporality, meaning and the scope of heritage studies. *Int. J. Herit. Stud.* **2001**, *7*, 319–338. [CrossRef]

34. Smith, L. *Uses of Heritage*, 1st ed.; Routledge: London, UK, 2006; p. 368.

35. Bala, S. *Digital inventories on Cultural Memories and Intangible Cultural Heritage: Case study of Yadav community of Haryana, India*; ICOM International Committee for Documentation: Delhi, India, 2012.

36. Mukherjee, B. *India's Intangible Cultural Heritage: A Civilisational Legacy to The World*; Government of India: New Delhi, India, 2015.

37. Lowenthal, D. *The Heritage Crusade and the Spoils of History*; Cambridge University Press: Cambridge, UK, 1998.

38. Quiroz, H.; Astrid, L. *Historic Urban Landscape. Kick-Off Meeting of the Project "Mapping Controversial Memories in the Historic Urban Landscape"*; Istituto Svizzero: Rome, Italy, 2015.

39. Throsby, D. Culture in sustainable development: Insights for the future implementation of art. 13. *Econ. Della Cult.* **2008**, *18*, 389–396.

40. Hitchcock, M.; King, V.; Parnwell, M. *Heritage Tourism in Southeast Asia*; NIAS Press: Copenhagen, Denmark, 2010.

41. Chalcraft, J.; Delanty, G. *Can Heritage be Transnationalised? The Implications of Transnationalism for Memory and Heritage in Europe and Beyond. AXIS 1. CULTURAL MEMORY TF1. Memory and Heritage*; University of Sussex: Falmer, UK, 2015.

42. Viejo-Rose, D. Cultural Heritage and Memory: Untangling the Ties that Bind. *Cult. Hist. Digit. J.* **2015**, *4*. [CrossRef]

43. Yu Park, H. HERITAGE TOURISM: Emotional Journeys into Nationhood. *Ann. Tour. Res.* **2010**, *37*, 116–135. [CrossRef]

44. Heritage, E. *Research Agenda: An Introduction to English Heritage's Research Themes and Programmes*; English Heritage: London, UK, 2005.

45. Jimura, T. The Relationship between World Heritage Designation and Local Identity. In *World Heritage, Tourism and Identity: Inscription and Co-Production*; Bourdeau, L., Ed.; Routledge: London, UK, 2015.

46. Sabat, V. *Surat ni Uttrayan*; Kapadiya, N., Ed.; Book World: Surat, India, 2020.

47. Stephenson, J. The Cultural Values Model: An Integrated Approach to Values in Landscapes. *Landsc. Urban Plan.* **2008**, *84*, 127–139. [CrossRef]

48. Burra Charter. *The Burra Charter: The Australia ICOMOS Charter for Places of Cultural Significance*; Australia ICOMOS: Burwood, Australia, 2013.

49. Mason, R. Assessing Values in Conservation Planning: Methodological Issues and Choices. In *Assessing the Values of Cultural Heritage*; Torre, M.d.l., Ed.; Getty Conservation Institute: Los Angeles, CA, USA, 2002.

50. Council of Europe. *The Role of Cultural Heritage in Enhancing Community Cohesion: Participatory Mapping of Diverse Cultural Heritage*; Council of Europe and the European Union: Paris, France, 2018.

51. Graham, B.; Ashworth, G.; Tunbridge, J. *A Geography of Heritage: Power, Culture and Economy*, 1st ed.; Routledge: London, UK, 2000.

52. Delanty, G. *The European Heritage from a Critical Cosmopolitan Perspective*; LSE 'Europe in Question' Discussion Paper Series; London School of Economics (LSE): London, UK, 2010. [CrossRef]

53. Delanty, G. *Europe and Asia beyond East and West*; Routledge: London, UK, 2006.

54. Gaur, R. Digital Preservation in India. *Desidoc J. Libr. Inf. Technol.* **2012**, *32*, 291–292. [CrossRef]

55. Udeaja, C. Scientometric Analysis and Mapping of Digital Technologies used in Cultural Heritage Field. In *ARCOM 2019*; ARCOM: London, UK, 2019.

56. Mallik, A. *Digital Hampi: Preserving Indian Cultural Heritage*; Springer Nature Singapore Pte Ltd.: Singapore, 2017.

57. Mallik, A. An Intellectual Journey in History: Preserving Indian Cultural Heritage. In *New Trends in Image Analysis and Processing—ICIAP 2013*; Petrosino, A., Maddalena, L., Eds.; Springer: Berlin/Heidelberg, Germany, 2013.

58. Girard, L.F.; Nocca, F. Integrating cultural heritage in urban territorial sustainable development. In *ICOMOS 19th General Assembly and Scientific Symposium "Heritage and Democracy"*; ICOMOS: New Delhi, India, 2018.

59. Leonard, C. Getting the city back to the people: Community heritage conservation in Ahmedabad. How the heritage trails have transformed Indian cities. In *Managing Heritage Cities in Asia and Europe: The Role of Public-Private Partnerships. Experts' Meeting & Public Forum in preparation for the 5th ASEM Culture Ministers' Meeting*; Asia-Europe Foundation: Yogyakarta, Indonesia, 2017.

60. Bumbaru, D. Initiatives of ICOMOS to Improve the Protection and Conservation of Heritage Sites Facing Natural Disasters and Climate Change. In *Cultural Heritage and Natural Disasters Risk Preparedness and the Limits of Prevention*; Meier, H.-R., Petzet, M., Will, T., Eds.; ICOMOS—International Council on Monuments and Sites: Charenton-le-Pont, France, 2007; pp. 203–213.

61. Dastgerdi, A.S. Climate Change and Sustaining Heritage Resources: A Framework for Boosting Cultural and Natural Heritage Conservation in Central Italy. *Climate* **2020**, *8*, 26. [CrossRef]

62. Serrainoa, M.; Lucchi, E. Energy Efficiency, Heritage Conservation, and Landscape Integration: The Case Study of the San Martino Castle in Parella (Turin, Italy). In *Proceedings of the Climamed 2017—Mediterranean Conference of HVAC; Historical Buildings Retrofit in the Mediterranean Area, Matera, Italy, 12–13 May 2017*; Elsevier Ltd.: Matera, Italy, 2017.

63. Riganti, P. Rapid urbanization and heritage conservation in Indian cities. *Semant. Sch.* **2017**, *17*, 1–16.

64. Pendlebury, J. Conservation values, the authorised heritage discourse and the conservation-planning assemblage. *Int. J. Herit. Stud.* **2013**, *19*, 709–727. [CrossRef]

65. Pendlebury, J.; Townshend, T.; Gilroy, R. The Conservation of English Cultural Built Heritage: A Force for Social Inclusion? *Int. J. Herit. Stud.* **2004**, *10*, 11–31. [CrossRef]

66. Labadi, S. *UNESCO, Cultural Heritage, and Outstanding Universal Value Value-based Analyses of the World Heritage and Intangible Cultural Heritage Conventions*; AltaMiraPress: Lanham, MD, USA, 2013.

67. INTACH. INTACH Charter. 2016. Available online: http://www.intach.org/about-charter-guidelines.php (accessed on 2 February 2019).

68. INTACH. *Cultural Mapping*; 2018; Available online: http://intangibleheritage.intach.org/projects/cultural-mapping/ (accessed on 2 February 2019).

Article

Toward Sustainability of South African Small-Scale Fisheries Leveraging ICT Transformation Pathways

Tsele T. Nthane [1,*], Fred Saunders [2], Gloria L. Gallardo Fernández [2] and Serge Raemaekers [1,3]

[1] Environmental and Geographical Science Department, University of Cape Town, Cape Town 7701, South Africa; serge@abalobi.org
[2] School of Natural Sciences, Technology, and Environmental Studies, University of Södertön, 141 89 Huddinge, Sweden; fred.saunders@sh.se (F.S.); gloria.l.gallardo.fernandez@sh.se (G.L.G.F.)
[3] Abalobi NPO,1 Westlake Dr, Cape Town 7945, South Africa
* Correspondence: nthtse001@myuct.ac.za

Received: 14 October 2019; Accepted: 7 January 2020; Published: 20 January 2020

Abstract: Though Internet and Communication Technologies (ICTs) have been employed in small-scale fisheries (SSFs) globally, they are seldom systematically explored for the ways in which they facilitate equality, democracy and sustainability. Our study explored how ICTs in South African small-scale fisheries are leveraged towards value chain upgrading, collective action and institutional sustainability—key issues that influence small-scale fishery contributions to marine resource sustainability. We held a participatory workshop as part of ongoing research in the town of Lambert's Bay, South Africa, in collaboration with small-scale fishers and the Abalobi ICT project. We mapped fisher value chain challenges and explored the role of ICT-driven transformation pathways, adopting Wright's 'Real Utopian' framework as the lens through which to explore equality, democracy and institutional sustainability. We found Abalobi's ICT platform had the potential to facilitate deeper meanings of democracy that incorporate socio-economic reform, collective action and institutional sustainability in South Africa's small-scale fisheries. Where fishers are not engaged beyond passive generators of data, this had the potential to undermine the goals of increasing power parity between small-scale fisheries and other stakeholders.

Keywords: small-scale fisheries; sustainability; ICT4F; South Africa; value chains; Real Utopias; technology; co-design

---

1. Introduction

Small-scale fisheries are essential to the livelihoods of rural coastal communities by providing both food security and employment [1,2]. Yet, South African small-scale fisheries remain marginalized with the state chronically under-resourced and unable to adequately cater to their needs. Key reasons are that fishers commonly target low value species [3,4], fisher landing sites are remote and spread across multiple actors [5] making them difficult to manage, and South African fisheries is saddled with the legacies of Apartheid in which traditional fishers remain socio-economically marginalized [6–8].

Beyond employment and food security, small-scale fishing activities are deeply interwoven with local cultural practices and traditions [9], where retention of subsistence catch is commonly used in non-profit exchanges for help with landing tasks or shared amongst friends and family members [10]. Fishing practices have historically played a crucial role in the economic and socio-cultural development of rural coastal communities, thus raising important questions of the consequences when these practices are altered through private rights regimes, external actors and marine resource depletion [11]. Small-scale practices have been shown to exhibit traits antithetical to behavioral assumptions of self-interested actors, with actors willing to change the distribution of material outcomes at personal cost and displaying concern for fairness and reciprocity [12]. For the fishers of Lambert's Bay, South Africa, fishing practices around 'Snoek' (*Thyrites atun*) and other linefish species have shaped the community's livelihood and cultural practices over centuries [13], which thus bear importance on collective action possibilities and the design of interventions [12].

Research has commonly focused on small-scale fishery policy and governance, value chain upgrading and the application of Internet and Communication Technologies (ICTs) [14–18]. Value chain research is primarily interested in revealing dominant power relations between actors, the terms of inclusion and exclusion, and more recently, incorporating 'horizontal' dimensions such as gender, the environment and sustainability [18–21]. Despite the proliferation of ICT for Fisheries (ICT4F) interventions, little work has been conducted on ICT-supported transformation possibilities that shift power relations within existing governance and market structures [22,23] within particular SSF settings or sought closer alignment with the ambitions of fisheries-related Sustainable Development Goals (SDGs) including: no poverty, zero hunger, gender equality, climate action and sustainability of life below water [24,25].

The urgency for systematically developed and contextually refined small-scale fisher transformation pathways is further apparent when understood in the context of decades of Apartheid and subsequent neoliberal fisheries policy that disenfranchised South African small-scale fishers [6–8,26–28]. We put forward that sustainability of marine resources must be facilitated by the design of institutions and governance arrangements that benefit small-scale fishers. We thus explore how small-scale fishers leverage ICT4F towards more democratic, equal and ultimately sustainable pathways.

1.1. Struggles of the Marginalized in South Africa Post-1994

South Africa held its first democratic elections in 1994, yet despite the promises and expectations of socio-economic redress, and the myriad programs directly and indirectly tethered to ideals of affirmative action, the patterns of inequality inherited from Apartheid remain [6,7,29,30]. Though progressive legislation such as the Bill of Rights and South Africa's much lauded Constitution have been enacted, the dire state of redress has precipitated a 'waiting' where the marginalized continue to wait on the delivery of the promises of socio-economic redress [31]. The majority of South African small-scale fishers live and harvest on rural coasts, while their struggles for access rights are fought in the large metropolis of Cape Town, where the headquarters of the State fisheries department is located.

Fisher protest action over the preceding two decades took place amidst the groundswell of public discontent, with fisher movements aligning themselves with trade unions and the struggles of the urban poor attempting to transform political spaces [32]. At the heart of small-scale fisher protests and broader social movements was dissatisfaction with the lack of socio-economic transformation despite the achievements of post-1994 political democratization [29].

1.2. South African Small-Scale Fisheries

South African societies are specific in the enormous expectations residents hold of the post-Apartheid state; this includes demands from artisanal fishers for universal access to marine resources [33]. South Africa's first democratic elections ushered in hopes of radical social, political and economic reforms in the fisheries sector. Industrialization of the sector in the 1950s had concentrated fishing rights in the hands of a few large industrial companies whilst systematically excluding artisanal fishers—restricting them to low-wage labor as crew and fish factory workers. Thus, the democratic government's first attempt at transformation of fisheries occurred between 1994–2007 with the Marine Living Resources Act of 1998 (MLRA), that sought to balance the needs of the industrialized commercial fishing sector with broadening access for artisanal fishers [6,34]. To retain their individual transferrable quotas (ITQs) under the MLRA, established companies were required to satisfy diversity requirements with the goal that ITQ redistribution to artisanal fishers would broaden access [8,35].

However, without infrastructural, financial and business skills to manage quota, artisanal fishers were often reduced to catching, processing and marketing agreements with larger industrial companies that reduced artisanal fishers to 'paper' quota holders without any real participation [36]. In addition, the MLRA recognized only a small-scale commercial fishing sector and made no provision for the men and women who derived a subsistence livelihood from marine resources. In effect, this excluded the majority of small-scale fishers who had, for generations, harvested marine resources [7].

Through a coalition of small-scale fishers, NGOs and university researchers, in 2004, the MLRA was challenged in the South African courts in "Kenneth George and Others vs. the Minister" [26]. Before a decision was reached by the Courts in 2007, the Minister settled, out of court, in an agreement that mandated the department to develop a policy, specific to the inclusion of small-scale fishers, in recognition of their historical livelihood dependence on marine resources [6]. Further, interim fishing licenses were immediately established that allowed small-scale fishers to harvest a limited amount of marine resources until the eventual promulgation of a new inclusive policy [37]. The result was the Small-Scale Fisher Policy of 2012 (SSFP) which provided the guidelines for the formal recognition and provision of South Africa's small-scale fishers.

The purpose of the new Small-Scale Fisheries Policy was to provide a framework under which small-scale fishers across all four of South Africa's coastal provinces could be granted collective community fishing rights with access to a 'basket' of resources [38]. As the legislative framework was finalized in 2016, the Department of Agriculture, Forestry and Fisheries (DAFF) spent the intervening period addressing contentious issues across South Africa's small-scale fisher communities, such as which men and women qualified as 'bona fide' fishers and thus could be provided for under the new policy. Simultaneously, this slow process forced small-scale fishers to organize, independently from the state, starting cooperatives and partnering with NGOs. One such NGO was Abalobi: a hallmark project that sought to leverage small-scale fisher data to facilitate value chain upgrading as one of multiple strategies to transform South African small-scale fisheries.

1.3. Value Chains of Small-Scale Fisheries of the Western Cape, South Africa

With 90% of South Africa's linefish caught within the Western Cape Province, small-scale fishers from Lambert's Bay contribute substantially to linefish harvests both through catches and as fish workers. Low-value species make up the majority of small-scale fisher catches with Snoek (Thyrsites atun), the largest contributor to fisher income and catch volumes [13]. Where Lambert's Bay fishers do harvest high value species such as Lobster (Jasus lalandii), these value chains remain linked to international Asian markets, with fishers unable to renegotiate their terms of incorporation and unable to influence the terms of trade [39].

Despite the importance of value chains to the sustainability of small-scale fisher livelihoods and the marine resource, state fisheries governance has not provided enough support to develop new, more viable small-scale fisher value chains nor the platforms for fishers to participate equitably in their existing value chains. The 2012 Small-Scale Fisher Policy represented the State's latest attempt at providing for small-scale fishers, promising economically viable access rights to all bona fide small-scale fishers. However, by 2019, the state had still not comprehensively provided small-scale fishers with viable access rights nor the platform to advance more equitable value chain participation. Consequently, small-scale fishers explored alternative value chain upgrading pathways with the assistance of non-state actors, an increasingly pivotal trend largely directed at facilitating value chain upgrading through ICTs and direct marketing arrangements.

1.4. Abalobi ICT4F Project

Through collaboration between small-scale fishers, University of Cape Town researchers, and with support from telecom companies, the Abalobi project initiated an Internet and Communication Technology for Fisheries (ICT4F) platform in 2015. The project sought to address 'social justice and poverty alleviation' in South African small-scale fisheries, aiming to transform fisheries science to include fisher experiential knowledge, develop locally-based systems of marine resource stewardship and build socio-ecological resilience primarily through collecting reliable small-scale fisheries data [40].

ICT4F applications have grown more complex in their aims since the earliest landmark studies showed how mobile phones reduced information search costs and price dispersion for small-scale fishers of Kerala [17,41]. Where ICT applications were implemented, they commonly provided extension services for price, weather, pest and technical information, though they are increasingly offering more complex services linking buyers to sellers and offering financial services [18,42,43]. The rise of more comprehensive ICT offerings for small-scale fishers has occurred simultaneously with the increased focus on ICT applications that better meet the local context of participants [43]. This focus on improving ICT practice is intended to achieve higher success rates by including participants early in the design of the ICT intervention to encourage buy-in, legitimacy of the programs, as well as more accurately directed practice [44,45].

The Abalobi platform consisted of five modules, namely: 'Fisher', 'Monitor', 'Manager', 'Co-op' and 'Marketplace'. Abalobi 'Fisher' was the core module and collected small-scale fisher logbook data. 'Monitor' was utilized by South African fisheries staff for catch monitoring. 'Manager' provided real-time fisheries data communication between small-scale fishers and Abalobi managers. 'Co-op' hosted collective accounting and asset management functions for small-scale fisher cooperatives, and 'Marketplace' captured small-scale fisher socio-cultural 'stories' on a traceability platform where fishers availed their catch for the Restaurant Supported Fishery (RSF). This 'storied' fish was modelled on the rapidly expanding arena of small-scale fisher direct marketing that facilitates shorter value chains between fishers and schools, hospitals, restaurants and homes [46–52]. At the time of writing, the most active modules were the Abalobi 'Fisher' and 'Marketplace' modules, with 'Monitor', 'Manager' and 'Co-op' in development.

Smartphones and ICTs are increasingly recognized as a tool for participatory fisheries data collection with the potential to increase the accuracy of small-scale fisheries data, facilitating communities in achieving sustainable development, an improved quality of life and supporting the poor and excluded [5,53,54]. Where small-scale fisheries data is collected, it is often irregularly so [54] or fails to fully capture the characteristics and essence of small-scale fisheries [55,56]. Whilst collecting reliable small-scale fisheries data remains imperative, the lack of public policies for the sector as well as the lack of skills in collecting this data continue to complicate the task [57,58].

In the absence of reliable small-scale fisher data, researchers risk masking small-scale fisher importance by under- or overreporting resource availability which ultimately influences fisheries legislation [5]. In order to leverage this data effectively in support of transformation pathways, it is imperative that a systematic approach is adopted that examines the potential of ICTs to shift power relations in governance and market structures.

Fisher livelihoods remain intricately connected with the health of the oceans, raising the importance of new approaches that encompass biological stock, socio-economic and socio-cultural assessments, ushering in a renewed focus on participatory and interdisciplinary approaches that incorporate fisher knowledge [58]. While stock assessments retain prime importance in fisheries management, this prominence is being challenged as calls to include social and economic livelihood data increase [59–62]. One avenue to achieving broader indicators of marine resource health is incorporating fisher knowledge [63–66].

Though novel, the Abalobi project is indicative of broader momentum within the field of Internet and Communication Technologies for Fisheries (ICT4F). This includes Fish Trax, an electronic fishery information system for small-scale fisheries; Norpac Fisheries Export, which provides traceability services for small-scale fisheries; ThisFish, which connects small-scale fishers directly to end markets; and Trace Register, which is a web-based traceability application facilitating direct marketing between fishers and their end markets.

1.5. ICT 4 Fisheries

ICTs were initially heralded as tools to 'leapfrog' underdeveloped communities by bridging the 'digital divide' towards more developed states, primarily through the proliferation of tele-centers located in rural areas. However, it quickly became apparent that tele-centers far underachieved their intended goals, and that ICT access often mirrored and thus, reinforced existing socio-economic inequalities. Yet the potential for ICTs to facilitate more robust rural livelihoods remains apparent, and it is thus critical to systematically investigate how and under what conditions ICTs are leveraged successfully towards the socio-economic development of rural communities, and in particular small-scale fisheries. We further need to build towards developing small-scale fisher criteria for selecting appropriate marine fishery ICT tools [66].

ICT interventions are extending beyond needs based design that focuses on short to medium term necessities, towards designing for long term aspirations [67]. Despite the proliferation of ICT studies, there remains a dearth in ICT research that provides positive evidence of successful outcomes of ICTs for environmental issues such as climate change and socio-ecological resilience [68]. Though collecting small-scale fisher data is a critical global priority, this can also have the effect of increasing the concentration of power in the hands of the few with the resources to access, analyze and extract meaning from it [69], raising questions of power between those who generate the data and those who effectively have control over it.

ICT interventions need to look beyond resolving issues of access and resources to that of wellbeing and fulfilment for the marginalized [70]. This entails defining how projects define development: whether development is defined as increased freedoms in the capabilities tradition, as expanded access to ICT artefacts for the disenfranchised, as increased economic productivity such as business profit through ICT use, or as improved wellbeing measured by satisfaction and fulfilment stemming from an ICT intervention [71]. Alternatively, development can also compellingly be defined as an increase in power parity between the stakeholders and beneficiaries [71], particularly because projects often privilege technical functionality over stakeholder engagement [72].

1.6. Neoliberalism in Small-Scale Fisheries

Neoliberal policies seek to control access to fisheries resources by strengthening property rights and reducing fishing effort in order to prevent marine resource overexploitation—typically through the introduction of Individual Transferrable Quotas (ITQs) [73]. With 'too many boats chasing too few fish', turning access into private property was intended to induce better stewardship of the resource whilst ensuring rights holders remained commercially vested in the health of the marine resource [74,75]. Though neoliberalism has increased profits generated through ITQs, these tend to remain concentrated in the hands of a few powerful actors at the exclusion of artisanal and indigenous fishers [76–78]. With assumptions rooted in the tragedy of the commons [78], small-scale fisheries were marginalized as industrial fisheries benefitted from the introduction of access rights and large quotas and subsidies [79]. For instance, of $35 billion worth of subsidies issued globally in 2009, small-scale fisheries received only 16 percent [80].

With the majority of South Africa's small-scale fishers residing along rural coasts, it remains an obvious problem that neoliberalism consolidated wealth and power into urban and industrial fishing operations and away from customarily governed open-access fishing on which small-scale fishers depend. These geographic and economic shifts led to detrimental effects undermining coastal livelihood identities, cultural values and artisanal indigenous fishery practices [12,76,81,82]. Urban concentration of fishing rights tends to empower privileged social groups, further increasing wealth and class disparities [78,83], in addition to the prevalent role of speculative finance in undermining and disrupting sustainable practices in small-scale fisher communities [73].

Inequitably distributed fishing rights have precipitated crises of legitimacy with artisanal fishers contesting the legitimacy of fishing rights bureaucracies [76,78] and their rights holders, particularly where these rights holders had no historical stake in the marine resource [83]; or when women have not been included [84]. Those impacted are commonly fishers closest to the resource such as small-scale fisher skipper-operators, their crew and those further along the fisheries value chain [76]. This dispossession can lead to a unique form of wildlife crime where previously accepted SSF subsistence practices become illegal in the absence of legislation that supports their livelihoods and cultural practices [85].

In response to these problems, new forms of fisheries governance emerged that encouraged sustaining marine resource diversity and coastal livelihoods by exploring indigenous modes of custodianship [86], or by institutionalizing collective management and action; for example, through territorial use rights for fisheries (TURFs) [87,88]. Inspired by indigenous small-scale fisher movements resisting overfishing, development projects in their territories and exploring pathways that bypass corporate fish processors [73], alternative pathways to capital intensive fisheries have sought to reposition fisherfolk as community actors integral to rural economies [89,90]. In addition, non-state actors are continually emerging as key partners of small-scale fisher resistance [90], as well as facilitators in the development of transformative fisher pathways. By undermining small-scale fisher livelihoods, neoliberalism jeopardizes the sustainability of community livelihoods and the resources on which they depend.

1.7. 'Real Utopias'

In order to explore the potentiality of a pathways approach to transformation for South African SSFs, here we engage with Wright's 'Real Utopias' framework as a means to support progressive social change. Wright argues that states have grown complex and incapable of addressing novel problems of contemporary society [91], and by considering alternative economic and social systems, utopian alternatives create better conditions for human flourishing than those under capitalism [92]. Using utopian framing, we can begin to imagine how things could be when the problems in contemporary society are resolved, thus providing us with immediate actionable tasks in pursuit of a better future [93]. At the heart of Wright's 'Real Utopias' is the fundamental goal to transform power relations and encourage equality, democracy and sustainability in the establishment of alternative transformation pathways.

Equality equates with living a 'flourishing life' through grass-roots political mobilization and collective action; democracy demands meaningful participation in decision making structures; and sustainability is ensured when future generations have access to the social and material conditions, at least at the same level as the present generation [94,95]. Utopian principles allow us to understand whether we are moving in the right direction, as well as the opportunity to determine how to increase the capacity for transformation towards our desired direction [96,97].

'Real Utopian' approaches consist of strategic alternatives existing within capitalism's niches from which varieties of non-market relations, egalitarian participation, democratic governance and collective action can take hold. Contemporary examples include Wikipedia, where anonymous collaborators contribute to the world's largest encyclopedia without compensation; the Quebec Social Economy, where day-care, elder-care and social housing services exist to meet community needs after deliberation through democratically elected councils; and Universal Basic Income (UBI) proposals that give every legal citizen income sufficient to live above the poverty line without any requirements related to work. Wright argues these strategies can be considered a broad socialist challenge to capitalism [97].

Under this framework, socialist inspired alternatives to capitalism, which adhere to Wright's three strategic 'logics of transformation', are shown in Table 1 below. 'Ruptural' transformations occur through a sharp break with existing institutions and social structures, 'interstitial' transformations build new modes of social empowerment in capitalist society's niches and peripheries, where they often do not seem to present an immediate threat to prevailing power structures, and 'symbiotic' pathways deepen and extend institutional forms of social empowerment that involve the state and civil society simultaneously, whilst solving practical challenges aligned with the interests of the powerful. Where 'ruptural' transformations reflect dramatic revolutionary coups, 'interstitial' and 'symbiotic' pathways facilitate growth that remains unmolested by powerful actors and thus, opportunities to successfully grow for the benefit of those who have produced them [94].

Table 1. Three Strategic Transformation Logics derived from [98].

Transformation	Political Tradition	Collective Actors	Relationship to the State	Relationship to the Powerful	Metaphors of Success
Ruptural	Revolutionary socialist/ Communist	Class-aligned political alliances	Attack the state	Confrontational	War (victories)
Interstitial Metamorphosis	Anarchist	Self-organized groups and social movements	Build alternatives outside of the state	Exclusive engagement with alternative social relations	Competition for resources
Symbiotic Metamorphosis	Social democratic	Coalitions of social forces and labor	Use the state: struggle on the terrain of the state	Collaborate with the bourgeoisie	Evolutionary adaptations

Rather than defined steps, utopian thinking encourages mechanisms that facilitate a move toward the desired direction, spurred on by incremental everyday actions that allow for adaptation, improvisation, and amendment [94,99]. In this study, we draw on Wright's (2013) 'Real Utopias' framework for (1) methodological guidance in developing and assessing the viability of desired pathways with small-scale fishers, and (2) to analyze and evaluate the pathways against the moral principles of equality, democracy and sustainability.

1.8. Lambert's Bay Case

Lambert's Bay is situated along South Africa's Cape coast (Figure 1) and remains a prominent fishing community due to the variety of commercially viable species including Snoek, Southern Mullet (*Liza richardsonnii*), Cape Bream (*Pachymetopon blochii*), White Steenbras (*Lithognathus lithognathus*), Galjoen (*Dichitius capenis*) and Dusky Kob (*Argyrosomus japonicas*) [13]. Artisanal fishers of Lambert's Bay mostly belong to the population group designated as Colored, (the South African state's racial stratification recognizes four racial classifications: African/Black, Indian/Asian, Colored and White. This reflects the racial stratification of colonial South Africa, allowing the state to marginalize those designated as Black, Indian and Colored. The State has continued with these classifications in order to redress past injustices perpetrated during colonialism and continued under Apartheid) [99], speak Afrikaans, and make up 74.53% of the town's population. Lambert's Bay, as nearly all South African towns, remains geographically split along Apartheid's historic racial zoning, with the White population living in the town, and the Colored and Black populations in the surrounding lower income areas. Overall, 76.5% of households are without home internet access, and only 26.4% of adults had graduated from secondary school [100].

Inconsistent application of fisher rights have led to divisions where ITQs have created differences in income, asset ownership and household food security [101] amongst artisanal fishers. Further, ITQs were allocated to a more educated elite as well as those in faraway metropoles—neither of whom had any historic dependence on marine resource harvesting. Artisanal fishers excluded from rights allocations thus had their livelihoods effectively criminalized [102].

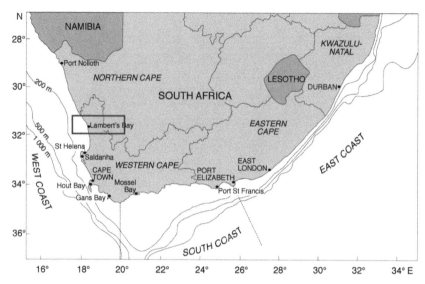

Figure 1. Lambert's Bay, South Africa (adapted from [103]).

2. Materials and Methods

This present study represents the ongoing research amongst the Lambert's Bay fisher community since 2013. Initial investigations focused on establishing household livelihood profiles of Lambert's Bay small-scale fisher families, exploring the role of quota allocations and the implications for small-scale fisher community dynamics [102]. Subsequent research (September 2014–March 2015) focused on co-design work conducted with support of the JUSTMAR Network (Global Marine Governance Network—Co-constructing a Sustainable Fisheries Future), financed both by the International Social Science Council, ISSC, and the Swedish Research Council/Swedish Research Links (VR/SR). JUSTMAR researchers delineated [104] fisher transformation pathways from Poland, Chile, South Africa and Vietnam. Whereas the JUSTMAR research mapped historical and current problems in the small-scale fisheries of the respective countries, as well as identifying possible pathways, the purpose of this workshop was to deepen the co-design of ICTs, where ICTs were increasingly seen as potential transformation pathways based on the future visioning of South African small-scale fishers. The intention was, moreover, not to envision and realize the 'ideal' but to work towards and facilitate conditions more in line with desired futures.

We held an evaluation and planning session in a community conferencing facility in Lambert's Bay, South Africa. We recruited participants through purposive sampling selection of Lambert's Bay fishers actively participating on the Abalobi platform, with a particular focus on skipper-boat owners, who were approached through telephone without the assistance of state agents or fisher associations. The focus on skipper-boat owners meant most participants were male and represented a relatively privileged group within the community both in terms of ownership over means of production and with regards to their association with the Abalobi project. While this represented a limitation to understanding power between diverse actors within the Lambert's Bay fishing community, our focus was primarily to explore the power dynamics between this particular group of fishers, middlemen and the State.

Though a larger number of fishers were expected to attend the workshop, some could not attend because the workshop coincided with particularly good fishing days. In total, six skipper-boat owners, 1 skipper, 1 fisherwoman and a local domestic violence counsellor participated. Our workshop was facilitated by two Abalobi project directors and four University researchers, of which two came from the University of Cape Town, South Africa and the remainder from Södertörn University,

Sweden, and primarily affiliated with JUSTMAR. All subjects gave their informed consent for inclusion before they participated in the workshop. The study was conducted in accordance with the Declaration of Helsinki, and the protocol was approved by the Ethics Committee of the University of Cape Town, South Africa.

Informed by the knowledge from prior JUSTMAR work, the researchers, fishers and NPO staff agreed in advance to exploring themes related to how ICTs can play a role in value chain participation, the associated challenges, opportunities and related governance concerns. Our format followed a mixed approach of plenary sessions and small group work where problems were presented with open discussion about solutions through presentations from fishers [105]. Our workshop's introduction highlighted how TURFS are widely seen as pathways for sustainable small-scale futures prominent in the fisheries of Vietnam and Chile. Our purpose was, thus, for South African small-scale fisheries to develop their own pathways, bearing in mind conflict, power and research constraints with a view to ultimately develop and enact selected ICT-based transformation strategies.

Though Lambert's Bay small-scale fishers primarily speak Afrikaans, our workshop was conducted in English, with translation assistance through a long term Lambert's Bay community member who was also an initiator of the Abalobi project. We conducted self-reporting strength and weakness exercises in order to understand a fisher's perceptions of challenges and opportunities within their practice. We subsequently explored future strategies, referencing how in Chile, fisherwomen developed a book as a means to raise their profile.

After mapping fisher problems in detail, participants were tasked with drawing up transformation pathways that would enable them to meet their value chain, upgrading goals through the ICT platform. This also required elucidating the roles and responsibilities of the various stakeholders towards fulfilling these opportunities, ensuring the viability, desirability and sustainability of a tangible action plan. In addition, our workshop sought to identify unintended consequences on both participating fishers and other actors. We took detailed field notes during the workshop and drew up fisher pathways on white boards in order to facilitate agreement on common issues. The outputs generated through the sessions directly informed the design of the Abalobi 'Marketplace' ICT module, which we subsequently piloted in two fishing communities.

Our data were coded for dominant themes continuously and iteratively [106,107] over a period of months, primarily guided by Real Utopian and Value Chain frameworks. Some constraints included the reality that fishers in different roles were not used to discussing issues openly together, for instance, crew and skippers, women amongst the men, and youth in front of their elders. The low scientific literacy levels may have led fishers to easily defer to those perceived to hold greater power or knowledge such as NPO staff and the researchers. However, the workshop approach facilitated in-depth discussion because of the relatively loose structure that left room for participants to grow in confidence.

3. Discussion

Through the workshop, fishers identified multiple short and longer term strategies to pursue, using the Abalobi ICT platform as presented in Table 2. It was apparent fishers connected the success of the Abalobi platform with the future subscription of all fishers in the Lambert's Bay community. Association with the Abalobi project represented a relatively privileged position amongst local fishers. This was evidenced through discussions on recruitment, where fishers involved with Abalobi saw themselves as singularly committed to fisher issues, and setting the terms of engagement for other fishers to join the platform:

Table 2. Fisher Strategy Workshop Outputs.

Workshop Output: Fisher Strategies
a. Recruit all fishers to the Abalobi ICT platform
b. Eliminate middleman
c. Use indigenous knowledge
d. Organize independently
e. Establish accident fund
f. Begin discussions with banks
g. Determine a salary structure
h. Keep commercial fleets out of inshore
i. Run local festivals and markets around fish
j. Use case studies to learn from other fishers

" . . . our fishers are also divided, the clique who are here [working with Abalobi] is [sic] very committed . . . but when we go to sea, there are no enemies." (FP1, young skipper, male, 03/10/2017)

"We want every fisher to be a part of Abalobi under our terms . . . if you don't want to listen to us then you must go." (FG1, skipper group, 03/10/2017)

Fishers further expressed the desire for the ICT platform to resolve issues related to eliminating the middleman in their value chains and developing local markets, participating effectively in fisheries management by leveraging indigenous knowledge, increased fisher self-organization, the establishment of an accident insurance fund, formal recognition by the banks, protecting their inshore fishing areas from commercial fleets, and learning from case studies in different country contexts.

The overarching challenge to the Lambert's Bay fishers was the ways in which their participation in value chains reinforced asymmetric power relations from which small-scale fishers derived little benefit. With fishers unable to renegotiate their terms of inclusion [39], deriving better returns from their value chains through strengthened bargaining positions thus guided our discussions around transformation pathways. Specifically, small-scale fishers identified the adverse nature of patron-client relationships as a key hindrance to receiving greater value for their harvests. In both their low value Snoek and high value Lobster chains, small-scale fishers derived the least benefit [13,39,108].

" . . . the main point is marketing, that's our biggest weakness in our fishing sector. With marketing, some middleman take our catch and go to Cape Town and spend its money there. And as I told you this type of period they come and let [sic] us some money and we must sign our debt by them [sic] and so it go." (FG1, skipper group, 03/10/2017)

"We are not organized, there is no organization to organize us. There is a caretaker who must speak with government for us . . . but the caretaker is also the marketing 'ou se boy' [sic]. There are no person to talk for us." (FG1, skipper group, 03/10/2017)

Although patron–client relationships have been noted to mediate fisher access in small-scale fisheries worldwide and though much derided, research is increasingly recognizing the ways in which middlemen in dense fisher networks may hold close kinship relationships with fishers and provide the operational capital for fishers to mobilize for the start of a new season [109–114]. Whilst this trend is observed across small-scale fisheries globally, fishers felt the South African state had effectively absolved themselves of the responsibility of supporting fisher marketing functions. While 'caretaker–marketers', as they are known in South Africa, provide loans to small-scale fishers, they also act with an impunity that may be detrimental to the economic welfare of fishers [39].

"Our relationship with the government, there are no relationship—the only relationship is with the caretaker and the caretaker take care of all of us." (FG1, skipper group, 03/10/2017)

Adversely included small-scale fisher value chain actors have overcome exploitation through the rise of local food movements that shorten small-scale fisher value chains by sourcing directly from fishers, bypassing the middleman and encouraging fisher post-harvest processing [48,49,115,116]. Despite the importance of effective governance to small-scale fisher value chains [13,117–120], small-scale fishers further bemoaned the lack of effective co-management structures through which they could exercise influence on policy.

Together with the Abalobi NPO, small-scale fishers synthesized three priority transformation pathways to pursue over the five-year period. First, fishers desired to strengthen independent fisher organizations to address the government's delay in implementing state supported fisher cooperatives and allocating their respective fishing rights. This was primarily in response to the delays in the implementation of the Small-scale Fisher Policy. Second, small-scale fishers desired to develop an alternative value chain that allowed fishers to participate more equitably by overcoming the asymmetric power relationships with other actors. Thirdly, fishers desired to collect fishing trip data through the Abalobi ICT platform in order to demonstrate their largely unrecognized contribution to South Africa's fishery sector, whilst also leveraging the data to advocate for greater participation in fisheries co-management.

"Abalobi gave us a much more way [sic] of thinking with what we are doing because now we can start monitoring our fishing, everything like that, and get it in the database . . . that is one of our strengths." (skipper, male, 03/10/2017)

"The indigenous knowledge is also a great strength because now we can use our knowledge, indigenous knowledge, somehow to be a part of Abalobi and there are some hope there . . . that's our strength in the room, we have a lot of knowledge here by us [sic]." (FG1, skipper group, 03/10/2017)

In order to support them more effectively, we examined these strategies through ICT and 'Real Utopian' lenses, relating them to other fisheries contexts in order to connect these actions of local fishers to global processes of anti-capitalist social change in small-scale fisheries.

4. ICTs, Real Utopias and Social Change in Small-Scale Fisheries

4.1. Addressing State Incapacity

Wright's definition of equality involves living a 'flourishing life' supported by grass-roots political mobilization and collective action [94]. Yet Lambert's Bay fishers endured a period of waiting due to State incapacity that only exacerbated their vulnerability from inequitable rights allocations. For instance, State-run fisher organization remained a laborious process, hamstrung by the necessity to ensure integrity of the verification and appeals process. Despite devolving verification to fisher communities, the extended delays and mistakes from the premature roll out of earlier years misallocated resources, and the subsequent delays continued to deny fishers access to state resources [7]. In this vacancy, fishers established formal and informal cooperatives to pursue alternatives to the state's incessant delays as a strategy of overcoming State incapacity

Fishers understood the potential of leveraging the Abalobi ICT platform towards greater independent mobilization, particularly in light of the perception the state had left them to the mercy of unscrupulous marketers. While the workshop represented one such arena to facilitate the actions of local fishers towards collective action, Abalobi-affiliated fishers further frequently met to resolve value chain challenges such as at whose house post-harvest processing should occur, how to compensate fish cleaners, as well as resolving perceptions of favoritism between fishers. In addition, the delays of the small-scale fisher policy implementation of community cooperatives left room for the growth of independent fisher organization, with the Abalobi platform operating as a de facto fisher cooperative, looking after the economic and political interests of small-scale fishers.

The wait of small-scale fishers for a policy that recognized them drew parallels with the wait for housing in South African urban spaces. In the urban space, this waiting shaped a politics of finding solutions in the grey spaces of informality and illegality [31]. Here, housing activists had sought sustainable alternatives to the hegemonic state discourse and housing policy towards inclusiveness for the homeless. While in the fisheries sector, the development of the policy marked citizens as legitimate wards of the state, their precarious existence throughout the policy's development and over the long term also required subversion of fisher regulations by poaching to support their livelihoods. Though South African state fisheries governance had scheduled completion of the verification process in 2020, this waiting had led fishers to start their own cooperatives.

In the urban housing literature, capacity was built through mobilizing local communities and demonstrating results through an alternative housing model [32], which highlights the importance for fishers of the viability of Abalobi's alternative value chain model. For small-scale fisher transformation pathways to embody equality, they will need to deliver on their promises promptly, where the perennial experiences of waiting result in exacerbated livelihood vulnerability.

The Abalobi ICT project facilitates collective action and mobilization by congregating small-scale fisher action toward value chain upgrading. Apart from the above 'co-design' workshops, Abalobi-affiliated small-scale fishers share marketplace information over 'Whatsapp' skipper groups, where this sort of collaborative skipper action is rare in Lambert's Bay. The Abalobi 'Fisher' module that captures logbook data from participating small-scale fisher trips further has the potential to be used for political mobilization through collective analysis of the data. Where development is understood as the increase in power parity [71], accurate small-scale fisheries data facilitates a pathway for fishers to leverage this important data towards gaining greater influence in fisheries management.

Collective action can enhance equality through the joint effort of pursuing political goals [91]. The Abalobi platform offers a space through which continued fisher cooperation can be leveraged towards achieving their political goals. A study in Nepal, linking ICT to fostering collective action, highlights the facilitative role of ICTs in generating and maintaining trust, acceptance, and alignment essential to cooperation through improved transparency and participation—where this collective action ultimately facilitates the expansion of individual freedoms [119].

4.2. Linking Actions of Local Fishers to Socio-Economic Outcomes

Post-Apartheid South Africa is marked by an increase in social inequality, particularly in the context of neo-liberal macroeconomic policies. Contesting neoliberalism has occurred across multiple civil society groups—here, we adopt understandings from South Africa's housing movement that draws important parallels for fisher struggles. Where Wright defines 'democracy' as meaningful participation, lessons from South African housing social movements highlight the inadequacy of basing democracy on free and fair elections, for example, mistaking institutional instruments with their democratic purpose [120]. The relationship between democracy and socio-economic rights remains complex, where too often liberal democracy is projected to lead to increased socio-economic rights, yet macroeconomic inequality has shown this to not be the case [29].

Participation processes need to result in real change. This includes more direct lines of accountability from local outcomes to local decision makers, as well as increased spaces from informal discussion concerning management issues [121]. These transformation pathways must, therefore, embody more than participation and transparency to also include socio-economic redress and gain in order to satisfy more comprehensive goals of democracy, especially in light of the ways in which democracy has not provided fishers with transformative socio-economic possibilities in post-Apartheid South Africa [8]. Though fisher rights are enshrined in a much-lauded Constitution, our experience in South African fisheries, as well as the literature, show they remain impoverished and marginalized with a significant schism between democracy and socio-economic redress.

One strategy to directly link democratic processes to socio-economic outcomes involved the Abalobi ICT platform facilitating local festivals and markets around their fish and developing a Restaurant Supported Fishery (RSF). While local fish markets are ubiquitous in fisheries globally, Apartheid laws denied small-scale fishers the opportunity to develop their own, creating highly centralized value chains under the control of a few large companies. Subsequent to the workshop, we thus established a small-scale fisher processing facility using the rudimentary facilities of a participating skipper. This processing facility conducted all the post-harvest fish processing of the new Abalobi-facilitated value chain, with fishers utilizing their traditional gutting methods and expertise. This decentralization of processing functions represented an example of increasing democratic freedoms with positive implications for socio-economic redress, as fishers were able to charge higher prices for their fish. Further, by facilitating small-scale fisher control of processing, fishers had the ability to bring to bear their local knowledge and cultural traditions in the post-harvest process. This further serves as a platform for fishers to leverage increased control of the post-harvesting process towards establishing independent local fish markets.

Transformative pathways linking democratic principles with real socio-economic benefits have been implemented in global fisher contexts, where in 2014, Scottish government reforms of Fixed Quota Allocations posited future scenarios for addressing moral economy issues in their fisheries. The Scottish government considered wellbeing above market precepts through six proposals: reducing fees on quota leases, facilitating youth entry in to fisheries, restricting quota allocation to active fishers, keeping quotas under Scottish ownership, keeping quotas within coastal communities, and supporting the traditions and practices of artisanal fishers [122]. These reforms can further serve as learnings and possibilities for South African fisher democratic ideals that place social and distributional objectives on equal importance as economic ones whilst recognizing coastal fishing communities, small-scale fisheries and communal traditions.

In considering transformation pathways, it is critical that small-scale fishers reflect on whether their existing practices contribute to economic welfare through operating in a democratic manner, allowing access and distributing that access justly. This is particularly important where communities have historically seen fractures brought on by the inequitable distribution of fishing rights. Fishers and the Abalobi staff will thus need to continuously reflect on how decisions are made and their implications—locating the processing facility in a particular skipper's home, for instance, resulted in perceptions of favoritism (pers. comm) requiring careful conflict management. Successful small-scale fisher-run processing facilities can serve as a rallying call for public support of their transformation pathways that demonstrate how fishers serve the public good by offering value and employment out of a publicly-owned resource, whilst also highlighting self-directed and self-supported fisher initiatives [123].

Case studies of Canadian Yukatat and Metlakatla fishing villages show how artisanal fisheries can be rebuilt on local and traditional knowledge, as well as incumbent subsistence technologies that open new opportunities within the existing neoliberal framework that retain local cultural traditions and support artisanal fisher identities. One of these strategies involves transferal of fish processing plant ownership rights to artisanal fishers who can then lease or operate the plant [123], which requires governments to provide subsidies towards value, adding capacity amongst artisanal fishers. These reforms also point to the importance of place-based identity and occupational stability, which South African fishers have struggled for through adhering to ideals of democracy and equality. Nonetheless, whether democracy and equality are conceptualized apart or highly integrated, fisheries governance needs to prioritize how fishers conceptualize traditional, moral and economic understandings of these ideals [124].

For small-scale fishers, democracy extends particularly to greater participation in the decisions that affect their livelihoods, such as state fisheries management decisions and policy prescriptions. Yet, democracy that has no direct benefit to the socio-economic improvement of livelihoods is inadequate. For South Africans and small-scale fishers for our purposes, the new democratic dispensation in 1994 and subsequent fisheries reform yielded no direct socio-economic benefit for small-scale fisheries. Rather, small-scale fishers remained marginalized and excluded from fisheries access under the neoliberalism-guided Marine Living Resource Act of 1998.

ICTs have the potential to facilitate democratic practice and improved socio-economic outcomes. By understanding development as increased economic productivity and power parity [71], ICT interventions should increasingly strive to pursue a welfare and fulfilment agenda beyond issues of access and resources [70]. Significantly, Abalobi's Marketplace platform offered fishers the opportunity to shorten their value chains by bypassing middlemen and selling directly to restaurants whilst receiving a higher price. Direct marketing programs in small-scale fisheries have shown to facilitate increased economic returns whilst increasing power parity between small-scale fishers and other value chain actors [46,49,115]. As demands on democracy are extended to meaningful participation in decision making, the urgency of concomitant economic benefits remain essential, particularly for fishers experiencing continued and historical marginalization.

4.3. Scaling Local Fisher Institutions

Reference [94] considers sustainability related to securing consumption of resources, as well as environmental conditions for future generations, at least at the same level as present, with scalable utopian institutions key to a sustainable future. Lessons from the alternative food movement highlight the importance of considering how transformation pathways can scale, as a key component of ensuring this sustainability. This involves exploring two related issues: the first concerns the capability for potential merging of multiple transformation pathways to facilitate innovative spillover and cross-fertilization—where one pathway may yield learnings for the growth of others. The second concerns the potential for transformation pathways to facilitate more interstitial space in which new pathways can grow, ultimately creating 'safe havens' of empowerment within existing institutional structures [125].

Fisher transformation pathways, such as Abalobi's Restaurant Supported Fishery value chain, needs to be flexible enough to navigate through the ways in which fishers will face resistance while leveraging the options to exploit resources within incumbent value chains. Transformation pathways begin with the risk of remaining just reactive alternatives to the mainstream, owing their dynamism to the 'oppositional status' created by the hegemonic neoliberal regime [126]. Fishers thus need to be careful to establish their transformation pathways beyond oppositional entities towards developing new governance modes based upon public priorities and fisher sovereignty [125].

Fisher desires to use the platform to engage with distant fishing communities was an example in which the ICT platform was leveraged to scale towards greater fisher collective action. Further, the 'Marketplace' module was piloted in two communities on opposite coasts of the Western Province of South Africa. Fishers used this collaboration to plan to offer diverse species on their future local markets, because certain species were exclusive to particular fishing areas. The increasing roll-out of the Abalobi platform across multiple communities, thus facilitates scaling of the platform with great potential for increasing the opportunities for the convergence of local movements, initiatives, ideas, and sustainable fishing practices.

In the alternative food movement, the twin crises of consumption and production and their impact on the urban consumer led to conditions that opened up more voids and spaces for new post-neoliberal institutional alternatives [126]. However, this conceptualization privileges the crisis of a wealthier consumer as the catalyst for change over that of the marginalized small producer who remains in a perennial state of vulnerability. Nonetheless, it is a useful way to understand how transformation pathways might develop; in South African small-scale fisheries the crisis amongst small-scale fishers has precipitated the development of alternative pathways more than a crisis of actors further along the value chain.

Convergence of food movements had been the key driver of food movements becoming major socio-political avenues for embedding and facilitating the transition to a post-neoliberal economy [126]. To leverage convergence, small-scale fishers of Lambert's Bay will need to develop closer relationships with other fisher communities and activities, a task made harder by the destructive role of ITQs.

However, in seeking convergence, fisher movements will need to be wary of appropriation and co-option, while remaining flexible enough to accommodate potentials for convergence [126,127]

Even where ICT projects are initially successful, when donor and state support is reduced or terminated, projects are prone to failure as demonstrated through a large study in India [71]. ICT- supported alternative pathways thus remain fundamentally tethered to the existence and scalability of the supporting ICT platform. Lessons from organizing within the social food movements show how convergence with other social movements may help solve the intractable existential crises of ICT projects whilst facilitating the maintenance of utopian alternatives [95,128]. ICT projects are commonly initiated by external partners with donor funding rather than by community members where the intervention occurs, reflecting power asymmetries related to resources. Consequently, money allotted to ICT projects seldom includes scope for exploring avenues for convergence with other ICT and analogue projects, which is crucial for sustainability, as lessons from alternative food movements indicate. Thus, explicitly pursuing areas of convergence with initiatives with similar aims might enhance the sustainability of utopian alternatives and ICT projects themselves.

5. Limitations

For small-scale fishers, individual access and subsequent use of marine resources is embedded in social relationships where neither 'property' nor 'rights' bear meaning other than as conditions imposed by external authorities [112], and where fishing practices continue to contribute to the unique cultural landscape of Lambert's Bay [13]. Thus, understanding their character and dynamics remains imperative for uncovering what matters to and for fishers, towards facilitating collective action [12], and the meanings of equality, democracy and sustainability as conceptualized by fishers.

These exchange relationships embedded in social and political conditions also determine the extent to which fishers capture a fair share of economic value of their resources which highlights the role of these relationships in disempowerment, exploitation and power inequalities [127]. Whilst emphasizing customary practices and traditions is essential for fisher transformation strategies, these practices may also represent and enable a tyranny of collective self-interest that encourage nepotism, ethnocentricity and gender inequalities—ultimately compromising the principles of transformative fisher pathways [128].

Though collecting small-scale fisheries data is critical to small-scale fisheries across the socio-ecological spectrum, it is often collected by non-state actors—raising questions of who generates the data, owns it, and has the skills and resources to analyze and use it. Power asymmetries exist inherently when non-state actors intervene in small-scale fisher communities. It is imperative ICT projects mitigate the ways in which power asymmetries within their projects undermine their stated goals. Small-scale fishers should not remain mere passive producers of data [69].

With co-design approaches at the forefront of greater stakeholder inclusion in ICT projects, project initiators must reflect on and actively mitigate the data inequalities present in their projects. This requires embedding participants in processes along the data chain beyond data generation, recognizing there are often different roles in different phases of ICT [129,130]. In pursuing greater power parity through ICT-based value chain upgrading, it would be a critical oversight if the Abalobi project were to neglect the power asymmetries embedded in ICT data chains.

For small-scale fishers, exploring new value chains supported through the Abalobi ICT platform necessarily requires redirecting catches that would ordinarily flow to the incumbent value chain and its associated actors. Where the strength of social networks are important determinants of benefit and credit flows in fisher communities [111,113], severing or even weakening the strength of these relationships may hold dire livelihood consequences, should the sustainability of the ICT-supported value chain be in financial peril.

Finally, Apartheid remains the central determinant of the life opportunities of South Africans from birth to death. The Apartheid government's inequitable distribution of resources remains stubbornly entrenched, affecting where people live, their education and subsequent career prospects, access to basic services and dignity. Lambert's Bay small-scale fishers remain marginalized, burdened with significant constraints for opportunities to lead flourishing lives. Directing social change where the trauma of the past continues to define the experiences of small-scale fishers poses complex and uncertain challenges for transformation pathways. It is imperative that fisher interventions remain attuned to this complexity where the real needs of fishers, including dignity, identity and cultural practice are not overlooked in favor of mainstream economic upgrading initiatives.

6. Conclusions

Whilst fisher participation is critical to ICT-related transformative small-scale fisheries interventions, it is imperative these pathways embody enhanced notions of democracy, equality and institutional and environmental sustainability. Through alternative small-scale fisher value chains, ICTs offered real material benefit for fishers whilst also facilitating democratic practices by leveraging fisher data towards increased power parity in management decisions. Early successes may help build confidence among fishers, authorities and market actors. This may be important to engender recognition that alternative pathways are indeed able to deliver on their promises. Lastly, sustaining transformative institutions requires ICTs to explore areas of convergence with other local fisher communities, particularly where fisher participation in transformation pathways necessitates a weakening of their long-held relationships with local stakeholders.

Author Contributions: S.R., F.S. and G.L.G.F. conceived and designed the project. F.S., G.L.G.F. and T.T.N. performed the analysis. T.T.N. wrote the paper with input from all authors. All authors have read and agreed to the published version of the manuscript.

Funding: This research was supported by the JUSTMAR Network (Global Marine Governance Network—Co-constructing a Sustainable Fisheries Future), financed both by the International Social Science Council, ISSC, and the Swedish Research Council/Swedish Research Links (VR/SR); and the APC was funded by the University of Cape Town, South Africa.

Conflicts of Interest: The authors declare no conflict of interest. The funders had no role in the design of the study; in the collection, analyses, or interpretation of data; in the writing of the manuscript, or in the decision to publish the results.

References

1. Béné, C.; Arthur, R.; Norbury, H.; Allison, E.H.; Beveridge, M.; Bush, S.; Campling, L.; Leschen, W.; Little, D.; Squires, D.; et al. Contribution of Fisheries and Aquaculture to Food Security and Poverty Reduction: Assessing the Current Evidence. *World Dev.* **2016**, *79*, 177–196. [CrossRef]
2. Grafeld, S.; Oleson, K.L.L.; Teneva, L.; Kittinger, J.N. Follow that fish: Uncovering the hidden blue economy in coral reef fisheries. *PLoS ONE* **2017**, *12*, e0182104. [CrossRef] [PubMed]
3. Bentley, N.; Stokes, K. Contrasting Paradigms for Fisheries Management Decision Making: How Well Do They Serve Data-Poor Fisheries? *Mar. Coast. Fish. Dyn. Manag. Ecosyst. Sci.* **2009**, *1*, 391–401. [CrossRef]
4. Geromont, H.F.; Butterworth, D.S. Generic management procedures for data-poor fisheries: Forecasting with few data. *ICES J. Mar. Sci.* **2015**, *72*, 251–261. [CrossRef]
5. Jeffers, V.F.; Humber, F.; Nohasiarivelo, T.; Botosoamananto, R.; Anderson, L.G. Trialling the use of smartphones as a tool to address gaps in small-scale fisheries catch data in southwest Madagascar. *Mar. Policy* **2019**, *99*, 267–274. [CrossRef]
6. Van Sittert, L.; Branch, G.; Hauck, M.; Sowman, M. Benchmarking the first decade of post-apartheid fisheries reform in South Africa. *Mar. Policy* **2006**, *30*, 96–110. [CrossRef]
7. Sowman, M. Subsistence and small-scale fisheries in South Africa: A ten-year review. *Mar. Policy* **2006**, *30*, 60–73. [CrossRef]
8. Isaacs, M. Small-scale fisheries reform: Expectations, hopes and dreams of "a better life for all". *Mar. Policy* **2006**, *30*, 51–59. [CrossRef]
9. Belhabib, D.; Sumaila, U.R.; Le Billon, P. The fisheries of Africa: Exploitation, policy, and maritime security trends. *Mar. Policy* **2019**, *101*, 80–92. [CrossRef]

10. Naranjo-Madrigal, H.; van Putten, I. The link between risk taking, fish catches, and social standing: Untangling a complex cultural landscape. *Mar. Policy* **2018**, *100*, 173–182. [CrossRef]

11. Antonova, A.S.; Rieser, A. Curating collapse: Performing maritime cultural heritage in Iceland's museums and tours. *Marit. Stud.* **2019**, *18*, 103–114. [CrossRef]

12. Henrich, J.; Boyd, R.; Bowles, S.; Camerer, C.; Fehr, E.; Gintis, H.; Mcelreath, R.; Alvard, M.; Barr, A.; Ensminger, J.; et al. "Economic man" in cross-cultural perspective: Behavioral experiments in 15 small-scale societies. *Behav. Brain Sci.* **2005**, *28*, 795–855. [CrossRef] [PubMed]

13. Isaacs, M. Small-scale fisheries governance and understanding the snoek (Thyrsites atun) supply chain in the ocean view fishing community, Western Cape, South Africa. *Ecol. Soc.* **2013**, *18*, 17–31. [CrossRef]

14. Eusebio, J.J. *A Research Framework on Value-Chain Analysis in Small Scale Fisheries*; Indiana University: Bloomington, IN, USA, 2004; pp. 1–27.

15. Jensen, R. The Digital Provide: Information (Technology), Market Performance, and Welfare in the South Indian Fisheries Sector. *Q. J. Econ.* **2007**, *122*, 879–924. [CrossRef]

16. Omar, S.; Chhachhar, A. A review on the roles of ICT tools towards the development of fishermen. *J. Basic Appl. Sci. Res.* **2012**, *2*, 9905–9911.

17. Srinivasan, J.; Burrell, J. *Revisiting the Fishers of Kerala, India*; Association for Computing Machinery: New York, NY, USA, 2013.

18. Aker, J.C.; Ghosh, I.; Burrell, J. The promise (and pitfalls) of ICT for agriculture initiatives. *Agric. Econ.* **2016**, *47*, 35–48. [CrossRef]

19. Riisgaard, L.; Bolwig, S.; Ponte, S.; Halberg, N.; Matose, F. Integrating Poverty and Environmental Concerns into Value-Chain Analysis: A Strategic Framework and Practical Guide. *Dev. Policy Rev.* **2010**, *28*, 195–216. [CrossRef]

20. Bush, S.R.; Oosterveer, P.; Bailey, M.; Mol, A.P.J. Sustainability governance of chains and networks: A review and future outlook. *J. Clean. Prod.* **2015**, *107*, 8–19. [CrossRef]

21. Bolwig, S.; Ponte, S.; Toit, A. Integrating Poverty and Environmental Concerns into Value-Chain Analysis: A Conceptual Framework. *Dev. Policy Rev.* **2010**, *28*, 173–194. [CrossRef]

22. Baudoin, M.-A.; Henly-Shepard, S.; Fernando, N.; Sitati, A.; Zommers, Z. From Top-Down to "Community-Centric" Approaches to Early Warning Systems: Exploring Pathways to Improve Disaster Risk Reduction Through Community Participation. *Int. J. Disaster Risk Sci.* **2016**, *7*, 163–174. [CrossRef]

23. Bremer, S.; Haque, M.M.; Haugen, A.S.; Kaiser, M. Inclusive governance of aquaculture value-chains: Co-producing sustainability standards for Bangladeshi shrimp and prawns. *Ocean Coast. Manag.* **2016**, *131*, 13–24. [CrossRef]

24. United Nations Development Programme. *Sustainable Development Goals|UNDP*; United Nations: New York, NY, USA, 2015.

25. Ntona, M.; Morgera, E. Connecting SDG 14 with the other Sustainable Development Goals through marine spatial planning. *Mar. Policy* **2018**, *93*, 214–222. [CrossRef]

26. Isaacs, M. Individual transferable quotas, poverty alleviation and challenges for small-country fisheries policy in South Africa. *MAST* **2011**, *10*, 63–84.

27. Isaacs, M.; Witbooi, E. Fisheries crime, human rights and small-scale fisheries in South Africa: A case of bigger fish to fry. *Mar. Policy* **2019**, *105*, 158–168. [CrossRef]

28. Sowman, M.; Cardoso, P.; Fielding, P.; Hauck, M.; Raemaekers, S.; Sunde, J.; Schultz, O. *Human Dimensions of Small-Scale Fisheries in the BCLME Region: An Overview Prepared by Prepared for Benguela Current Commission (BCC) and Food and Agricultural Organisation (FAO)*; Food and Agricultural Organisation of the United Nations: Rome, Italy, 2011.

29. Stokke, K.; Oldfield, S. Social Movements, Socio-economic Rights and Substantial Democratisation in South Africa. In *Politicising Democracy*; Harriss, J., Stokke, K., Törnquist, O., Eds.; Palgrave Macmillan: London, UK, 2005; pp. 127–147.

30. Friedman, S. The More Things Change . . . South Africa's Democracy and the Burden of the Past. *Soc. Res. Int. Q.* **2019**, *86*, 279–303.

31. Oldfield, S.; Greyling, S. Waiting for the state: A politics of housing in South Africa. *Environ. Plan. A* **2015**, *47*, 1100–1112. [CrossRef]

32. Millstein, M.; Oldfield, S.; Stokke, K. uTshani BuyaKhuluma-The Grass Speaks: The political space and capacity of the South African Homeless People's Federation. *Geoforum* **2003**, *34*, 457–568. [CrossRef]

33. Bénit-Gbaffou, C.; Oldfield, S. Accessing the State: Everyday Practices and Politics in the South. *J. Asian Afr. Stud.* **2011**, *46*, 445–453. [CrossRef]
34. Isaacs, M. Multi-stakeholder process of co-designing small-scale fisheries policy in South Africa. *Reg. Environ. Chang.* **2016**, *16*, 277–288. [CrossRef]
35. Witbooi, E. Subsistence Fishing in South Africa: Implementation of the Marine Living Resources Act. *Int. J. Mar. Coast. Law* **2002**, *17*, 431–440.
36. Ponte, S.; Sittert, L. Van The Chimera of redistribution in post-apartheid South Africa: "Black Economic Empowerment" (BEE) in Industrial Fisheries. *Afr. Aff.* **2007**, *106*, 437–462. [CrossRef]
37. Sowman, M.; Scott, D.; Green, L.J.F.; Hara, M.M.; Hauck, M.; Kirsten, K.; Paterson, B.; Raemaekers, S.; Jones, K.; Sunde, J.; et al. Shallow waters: Social science research in South Africa's marine environment. *Afr. J. Mar. Sci.* **2013**, *35*, 385–402. [CrossRef]
38. Department of Agriculture, Forestry and Fisheries. *Policy for the Small Scale Fisheries Sector in South Africa*; Department of Agriculture, Forestry, and Fisheries: Cape Town, South African, 2012.
39. Wentink, C.R.; Raemaekers, S.; Bush, S.R. Co-governance and upgrading in the South African small-scale fisheries value chain. *Marit. Stud.* **2017**, *16*, 5. [CrossRef]
40. Abalobi. About|ABALOBI. Available online: http://abalobi.info/about/ (accessed on 2 September 2017).
41. Srinivasan, J.; Burrell, J. Revisiting the Fishers of Kerala, India (Working Draft). In Proceedings of the Forthcoming at the International Conference on Information and Communication Technologies and Development, Cape Town, South Africa, 7–10 December 2013; pp. 56–66.
42. Jensen, R. Information, Efficiency and Welfare in Agricultural Markets. In Proceedings of the 27th International Association of Agricultural Economists Conference, Beijing, China, 16–22 August 2009; pp. 1–24.
43. Duncombe, R. Mobile Phones for Agricultural and Rural Development: A Literature Review and Suggestions for Future Research. *Eur. J. Dev. Res.* **2016**, *28*, 213–235. [CrossRef]
44. David, S.; Sabiescu, A.G.; Cantoni, L. Co-Design with Communities. A Reflection on the Literature. In Proceedings of the 7th International Development Informatics Association Conference, Lugano, Switzerland, 1–3 November 2013.
45. Smith, R.C.; Bossen, C.; Kanstrup, A.M. Participatory design in an era of participation. *CoDesign* **2017**, *13*, 65–69. [CrossRef]
46. Brinson, A.; Lee, M.-Y.; Rountree, B. Direct marketing strategies: The rise of community supported fishery programs. *Mar. Policy* **2011**, *35*, 542–548. [CrossRef]
47. Campbell, L.M.; Boucquey, N.N.; Stoll, J.; Coppola, H.; Smith, M.D. From Vegetable Box to Seafood Cooler: Applying the Community-Supported Agriculture Model to Fisheries. *Soc. Nat. Resour.* **2014**, *27*, 88–106. [CrossRef]
48. Bolton, A.E.; Dubik, B.A.; Stoll, J.S.; Basurto, X. Describing the diversity of community supported fishery programs in North America. *Mar. Policy* **2016**, *66*, 21–29. [CrossRef]
49. Witter, A.; Stoll, J. Participation and resistance: Alternative seafood marketing in a neoliberalera. *Mar. Policy* **2017**, *80*, 130–140. [CrossRef]
50. Bush, S.R.; Bailey, M.; van Zwieten, P.; Kochen, M.; Wiryawane, B.; Doddema, A.; Mangunsong, S.C. Private provision of public information in tuna fisheries. *Mar. Policy* **2017**, *77*, 130–135. [CrossRef]
51. Lewis, S.G.; Boyle, M. The Expanding Role of Traceability in Seafood: Tools and Key Initiatives. *J. Food Sci.* **2017**, *82*, A13–A21. [CrossRef] [PubMed]
52. Barclay, K.; Miller, A. The sustainable seafood movement is a Governance concert, with the audience playing a key role. *Sustainability* **2018**, *10*, 180. [CrossRef]
53. Morgera, E.; Ntona, M. Linking small-scale fisheries to international obligations on marine technology transfer. *Mar. Policy* **2018**, *93*, 295–306. [CrossRef]
54. Qureshi, S. Are we making a Better World with Information and Communication Technology for Development (ICT4D) Research? Findings from the Field and Theory Building. *Info. Tech. Dev.* **2015**, 511–522. [CrossRef]
55. Previero, M.; Gasalla, M.A. Mapping fishing grounds, resource and fleet patterns to enhance management units in data-poor fisheries: The case of snappers and groupers in the Abrolhos Bank coral-reefs (South Atlantic). *Ocean Coast. Manag.* **2018**, *154*, 83–95. [CrossRef]
56. Chuenpagdee, R.; Rocklin, D.; Bishop, D.; Hynes, M.; Greene, R.; Lorenzi, M.R.; Devillers, R. The global information system on small-scale fisheries (ISSF): A crowdsourced knowledge platform. *Mar. Policy* **2019**, *101*, 158–166. [CrossRef]

57. Hordyk, A.; Ono, K.; Valencia, S.; Loneragan, N.; Prince, J. A novel length-based empirical estimation method of spawning potential ratio (SPR), and tests of its performance, for small-scale, data-poor fisheries. *ICES J. Mar. Sci.* **2015**, *72*, 217–231. [CrossRef]

58. Pita, C.; Villasante, S.; Pascual-Fernández, J.J. Managing small-scale fisheries under data poor scenarios: Lessons from around the world. *Mar. Policy* **2019**, *101*, 154–157. [CrossRef]

59. Rubio-Cisneros, N.T.; Moreno-Báez, M.; Glover, J.; Rissolo, D.; Sáenz-Arroyo, A.; Götz, C.; Salas, S.; Andrews, A.; Marín, G.; Morales-Ojeda, S.; et al. Poor fisheries data, many fishers, and increasing tourism development: Interdisciplinary views on past and current small-scale fisheries exploitation on Holbox Island. *Mar. Policy* **2019**, *100*, 8–20. [CrossRef]

60. De la Barra, P.; Iribarne, O.; Narvarte, M. Combining fishers' perceptions, landings and an independent survey to evaluate trends in a swimming crab data-poor artisanal fishery. *Ocean Coast. Manag.* **2019**, *173*, 26–35. [CrossRef]

61. Goti-Aralucea, L. Assessing the social and economic impact of small scale fisheries management measures in a marine protected area with limited data. *Mar. Policy* **2019**, *101*, 246–256. [CrossRef]

62. Tallman, R.F.; Roux, M.-J.; Martin, Z.A. Governance and assessment of small-scale data-limited Arctic Charr fisheries using productivity-susceptibility analysis coupled with life history invariant models. *Mar. Policy* **2019**, *101*, 187–197. [CrossRef]

63. Fischer, J.; Jorgensen, J.; Josupeit, H.; Kalikoski, D.; Lucas, C.M. *Fishers' Knowledge and the Ecosystem Approach to Fisheries. Applications, Experiences and Lessons in Latin America*; Food and Agriculture Organisation of the United Nations: Rome, Italy, 2015.

64. Parsons, M.; Fisher, K.; Nalau, J. Alternative approaches to co-design: Insights from indigenous/academic research collaborations. *Curr. Opin. Environ. Sustain.* **2016**, *20*, 29–105. [CrossRef]

65. Barnes, M.L.; Mbaru, E.; Muthiga, N. Information access and knowledge exchange in co-managed coral reef fisheries. *Biol. Conserv.* **2019**, *238*, 108198. [CrossRef]

66. Rathwell, K.J.; Armitage, D.; Berkes, F. Bridging knowledge systems to enhance governance of the environmental commons: A typology of settings. *Int. J. Commons* **2015**, *9*, 851–880. [CrossRef]

67. Sabu, M.; Shaijumon, C.S.; Rajesh, R. Factors influencing the adoption of ICT tools in Kerala marine fisheries sector: An analytic hierarchy process approach. *Technol. Anal. Strateg. Manag.* **2018**, *30*, 866–880. [CrossRef]

68. Sein, M.K.; Thapa, D.; Hatakka, M.; Saebø, Ø. A holistic perspective on the theoretical foundations for ICT4D research. *Inf. Technol. Dev.* **2019**, *25*, 7–25. [CrossRef]

69. Brown, A.N.; Skelly, H.J. How Much Evidence Is There Really ? Mapping the Evidence Base for ICTD Interventions. *Inf. Technol. Int. Dev.* **2019**, *15*, 16–33.

70. Cinnamon, J. Data inequalities and why they matter for development. *Inf. Technol. Dev.* **2019**. [CrossRef]

71. Cibangu, S.K. Marginalization of indigenous voices in the information age: A case study of cell phones in the rural Congo. *Inf. Technol. Dev.* **2019**. [CrossRef]

72. Chipidza, W.; Leidner, D. A review of the ICT-enabled development literature: Towards a power parity theory of ICT4D. *J. Strateg. Inf. Syst.* **2019**, *28*, 145–174. [CrossRef]

73. Jacobs, C.; Rivett, U.; Chemisto, M. Developing capacity through co-design: The case of two municipalities in rural South Africa. *Inf. Technol. Dev.* **2018**, *25*, 204–226. [CrossRef]

74. Pinkerton, E. Hegemony and resistance: Disturbing patterns and hopeful signs in the impact of neoliberal policies on small-scale fisheries around the world. *Mar. Policy* **2017**, *80*, 1–9. [CrossRef]

75. Pinkerton, E.; Davis, R. Neoliberalism and the politics of enclosure in North American small-scale fisheries. *Mar. Policy* **2015**, *61*, 303–312. [CrossRef]

76. Bess, R. New Zealand's indigenous people and their claims to fisheries resources. *Mar. Policy* **2001**, *25*, 23–32. [CrossRef]

77. Olson, J. Understanding and contextualizing social impacts from the privatization of fisheries: An overview. *Ocean Coast. Manag.* **2011**, *54*, 353–363. [CrossRef]

78. Carothers, C. A survey of US halibut IFQ holders: Market participation, attitudes, and impacts. *Mar. Policy* **2013**, *38*, 515–522. [CrossRef]

79. Hardin, G. The Tragedy of the Commons. *Source Sci. New Ser.* **1968**, *162*, 1243–1248.

80. Mansfield, B. "Modern" industrial fisheries and the crisis of overfishing. In *Global Political Ecology*; Peet, R., Robbins, P., Watts, M., Eds.; Routledge: London, UK, 2010; pp. 98–113.

81. Barnett, A.J.; Messenger, R.A.; Wiber, M.G. Enacting and contesting neoliberalism in fisheries: The tragedy of commodifying lobster access rights in Southwest Nova Scotia. *Mar. Policy* **2017**, *80*, 60–68. [CrossRef]
82. Breslow, S.J. Accounting for neoliberalism: "Social drivers" in environmental management. *Mar. Policy* **2015**, *61*, 420–429. [CrossRef]
83. Davis, A.; Ruddle, K. Massaging the Misery: Recent Approaches to Fisheries Governance and the Betrayal of Small-Scale Fisheries. *Hum. Organ.* **2012**, *71*, 244–254. [CrossRef]
84. Said, A.; Tzanopoulos, J.; Macmillan, D. Bluefin tuna fishery policy in Malta: The plight of artisanal fishermen caught in the capitalist net Wildlife Trade View project Investigating tiger poaching in the Bangladesh Sundarbans View project Bluefin tuna fishery policy in Malta: The plight of art. *Mar. Policy* **2016**, *73*, 27–34. [CrossRef]
85. Gallardo-Fernández, G.L.; Saunders, F. "Before we asked for permission, now we only give notice": Women's entrance into artisanal fisheries in Chile. *Marit. Stud.* **2018**, *17*, 177–188. [CrossRef]
86. Peterson, M.N.; Von Essen, E.; Hansen, H.P.; Peterson, T.R. Illegal fishing and hunting as resistance to neoliberal colonialism. *Crimelaw. Soc. Chang.* **2017**, *67*, 401–413. [CrossRef]
87. Thornton, T.F.; Hebert, J. Neoliberal and neo-communal herring fisheries in Southeast Alaska: Reframing sustainability in marine ecosystems. *Mar. Policy* **2015**, *61*, 366–375. [CrossRef]
88. Gallardo Fernández, G.L.; Stotz, W.; Aburto, J.; Mondaca, C.; Vera, K. Emerging commons within artisanal fisheries. The Chilean territorial use rights in fisheries (TURFs) within a broader coastal landscape. *Int. J. Commons* **2011**, *5*, 459. [CrossRef]
89. Altieri, M.A.; Rojas, A. Ecological Impacts of Chile's Neoliberal Policies, with Special Emphasis on Agroecosystems. *Environ. Dev. Sustain.* **1999**, *1*, 55–72. [CrossRef]
90. St Martin, K. The Difference that Class Makes: Neoliberalization and Non-Capitalism in the Fishing Industry of New England. *Antipode* **2007**, *39*, 527–549. [CrossRef]
91. Altamirano-Jiménez, I. The sea is our bread: Interrupting green neoliberalism in Mexico. *Mar. Policy* **2017**, *80*, 28–34. [CrossRef]
92. Fung, A.; Wright, E.O. Deepening Democracy: Innovations in Empowered Participatory Governance. *Polit. Soc.* **2001**, *29*, 5–41. [CrossRef]
93. Emirbayer, M.; Noble, M. The peculiar convergence of Jeffrey Alexander and Erik Olin Wright. *Theory Soc.* **2013**, *42*, 617–645. [CrossRef]
94. Box, R.C. Progressive Utopias Marcuse, Rorty, and Wright. *Adm. Theory Prax.* **2012**, *34*, 60–84.
95. Wright, E.O. Transforming Capitalism through Real Utopias. *Ir. J. Sociol.* **2013**, *21*, 6–40. [CrossRef]
96. Brie, M. Review of Envisioning Real Utopias by Erik Olin Wright. *Int. Crit. Thought* **2011**, *1*, 462–468. [CrossRef]
97. Williamson, T. Emancipatory Politics, Emancipatory Political Science: On Erik Olin Wright's Envisioning Real Utopias. *New Polit. Sci.* **2012**, *34*, 386–395. [CrossRef]
98. Wright, E.O. Basic Income as a Socialist Project. *Basic Income Stud.* **2006**, *1*. [CrossRef]
99. Mccabe, C. Transforming capitalism through real utopias: A critical engagement. *Ir. J. Sociol.* **2013**, *21*, 51–61. [CrossRef]
100. Khalfani, K.A.; Zuberi, T. Racial classification and the modern census in South Africa, 1911–1996. *Race Soc.* **2001**, *4*, 161–176. [CrossRef]
101. Statistics South Africa. Statistics South Africa. Census. 2011. Available online: http://www.statssa.gov.za/?page_id=4286&id=14 (accessed on 31 May 2019).
102. Fabinyi, M.; Foale, S.; Macintyre, M. Managing inequality or managing stocks? An ethnographic perspective on the governance of small-scale fisheries. *Fish Fish.* **2015**, *16*, 471–485. [CrossRef]
103. Nthane, T.; Raemaekers, S.; Waldeck, N. New Policy Can Bring Unity to Lamberts Bay Fishers. Masifundise Development Trust, 20 May 2015. Available online: http://masifundise.org/new-policy-can-bring-unity-to-lamberts-bay-fishers/ (accessed on 31 May 2019).
104. FAO. *FAO Fishery Country Profile—The Republic of South Africa*; Food and Agricultural Organisation of the United Nations: Rome, Italy, 2003.
105. Saunders, F.P.; Gallardo-ferna, G.L.; Van Tuyen, T.; Raemaekers, S.; Marciniak, B.; Plá, R.D. Transformation of small-scale fisheries —Critical transdisciplinary challenges and possibilities. *Curr. Opin. Environ. Sustain.* **2016**, *20*, 26–31. [CrossRef]

106. Newing, H.S. *Conducting Research in Conservation: A Social Science Perspective*, 1st ed.; Newing, H.S., Ed.; Routledge: New York, NY, USA, 2011; ISBN 9780415457910.

107. Miles, M.B.; Huberman, A.M.; Saldaña, J. *Qualitative Data Analysis: A Methods Sourcebook*, 3rd ed.; Huberman, M., Miles, M.B., Saldanña, J., Eds.; SAGE Publications Ltd: New York, NY, USA, 2014; ISBN 148332379X.

108. Hara, M.M. Analysis of South African Commercial Traditional Linefish Snoek Value Chain. *Mar. Resour. Econ.* **2014**, *29*, 279–299. [CrossRef]

109. González-Mon, B.; Bodin, Ö.; Crona, B.; Nenadovic, M.; Basurto, X. Small-scale fish buyers' trade networks reveal diverse actor types and differential adaptive capacities. *Ecol. Econ.* **2019**, *164*, 106338. [CrossRef]

110. Crona, B.I.; Basurto, X.; Squires, D.; Gelcich, S.; Daw, T.M.; Khan, A.; Havice, E.; Chomo, V.; Troell, M.; Buchary, E.A.; et al. Towards a typology of interactions between small-scale fisheries and global seafood trade. *Mar. Policy* **2016**, *65*, 1–10. [CrossRef]

111. Kininmonth, S.; Crona, B.; Bodin, Ö.; Vaccaro, I.; Chapman, L.J.; Chapman, C.A. Microeconomic relationships between and among fishers and traders influence the ability to respond to social-ecological changes in a small-scale fishery. *Ecol. Soc.* **2017**, *22*. [CrossRef]

112. Wamukota, A.; Brewer, T.D.; Crona, B. Market integration and its relation to income distribution and inequality among fishers and traders: The case of two small-scale Kenyan reef fisheries. *Mar. Policy* **2014**, *48*, 93–101. [CrossRef]

113. Drury O'Neill, E.; Crona, B. Assistance networks in seafood trade – A means to assess benefit distribution in small-scale fisheries. *Mar. Policy* **2017**, *78*, 196–205. [CrossRef]

114. Drury O'Neill, E.; Crona, B.; Ferrer, A.J.G.; Pomeroy, R.; Jiddawi, N.S. Who benefits from seafood trade? A comparison of social and market structures in small-scale fisheries. *Ecol. Soc.* **2018**, *23*. [CrossRef]

115. Stoll, J.S.; Dubik, B.A.; Campbell, L.M. Local seafood: Rethinking the direct marketing paradigm. *Ecol. Soc.* **2015**, *20*. [CrossRef]

116. Godwin, S.C.; Francis, F.T.; Howard, B.R.; Malpica-Cruz, L.; Witter, A.L. Towards the economic viability of local seafood programs: Key features for the financial performance of community supported fisheries. *Mar. Policy* **2017**, *81*, 375–380. [CrossRef]

117. Hamilton-Hart, N.; Stringer, C. Upgrading and exploitation in the fishing industry: Contributions of value chain analysis. *Mar. Policy* **2016**, *63*, 166–171. [CrossRef]

118. Bair, J. Analysing economic organization: Embedded networks and global chains compared. *Econ. Soc.* **2008**, *37*, 339–364. [CrossRef]

119. McCay, B.J.; Micheli, F.; Ponce-Díaz, G.; Murray, G.; Shester, G.; Ramirez-Sanchez, S.; Weisman, W. Cooperatives, concessions, and co-management on the Pacific coast of Mexico. *Mar. Policy* **2014**, *44*, 49–59. [CrossRef]

120. Léopold, M.; Thébaud, O.; Charles, A. The dynamics of institutional innovation: Crafting co-management in small-scale fisheries through action research. *J. Environ. Manag.* **2019**, *237*, 187–199.

121. Thapa, D.; Sein, M.K.; Sæbø, Ø. Building collective capabilities through ICT in a mountain region of Nepal: Where social capital leads to collective action. *Inf. Technol. Dev.* **2012**, *18*, 5–22. [CrossRef]

122. Beetham, D. Conditions for democratic consolidation. *Rev. Afr. Polit. Econ.* **1994**, *21*, 157–172. [CrossRef]

123. Brewer, J.F.; Molton, K.; Alden, R.; Guenther, C. Accountability, transformative learning, and alternate futures for New England groundfish catch shares. *Mar. Policy* **2017**, *80*, 113–122. [CrossRef]

124. Pinkerton, E. The role of moral economy in two British Columbia fisheries: Confronting neoliberal policies. *Mar. Policy* **2015**, *61*, 410–419. [CrossRef]

125. Langdon, S.J. Foregone harvests and neoliberal policies: Creating opportunities for rural, small-scale, community-based fisheries in southern Alaskan coastal villages. *Mar. Policy* **2015**, *61*, 347–355. [CrossRef]

126. Hursh, D.W.; Henderson, J.A. Contesting global neoliberalism and creating alternative futures. *Discourse Stud. Cult. Polit. Educ.* **2011**, *32*, 171–185. [CrossRef]

127. Marsden, T.; Franklin, A. Replacing neoliberalism: Theoretical implications of the rise of local food movements. *Local Environ.* **2013**, *18*, 636–641. [CrossRef]

128. Adhuri, D.S.; Rachmawati, L.; Sofyanto, H.; Hamilton-Hart, N. Green market for small people: Markets and opportunities for upgrading in small-scale fisheries in Indonesia. *Mar. Policy* **2016**, *63*, 198–205. [CrossRef]

129. Ruddle, K.; Davis, A.; Zone, P.U.; District, H.D.; Noi, H.; Nam, V. Human rights and neo-liberalism in small-scale fisheries: Conjoined priorities and processes. *Mar. Policy* **2013**, *39*, 87–93. [CrossRef]

130. Sein, M.K.; Thapa, D.; Hatakka, M.; Saebo, O. *What Theories do We Need to Know to Conduct ICT4D Research?* SIG GlobDev: Dublin, Ireland, 2016.

sustainability

Brief Report

Spatial Distribution Patterns and Ethnobotanical Knowledge of Farmland Demarcation Tree Species: A Case Study in the Niyodo River Area, Japan

Yoshinori Tokuoka [1,*], Fukuhiro Yamasaki [2], Kenichiro Kimura [3], Kiyokazu Hashigoe [4] and Mitsunori Oka [5]

[1] Division of Biodiversity, Institute for Agro-Environmental Sciences, National Agriculture and Food Research Organization 3-1-3, Kannondai, Tsukuba, Ibaraki 305-8517, Japan
[2] Genetic Resources Center, National Agriculture and Food Research Organization, 2-1-2, Kannondai, Tsukuba, Ibaraki 305-0856, Japan; fukuhiro.y@affrc.go.jp
[3] Rural Development Division, Japan International Research Center for Agricultural Sciences, 1-1, Ohwashi, Tsukuba, Ibaraki 305-8686, Japan; cxx02377@affrc.go.jp
[4] 1-8-37-401, Iwasaki, Matsuyama, Ehime 790-0854, Japan; ji5ost.k-hashigoe@abelia.ocn.ne.jp
[5] Research Institute, Tokyo University of Agriculture, 1-1-1, Sakuragaoka, Setagaya, Tokyo 156-8502, Japan; oka.mitsunori@gmail.com
* Correspondence: tokuoka@affrc.go.jp

Received: 28 November 2019; Accepted: 27 December 2019; Published: 1 January 2020

Abstract: Isolated trees in farmlands serve various ecological functions, but their distribution patterns and planting history are often unknown. Here, we examined the spatial distribution, uses, and folk nomenclature of farmland demarcation trees planted in the Niyodo River area in Japan. Hierarchical clustering using the data from 33 locations distinguished four tree composition groups characterized by the combination of *Euonymus japonicus*, *Ligustrum obtusifolium*, *Deutzia crenata*, and *Celtis sinensis*. Near the upper to middle reaches of the river, the group characterized by *E. japonicus* dominated. Near the middle to lower reaches, the group characterized by *L. obtusifolium* occurred relatively frequently. The other two groups were found sporadically near the upper to lower reaches. The locally unique plant name *nezu*, used for *L. obtusifolium*, seems to have originated from a word meaning "the tree does not sleep and keeps the watch" in Japanese. In the study area, *D. crenata* was one of the plant species utilized for the sticks (*magozue*) used in traditional funeral ceremonies, which might help to explain why local people maintain *D. crenata* around homesteads as a demarcation tree. These findings highlight both the commonalities and uniqueness of demarcation tree culture in different regions of Japan and contribute to deepening our understanding of agricultural heritage.

Keywords: agricultural heritage; cultural landscape; folk nomenclature; floristic composition; traditional knowledge

1. Introduction

The expansion of intensive agriculture and farmland abandonment threatens traditional agricultural landscapes and the intangible knowledge assets supporting such landscapes [1,2]. The utilization of traditional landscapes and local knowledge in other forms, for instance, as a destination for tourists seeking traditional agricultural experiences, has been proposed as a means for their conservation. Identifying the indigenous cultural values in traditional agricultural landscapes is often difficult, however, because written historical documents are insufficient or lacking and no comparison among similar production landscapes has been performed.

Isolated trees in farmland serve multiple functions. For example, they support faunal biodiversity [3] and local food security [4], provide various collective goods [5], and maintain farmland demarcation. However, how the locally specific composition of isolated trees was historically shaped remains unknown in most regions, because of the lack of historical information. Under such limitations, ethnobiological approaches that analyze the ecological, botanical, and linguistic perceptions of local people can provide clues to the history of isolated trees in farmland landscapes.

In the Musashino region of eastern Japan, isolated *Deutzia crenata* shrubs were originally planted for farmland demarcation before the 17th century, and the species began to be used as a windbreak after settlement expansion [6]. Since then, *D. crenata* demarcation trees and windbreaks in the region were replaced with various beneficial plants used for hedgerow windbreaks, reflecting the changing needs of the local economy and the horticultural preferences of the local people. For example, the tea plant *Camellia sinensis* was planted from the mid-19th to early 20th century, and many gardening plants such as *Rhododendron* spp., *Euonymus japonicus*, and *Juniperus chinensis* were planted from the early to mid-20th century, accompanying the increasing number of florists in rural areas. In recent years, spatial distribution patterns and some planting background information on demarcation tree species were reported for upland fields in eastern Japan [7] and areas along the Hijikawa River in southwestern Japan [8]. However, because of the lack of comparison studies on farmland demarcation trees, how the findings from these localities are common or unique remains difficult to evaluate.

In this study, we investigated the spatial distribution patterns, folk nomenclature, and uses of farmland demarcation trees on the alluvial plains along the Niyodo River in the Kochi Prefecture of Shikoku Island, southwestern Japan. Based on the results, we discuss the commonality and uniqueness of demarcation tree culture at different localities and share ethnobotanical knowledge of specific tree species maintained around homesteads in Japan.

2. Material and Methods

2.1. Study Area

Farmland demarcation trees were investigated in areas along the Niyodo River in Kochi Prefecture (Figure 1). In the present and our past studies [7,8], "demarcation" was visually assessed at each study site depending on the kind of crops and the differences in soil and weed management conditions of adjacent farm parcels with isolated trees along their boundaries. Therefore, demarcation may not be directly relevant to the land ownership.

According to the climate data obtained at the Susaki weather station (1981–2010), the mean annual temperature was 16.7 °C and mean annual rainfall was 2604.3 mm. Upland fields and paddy fields predominate the alluvial plains, and steep mountains surround the river basin. Therefore, flooding has repeatedly damaged the farmland and residential areas for centuries [9].

2.2. Marker Sampling

We mostly followed the protocol of our previous study [8] for marker sampling and interviews. We recorded the demarcation trees on the alluvial plains between the Yokobatake area at the upper reach and the Harunochou Saibata area at the lower reach of the Niyodo River. The names of all of the districts investigated here were recorded in farm production statistics collected in the 16th century [10]. Depending on the size of the farmland area and the abundance of demarcation trees remaining at each village, we randomly selected one to three sampling points in each small village area. As an exception, however, the sampling point including densely resprouted and possibly older *D. crenata* individuals at the Imanari district was arbitrarily added, because we considered it important to discuss the diversity and traditional state of tree composition. At each selected sampling point, up to five farmland boundaries marked with isolated woody plants were explored and the species of each marker plant was identified. At a location in the Oochi area, only one individual of *Ligustrum obtusifolium* on a farmland boundary was recorded. This datum was omitted from the following analysis on tree composition. For the other sampling points, at least two farmland boundaries marked with demarcation trees were included in the analysis. Consequently, a total of 33 locations were surveyed from 22 August to 12 September 2019. In this paper, the plant nomenclature follows the YList [11] and the plant species origin and leaf types follow Miyawaki et al. [12].

Figure 1. Study site location, scenery, and examples of four demarcation trees. (**a**) Maps of the study site. (**b**) Demarcation trees in upland fields in the Imanari area, Ochi town. (**c**) *Euonymus japonicus*, (**d**) *Ligustrum obtusifolium*, (**e**) *Deutzia crenata*, and (**f**) *Celtis sinensis*.

Sustainability **2020**, *12*, 348

The longitude and latitude of the boundary markers were recorded by comparing their positions in the farmland with that on photographs available on the internet (cyberjapandata.gsi.go.jp) taken by the Geospatial Information Authority of Japan [13]. For the map created using Quantum GIS software version 2.18.16 [14], the mean longitude and latitude of surveyed markers around each chosen point were used as the representative point. The village section data were obtained from [15].

2.3. Interviews

Semi-structured interviews were conducted at 32 locations close to the marker sampling points. The informants were composed of 22 males, 8 females, and two couples in the following age groups: 50 s ($n = 1$), 60 s ($n = 6$), 70 s ($n = 17$), 80 s ($n = 6$), 90 s ($n = 1$), unknown ($n = 1$). Because the collection of information on the demarcation tree history was our priority, when we met several family members at a focal field, we chose the oldest informant (s) for interviewing. In each interview, we asked about the following items: the local name of the marker species that was present at the informant's or neighboring fields, the reason that species was chosen, multiple uses of plants, introducer and introduction period, means of planting, and management method. We also recorded the additional information provided during the interviews. Hereafter, the local plant names and folk habit are written in italic type.

3. Statistical Analysis

The compositional similarities of demarcation tree species among the 33 locations were compared by hierarchical clustering [16]. The matrix used in this analysis consists of 33 rows (locations) and 11 columns (marker species), and each cell is filled with the presence/absence of each species at each study location. Using this matrix, dissimilarity indices based on the Jaccard index were calculated and hierarchical clustering was performed with complete linkage. In this study area, the number of *E. japonicus* individuals was especially high from the upper to middle reaches, which made it difficult to examine the compositional variation characterized by some minor accompanying species when the abundance-based distance measures were used. Therefore, the Jaccard index was adopted. This analysis was conducted with the *vegan* and *gplots* packages using R software version 3.3.3 [17].

4. Results

4.1. Spatial Distribution of the Marker Plants

As shown in Table 1, we recorded a total of 250 individuals of 11 woody species (including formerly cultivated crop tree varieties of *Morus* sp.). The cluster analysis distinguished four marker composition groups at the minimum level of pruning of the dendrogram (Figure 2). Group 1 ($n = 5$) was characterized by the presence of *L. obtusifolium* and partly *Ce. sinensis* and the absence of *E. japonicus*. Group 2 ($n = 5$) was characterized by the presence of *D. crenata*. Group 3 ($n = 4$) was characterized by the co-presence of *L. obtusifolium*, *Ce. sinensis*, and *E. japonicus*. Group 4 ($n = 19$) was characterized by the presence of *E. japonicus* and the absence of most other species. Group 1 was located in areas along the right bank of the middle to lower reaches (Figure 3). Groups 2 and 3 occurred sporadically between the upper and lower reaches. Group 4 was located at the upper reaches and in areas along the left bank of the middle reaches.

Table 1. Observed number of individuals, local name, species origin, and morphology of demarcation tree species in areas along the Niyodo River.

Species	Observed Number of Individuals (% of Total)	Local Name (Response/Total Answers) [a]	Species Origin [b]	Morphology [c]
Euonymus japonicus	174 (69.6)	*masaki* * (12/18), DN (5/18), F (1/18)	N	E
Ligustrum obtusifolium	36 (14.4)	*nezu* (3/6), DN (2/6), *nezunoki* (1/6)	N	D
Deutzia crenata	13 (5.2)	DN (7/10), *tsuge* (1/10), *utsuge* (1/10), *utsugi* * (1/10)	N	D
Celtis sinensis	7 (2.8)	*enoki* * (2/2)	N	D
Morus sp.	6 (2.4)		A or N	D
Salix chaenomeloides	4 (1.6)		N	D
Cleyera japonica	4 (1.6)		N	E
Eurya emarginata	2 (0.8)	DN (1/1)	N	E
Euonymus sieboldianus	2 (0.8)	DN (1/1)	N	D
Gardenia jasminoides	1 (0.4)		N	E
Laurocerasus spinulosa	1 (0.4)		N	E

[a] Asterisk indicates the standard Japanese name of the plant species. DN indicates that the informant did not know the plant name. F indicates that the informant forgot the plant name; [b] A, alien species; N, native species; [c] D, deciduous; E, evergreen.

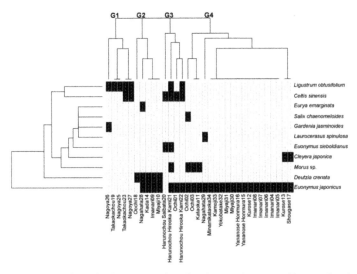

Figure 2. Dendrogram of woody plant species planted as demarcation trees at 33 survey locations along the Niyodo River, Kochi Prefecture. The four marker composition groups, which were determined by pruning this dendrogram at the minimum level of branching, correspond to the labeled numbers in Figure 3.

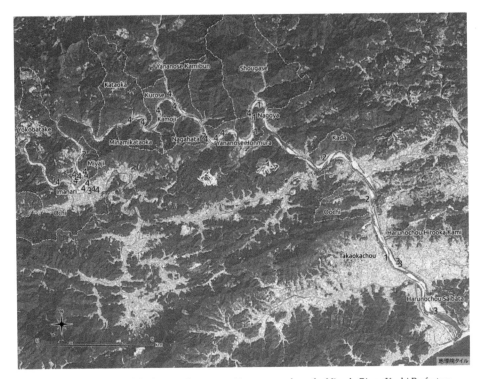

Figure 3. Distribution of the four marker composition groups along the Niyodo River, Kochi Prefecture. In the surveyed area, the Niyodo River follows a southeastern course. The numbers (1–4, highlighted in cyan) indicate the marker composition groups shown in Figure 2. Village names are highlighted in yellow. The dash-dotted yellow lines show the village boundaries. The recent aerial photograph was obtained from the Geospatial Information Authority of Japan (Chiriin-chizu; https://maps.gsi.go.jp/).

4.2. Folk Nomenclature

The folk nomenclature recorded for the four dominant species indicated simple naming systems, and for the most part scientific names of trees correspond well to a specific local name (e.g., *masaki* used for *E. japonicus*, *enoki* used for *Ce. sinensis*) or similar local names (e.g., *nezu* and *nezunoki* [*-noki* means tree in Japanese] used for *L. obtusifolium*). The origin and meaning of the name *tsuge* for *D. crenata* was unknown, but it may be a slight modification or misremembering of *utsuge*. According to a male informant at Takaokachou, the Japanese word *nezu* means "do not sleep and keep the watch" (*nezu-ni-ban-wo-suru*). Another male informant at Nagoya stated that there were at least two other species named *nezu*. These statements indicate that the planting practices for demarcation trees are linked strongly to the folk generic name *nezu* in some localities.

4.3. Introduction, Management, and Multiple Usage

According to the interviews, nine informants planted their demarcation trees using cuttings of *E. japonicus*. One informant at Nagahata planted *Eurya emarginata*, the source of which was unknown. A male informant at Takaokachou stated that the *Morus* individuals remaining on his neighboring farmland boundaries were left after the cessation of silkworm raising in the area and may not be demarcation trees. In the study region, sericulture drastically declined during WWII [18]. Except for these answers, 31 informants stated that they did not know the origin of old demarcation trees, because most of them were planted before they were born.

Some informants proposed the reason of species choice as the durability of trees against repetitive cutting in the cropping environments for *E. japonicus* (*n* = 3) and *L. obtusifolium* (*n* = 1). Another reason given was the ease of management because of slow growth and the size at maturity for *D. crenata* (*n* = 1) and *L. obtusifolium* (n = 1) and the rigorous resprouting ability of *D. crenata* (*n* = 1). The informants reported no strict season or way of cutting for the marker individuals of *E. japonicus* (*n* = 16), *D. crenata* (*n* = 8), *L. obtusifolium* (*n* = 5), *Ce. sinensis* (*n* = 2), *Euonymus sieboldianus* (*n* = 1), or *E. emarginata* (*n* = 1).

The shaded areas around *E. japonicus* (*n* = 2), *Ce. sinensis*, and *E. sieboldianus* (*n* = 1) trees were used for resting places. After WWII, local florists collected the branches of *E. japonicus* in the Imanari area (*n* = 1). Although not always acquired from demarcation trees, long and straight stems of *D. crenata* were used for the sticks, called *magozue*, which were used in traditional funeral ceremonies in the Nagahata area (*n* = 1). Bamboo species (*n* = 2, in the Nagahata and Minamikataoka areas) and the stems of a fern locally called *onishida* (*n* = 1 in the Miyaji area) were also used as *magozue*.

4.4. Marker Replacement and Removal

One informant in the Miyaji area stated that although, at present, many tree individuals used for demarcation are *E. japonicus*, there were more *D. crenata* individuals in the past. Some informants stated that some trees were removed because they became obstacles for machine plowing or agricultural production in the Imanari area (*n* = 2). Others noted that there used to be more individual demarcation trees in the past in the Yananose Honmura and Takaokachou areas (*n* = 2). Moreover, another informant in the Kada area stated that after public farmland surveys were conducted and artificial pillar markers were placed, there was no need to maintain the traditional tree markers.

5. Discussion

5.1. Commonality and Uniqueness of Tree Species Choice

A comparison of the present study with previous reports [6–8] shows that the dominance of *E. japonicus* and *D. crenata* in some districts and *Morus* individuals, perhaps left after their use for commercial sericulture, were common in those places. In contrast, the dominance or sub-dominance of *L. obtusifolium* was unique in the study region. Moreover, the dominance of *Chaenomeles speciosa* and the relatively frequent *Salix* usage (including a fiber crop variety *Salix koriyanagi* and wild species *Salix pierotii* and *Salix chaenomeloides*) along the Hijikawa River areas in Ozu city [8] and the dominance of *D. crenata* accompanying sub-dominant *Pourthiaea villosa* and the tea plant Ca. *sinensis* in some parts of the Ibaraki Prefecture, Eastern Japan [7] were also respectively unique. Although the reason why such distinct variation of dominant tree abundance was observed among those places remains unclear, the use of *Morus*, *S. koriyanagi*, and Ca. *sinensis* highlights that maintaining the former crops for demarcation is a common behavior of local farmers.

5.2. Implications from Folk Nomenclature

The local name *nezu* or *nezunoki* and its etymological meaning ("the trees do not sleep and keep the watch") for *L. obtusifolium* seem indigenous. The statement of folk generic use of *nezu* for several different species also highlights that the wording reflects the importance of demarcation tree planting along the right bank of the middle to lower reaches of the Niyodo River. In this study region, many local people (12 out of 18 informants, Table 1) recognized the name of *E. japonicus* as *masaki*, which is also the scientific name in Japanese. In contrast, the local people living along the Hijikawa River hardly recognized the correct plant name and often called *E. japonicus boke*, which is the local and scientific name of another dominant demarcation tree species, *C. speciosa* [8]. These differing degrees of recognition of *E. japonicus* in the two relatively close localities indicate that, although the species used and landscapes are similar, their planting background seems independent of each other. As noted in previous ethnobiological research [19–22], the comparison of the folk nomenclature gathered in this study with that from previous works, such as [8], can deepen the ethnobiological understanding of local plant use.

5.3. Folk Plant Usages

Although the demarcation trees were seldom used for multiple purposes, some statements provided important insights. These examples were the use of *D. crenata* sticks at home funerals and the historical commercial use of *E. japonicus* in the florist trade. A previous report from Hidaka village in Kochi Prefecture [23] and our results show that using small sticks made of various plant materials, known as *magozue*, at home funerals was part of the traditional ritual in at least several villages in the Niyodo River basin. Similarly, a single stick of *D. crenata* was laid beside the dead body at home funerals in central Japan [24] and eastern Japan [7]. As frequently mentioned in the old Waka poems [25] and represented in the naming of the month *uzuki* in the Japanese calendar, *D. crenata* is a symbolic plant in Japanese culture. In addition to these features, the use of *D. crenata* sticks at home funerals may be another motivation for local people to maintain this species around their homesteads in wide areas across Japan. The use of *E. japonicus* by florists was also recorded in the Hijikawa area after WWII [8] and in eastern Japan before the war [6]. These ornamental needs in the early to mid-20th century and the ease of planting by cuttings may be additional reasons for the increase of *E. japonicus* markers in recent decades in different rural areas.

6. Conclusions

This study elucidated the unique tree composition and some background of the trees' planting based on folk nomenclature and usage in areas along the Niyodo River. Our findings shed light on the commonalities and uniqueness of demarcation tree culture across various landscapes. Moreover, this study provides insight into the ritual importance of *D. crenata* at home funerals across Japan, perhaps since olden times. However, local people in the study region have been gradually removing the demarcation trees because they are becoming obstacles as agricultural practices are modernized and the trees are being replaced with artificial pillars after public land surveying. Although the trees may have few practical merits in present-day food production and land management systems, the different composition of demarcation tree species and local peoples' intangible ethnobotanical knowledge of these trees in each landscape represent an invaluable cultural heritage and should be conserved. In further research, the regional variation of demarcation tree landscapes and their ethnobotanical history should be examined using similar multidisciplinary approaches. In such projects, the collection of information from local elderly informants must be prioritized, because such information is irretrievable after generational change in rural communities.

Author Contributions: Y.T. conceived of and designed the study and K.H. supported data collection. F.Y., K.K., and M.O. provided comments on local plant use and folk habits. All authors read and approved the final manuscript.

Funding: This study was supported by a Japan Society for the Promotion of Science KAKENHI Grant (18K05696).

Acknowledgments: We thank all the villagers who consented to be interviewed.

Conflicts of Interest: The authors declare that they have no competing interests.

References

1. Bouchenaki, M. The interdependency of the tangible and intangible cultural heritage. In Proceedings of the ICOMOS 14th General Assembly and Scientific Symposium, Victoria Falls, Zimbabwe, 27–31 October 2003; pp. 1–5.
2. Daugstad, K.; Rønningen, K.; Skar, B. Agriculture as an upholder of cultural heritage? Conceptualizations and value judgements: A Norwegian perspective in international context. *J. Rural Stud.* **2006**, *22*, 67–81. [CrossRef]

3.   Fischer, J.; Lindenmayer, D.B. The conservation value of paddock trees for birds in a variegated landscape in southern New South Wales. 1. Species composition and site occupancy patterns. *Biodivers. Conserv.* **2002**, *11*, 807–832. [CrossRef]

4.   Vityakon, P. The traditional trees-in-paddy-fields agroecosystem of Northeast Thailand: Its potential for agroforestry development. *Reg. Dev. Dialogue* **1993**, *14*, 125–148.

5.   Dewees, P.A. Trees and farm boundaries: Farm forestry, land tenure and reform in Kenya. *Africa* **1995**, *65*, 217–235. [CrossRef]

6.   Yamamoto, R. Musashino-no-kaihatsu-to-koutiboufugaki-no-hattatsu [Agricultural land development and the history of windbreak in the Musashino region]. *Noukou-no-gijyutsu* **1981**, *4*, 1–24. (In Japanese)

7.   Tokuoka, Y.; Hosogi, D. Spatial distribution and management of isolated woody plants traditionally used as farmland boundary markers in Ibaraki Prefecture, Japan. *SpringerPlus* **2012**, *1*, 57. [CrossRef] [PubMed]

8.   Tokuoka, Y.; Yamasaki, F.; Kimura, K.; Hashigoe, K.; Oka, M. Tracing chronological shifts in farmland demarcation trees in southwestern Japan: Implications from species distribution patterns, folk nomenclature, and multiple usage. *J. Ethnobiol. Ethnomed.* **2019**, *15*, 21. [CrossRef] [PubMed]

9.   Ministry of Land, Infrastructure, Transport and Tourism. *Niyodogawasuikei-No-Ryuuiki-Oyobi-Kasen-No-Gaiyou [Basin and Rivers of Niyodo Riverrine System]*; Ministry of Land, Infrastructure, Transport and Tourism: Tokyo, Japan, 2007; pp. 1–88. (In Japanese)

10.   Shimonaka, K. (Ed.) *Kouchiken-No-Chimei [Place Names in Kochi Prefecture]*; Heibonsha: Tokyo, Japan, 1983; pp. 1–755. (In Japanese)

11.   Yonekura, K.; Kajita, T. *BG Plants Wamei-Gakumei (Japanese-Latin) Index (YList)*. 2003. Available online: https://http://ylist.info/ylist_simple_search.html (accessed on 31 December 2019).

12.   Miyawaki, A.; Okuda, S.; Fujiwara, R. *Handbook of Japanese Vegetation*; Shibundo: Tokyo, Japan, 1994; pp. 1–910. (In Japanese)

13.   Geospatial Information Authority of Japan. Chiriin-Chizu. 2019. Available online: https://maps.gsi.go.jp (accessed on 28 November 2019).

14.   QGIS Development Team. *QGIS Geographic Information System*; Open Source Geospatial Foundation Project: Chicago, IL, USA, 2016.

15.   Shobunsha. *Mapple Minisiparity Polygon Data: Rel. 12_1411*; Shobunsha Corp: Tokyo, Japan, 2015.

16.   Anderson, M.J. A new method for non-parametric multivariate analysis of variance. *Austral. Ecol.* **2001**, *26*, 32–46.

17.   R Development Core Team. *R: A Language and Environment for Statistical Computing*; R Foundation for Statistical Computing: Vienna, Austria, 2017.

18.   Haruno-Choushi-Hennsann-Iinnkai. *Haruno-Chousi [History of Haruno Town]*; Kochi-Insatsu Co., Ltd.: Kochi, India, 1971; pp. 1–780. (In Japanese)

19.   Berlin, B. *Ethnobiological Classification: Principles of Categorization of Plants and Animals in Traditional Societies*; Princeton University Press: Princeton, NJ, USA, 1992; pp. 1–335.

20.   Johnson, L.M.; Hunn, E.S. *Landscape Ethnoecology: Concepts of Biotic and Physical Space*; Berghahn Books: New York, NY, USA, 2010; pp. 1–319.

21.   Martin, G.J. *Ethnobotany: A People and Plants Conservation Manual*; Chapman and Hall: London, UK, 1995; pp. 1–268.

22.   Turner, N. *Ancient Pathways, Ancestral Knowledge: Ethnobotany and Ecological Wisdom of Indigenous Peoples of Northwestern North America*; McGill-Queen's University Press: Montreal, QC, Canada, 2015; pp. 1–554.

23.   Umeno, M. The realities and changes of funeral services in Owada, Hidawa village, Kochi Prefecture. *Bull. Nat. Mus. Jpn. Hist.* **2015**, *191*, 483–509. (In Japanese)

24.   Uehara, K. *Illustrated Encyclopedia of Trees*; Ariake Shobou: Tokyo, Japan, 1961; pp. 1–1203. (In Japanese)

25.   Hida, N. *Nihon Teien no Shokusaishi [History of Japanese Garden Plantings]*; Kyoto University Press: Kyoto, Japan, 2002; pp. 1–435. (In Japanese)

Article

What Difference Does Public Participation Make? An Alternative Futures Assessment Based on the Development Preferences for Cultural Landscape Corridor Planning in the Silk Roads Area, China

Haiyun Xu [1,*], Tobias Plieninger [2,3] , Guohan Zhao [1] and Jørgen Primdahl [1]

1   Department of Geosciences and Natural Resource Management, University of Copenhagen, 1958
    Frederiksberg, Denmark; gz@ign.ku.dk (G.Z.); jpr@ign.ku.dk (J.P.)
2   Faculty of Organic Agricultural Sciences, University of Kassel, 37213 Witzenhausen, Germany;
    plieninger@uni-kassel.de
3   Department of Agricultural Economics and Rural Development, University of Göttingen, 37073 Göttingen,
    Germany
*   Correspondence: hx@ign.ku.dk

Received: 12 October 2019; Accepted: 12 November 2019; Published: 19 November 2019

check for updates

Abstract: Landscape corridor planning (LCP) has become a widespread practice for promoting sustainable regional development. This highly complex planning process covers many policy and planning issues concerning the local landscape, and ideally involves the people who live in the area to be developed. In China, regional planners and administrators encourage the development of landscape corridor planning. However, the current LCP process rarely considers ideas from local residents, and public participation is not recognized as beneficial to planning outcomes. We use a specific Chinese case of LCP to analyze how citizen involvement may enrich sustainable spatial planning in respect to ideas considered and solutions developed. To this end, we compare a recently approved landscape corridor plan that was created without public participation with alternative solutions for the same landscape corridor, developed with the involvement of local residents. These alternatives were then evaluated by professional planners who had been involved in the initial planning process. We demonstrate concrete differences between planning solutions developed with and without public participation. Further, we show that collaborative processes can minimize spatial conflicts. Finally, we demonstrate that public participation does indeed contribute to innovations that could enrich the corridor plan that had been produced exclusively by the decision-makers. The paper closes with a discussion of difficulties that might accompany the involvement of local residents during sustainable LCP in China.

Keywords: cultural landscape corridor planning; participation; conflicts; development preferences; alternative future assessments; scenario planning

1. Introduction

Landscape corridor planning (LCP) has become a widespread practice in recent decades, partly driven by a collective shift toward protecting against fragmentation caused by infrastructure and other urban and agricultural development, and partly as a way to promote regional sustainable development and to enhance ecological and cultural values. LCP typically includes traditional approaches to landscape planning including surveys and an analysis of spatial patterns, functions, and changes, combined with various forms of stakeholder involvement in the plan-making process [1]. Throughout the paper, LCP is understood to be a form of spatial planning through which future developments of the landscape corridor in question are both envisioned and controlled through different measures. Land use regulations, management incentives, habitat restoration and tourist facility investments are examples of such measures. In the next section, we discuss the field of planning with reference to a traditional definition of planning and a more strategic form of spatial planning.

Landscape corridors, or "greenways" [2], are linear patterns that provide connectivity across landscapes and regions and are thus subject to public spatial planning processes. LCP represents a complex form of planning that includes an array of policy and planning issues and involves many types of expertise decision-makers [3,4]. Whereas early approaches to landscape corridor planning were based almost exclusively on landscape analysis and judgements made by planning experts [5,6], these planning processes have become common practice in many countries.

Although the value and role of public participation in spatial planning have been discussed for a long time, the development and the current state of the art in terms of participatory approaches to planning vary significantly from country to country. In the West, there is a long tradition for public participation, whereas developing countries like China, participation represents a relatively new dimension of planning.

For instance, as early as the 1960s, Davidoff claimed that an inclusive planning process encourages a more democratic form of urban planning and management [7]. Since then, planning scholars have emphasized the increasing importance of public participation during regional plan-making. Over time, public participation came to be regarded as a means of reflecting democratic ideals within local planning and development [8–10], and multiple case studies have since indicated that broader public participation and collaboration can improve the process of planning and managing landscapes [11,12]. Despite these benefits, there are to our knowledge no studies of the concrete differences in planning solutions between planning with or without participation to indicate potential enrichments through participation, and there surely are no such studies with reference to China. From a planning solution point of view, participation appears to under-researched.

Despite this scholarly shortcoming, there been a long-standing debate about the value of public participation, including whether it is necessary or worthwhile to incorporate public input into decision-making [13]. For example, the process of collecting local resident feedback requires more resources because municipalities have to satisfy more stakeholders, resulting in more lengthy and costly planning and construction processes [14], or because the collaborative process sometimes could be deliberately designed to slow down environmental decision-making to favour the status quo instead of promoting the new project [15].

On this background, we will be focusing on how public participation may contribute to planning solutions in LCP using one specific case study: the landscape corridor plan in Zhangye Municipality located within the historic Silk Roads region in Western China. Our focus was the potential for the participation of local residents to enrich the overall planning content.

Landscape corridor planning in China has received increased policy attention, especially after the Silk Roads areas and the Great Canal were included as cultural routes in the list of UNESCO World Heritage Sites [16,17]. Several local and regional corridors have been or currently are being planned along these sites, and the typical planning approach is mainly based on expert analysis and judgement, supplemented only with the preferences of high-level municipal decision-makers. Here, public participation in the planning process has not been recognized as beneficial to planning outcomes in LCP.

Given this background, and with reference to the Zhangye planning case, we thus addressed the following questions: (1) To what extent does the planning content related to land use zoning and proposed major landscape projects differ from local stakeholder development preferences? and (2) What differences would have been made to the planning content if the planning process had included participation from local residents? (3) How do professional planners involved in the current plan perceive alternative plan solutions based on after-the-fact inputs from local residents?

2. Landscape Corridor Planning—Approaches to Analysis and Public Participation

Upon review of the dominant definitions of planning in the 1970s, Lundquist [18] suggested the following (translated from Swedish): "Planning is a future-oriented process through which the actor seeks control over the environment in order to be able to pursue his intentions." This definition is about pursuing the intentions of an "actor", and gaining control to make the future more certain. Landscape planning at that time was about surveying and analyzing the conditions and potential of a given landscape based on proposed planning solutions, which basically were about protecting what should not be changed, and locating new developments based on the best available areas and sites outside these protected areas. In Ian McHarg's book Design with Nature, such an approach formed the backbone of these highly innovative examples, showing how to integrate ecological dimensions into spatial planning for different types of landscapes and planning problems, and on different scales [5].

Since those early interpretations, this approach to spatial planning developed further in Europe and North America and has become a practice that includes stakeholders of various kinds. There are several reasons for this, including some related to politics and economics, but the bottom line is that over time, it has simply become increasingly clear that spatial planning and subsequent implementation does not function well without the involvement of key stakeholders [19,20]. Collaborations, co-design, and co-creations have become common terms for such approaches. According to Healey [21], strategic spatial planning may be understood as "a self-conscious collective effort to re-imagine a city, urban region or wider territory, and to translate the result into priorities for the area and strategic infrastructure investments, conservation measures, and principles of land use regulation."

Public participation and stakeholder involvement has also gained importance in landscape planning [22], including LCP [2]. It may, however, still be an exaggeration to claim that public participation is mainstream in LCP practice. In a review of LCP cases in Europe, it was found that public stakeholders and their approaches to involving public participation were rarely considered in current landscape corridor planning practices [1]. In traditional landscape planning, including LCP, various surveys and analyses of the landscape in question are carried out to support the implementation of the plan. Such work includes historical map-based analyses, sustainability analyses of various kinds, ecological analyses, several analyses of climate conditions, visual analyses, and many others [23]. Turner [24,25] has, for instance, criticized this Survey-Analysis-Design approach for being too mechanistic, and for being based on the assumption that, if done properly, such surveys and analyses may, more or less, automatically, lead to complete planning solutions. Turner terms it the SAD approach because it often leads to "sad results". Stiles [26], on the other hand, disagrees with this view, arguing instead that the SAD approach should not be abandoned because, despite its limitations, it nonetheless supports rational solutions and remains open to criticism.

Whereas it is difficult to imagine that an LCP process can be carried out without surveys and analyses, we can nevertheless agree with Turner that in practice, the SAD method may not be sufficient to deal efficiently or effectively with most landscape planning problems. It is basically based on expert judgements alone and, as such, may hamper collaboration on priority tasks and in creating design solutions.

3. Public Participation in Planning Process in China

Spatial planning in China has developed mainly as an expert-driven technical process that focuses on social, economic, and environment objectives formulated in advance by governmental bodies, as well as on the physical organization of space as understood and designed mainly by professional planners [27]. Even though balancing different demands of public and private interests is a key element in spatial planning process [28], the current top-down planning process in China has a large effect on which public interests are given a voice in spatial planning, including LCP.

In China, although public participation in spatial planning is far from widespread, it is evolving, and a fast-growing body of literature arguing for more participation in Chinese planning is emerging. Still, more research about the effects of community participation, public hearings, social impact assessments, and user discussions in regional sustainable development are needed [29–31], as Chinese political and economic systems are different from some modelled Western capitalist systems. It is therefore likely that including public involvement in Chinese landscape planning practices will evolve in other ways. These differences are due largely to weakly developed state-civil society relationships and environmental legislation [30].

According to Van der Ploeg, ignoring local resident interests on local resources and land use can lead to conflicts around rural development in China [32]. Conflicts in the past decades have been mainly due to the lack of a clear set of common interests among the highly diverse populations of rural China [33]. These conflicts are thus caused by diverse interests among differing groups of people, a classic situation in planning everywhere in the world. For instance, residents, the committees of each village, and the municipality would have different interests in development directions for local land use and rural development planning. In this context, the idea of enabling local populations to present their common interests during planning and decision-making has therefore received increasing attention from Chinese authorities.

Given this history, our study analyzes how the involvement of public citizens in a Chinese context may enrich spatial planning with respect to ideas and solutions. We accomplish this by selecting a recently approved landscape corridor plan, investigating how alternative plan solutions could have been developed through local resident involvement, and further reviewing these alternatives with a group of professional planners who were involved in the original planning process.

4. Material and Methods

4.1. Study Area and Current Corridor Plan

A Silk Roads cultural landscape corridor plan in a suburb area of Zhangye Municipality (Figure 1) was used to analyze a local, recently approved landscape corridor plan, and to investigate the implications of public participation for innovating planning content. The corridor plan was first proposed by the local municipality as a part of a conservation project along the Silk Roads region in Western China in 2015. Since the Silk Roads were added onto the list of World Heritage Sites in 2014, the municipalities located in these regions began to promote various local cultural landscape conservation and development projects. There exists as well an overall regional plan that aims to protect local natural resources and heritage sites, improve the local agricultural food production industry system, and promote regional cultural and ecological tourism development. The regional development of the whole Silk Roads cultural landscape corridor will be guided by a number of local corridor plans. The case analyzed in this paper is one such local plan.

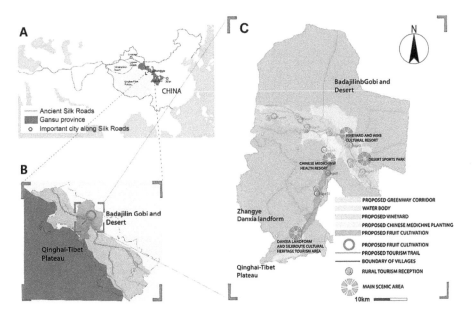

Figure 1. The location of the Zhangye cultural landscape corridor plan along the Silk Roads in Northwestern China. (**A**). General map of the Silk Roads area; (**B**). Our study area within the Silk Roads region, surrounded by the Badajilin Desert in the north and the Qinghai-Tibet Plateau in the South; (**C**). The current Zhangye cultural landscape corridor plan, including the proposed tourism trail and proposed development package projects (approved by local decision-makers).

Our study area is located in the narrow central part of the Heixi Corridor, historically known as the sole route used by ancient Silk Roads traders to pass between the northern Badajilin Desert and the Qinghai-Tibet Plateau in the South (Figure 1A,B). The existing Silk Roads linear heritage within Zhangye Municipality includes stretches of the Great Wall of China along with numerous temples, heritage trading sites, cave sculptures and other architectural remnants of historical civilizations and their early cultural exchanges with Europe.

In 2015, Zhangye municipality's corridor project started with the idea of building a recreational trail that would revitalize the region by connecting these Silk Roads cultural sites. Prior to plan implementation, pedestrian and tourist access to these cultural sites, or even to the scenic landscapes between them, was characterized by discontinuity. Hence, decision-makers from local municipalities began looking for a way to connect existing Silk Roads cultural resources while also enhancing opportunities for recreational and tourist activities via a connected corridor plan. Developed by China Agricultural University and with the support of local planning departments in 2017, one such corridor plan was then adopted by Zhangye municipality. The aim of that plan was to preserve and strengthen the existing landscape patterns and vegetation, while at the same time emphasizing the aforementioned historical and cultural areas. In short, the plan would combine the local agricultural and ecotourism industries for the betterment of regional sustainable development. With a positive outlook, this proposal for the cultural landscape corridor was approved. Implementation spanning the area between ten different villages began in Summer 2018. The current cultural landscape plan, presented in Figure 1C, includes: (1) a greenway corridor that contains a newly planted vineyard along the river and a recreational trail, as well as plans for a wine cultural resort; (2) a desert sports park; (3) a Chinese medicine garden; (4) a Chinese medicine-focused health and culture resort; (5) a rural recreation system with access to the river tributary, including access to eight tourism service centers, reception centers, and homestay dwellings along the recreation trail; (6) a fruit cultivation area; and (7) a central Danxia landform and Silk Roads cultural tourist area. The tourist area will be built in the pseudo-classic style of a tourism village, with the Silk Roads cultural features being the focus of the outside reception areas and the Danxia landform scenic area visible to the West.

4.2. Methods

Based on our analysis of the current plan, a two-step framework was designed to study the potential enrichments that could be gained by including local resident participation in building the final plan: (1) Exploring the conflict area through different development preferences between local residents and decision-makers; and (2) Assessing alternative future development through scenario planning. Our final study applied a variety of different methods and research techniques, including local resident participatory mapping surveys, spatial analyses (Kernel Density analysis), semi-structured interviews, group meetings and workshops, and scenario planning.

4.2.1. Conflict Areas with Different Development Preferences Between Local Stakeholders and Decision Makers

Different stakeholders—from farmers, to local tourist operators, village residents, political decision makers, or professional planners—are likely to have different perspectives and preferences concerning spatial development [19]. Conflicts may therefore appear during the planning process, specifically concerning the promotion and development of different land use areas. In this study, we explore such conflict areas by comparing elements of the current plan with the spatial implications we create based on local stakeholder preferences.

4.2.2. Building the Map of Local Resident Preferences

The spatial patterns that arose from an analysis of local resident views were built via a participatory mapping practice using local residents' development preferences. We performed this participatory mapping survey in January 2018, when the research team, with the help of the municipality's planning department, was introduced to the committees of ten villages along the Silk Roads corridor. Our researchers went to these 10 villages and informed local residents about our roles, the surveys, and our main aims thanks to the committees in each village. We then selected our participants randomly and independently from this committee sample. Finally, 20 participants from each village were invited by researchers, through snowball sampling ($n = 200$), to participate in small group interviews. We began these interviews by showing each two- or three-member group a satellite image of the village in question and a paper map covering the ten villages, thereby displaying current land cover, as well as the locations of current residential settlements. These steps were used to help respondents identify the locations of their familiar landmarks, as well as the location of each village within our study area.

Next, we used the corridor planning proposal to explain future area plans to participant-residents. Then, in response to our questions, these individuals were asked to mark which areas they thought were the most suitable toward the future planning of each kind of development preference, using questions like, "Which place do you think is most suitable for planning a desert sports park within this area?." Respondents' development preferences were then mapped using a pencil to draw points on the ten-village paper map mentioned above. To gain a deeper understanding of these preferences, our researchers then asked participants about the reasoning behind their chosen preferences.

From there, we digitalized the mapped points into ArcGIS (version 10.6, ESRI, Redlands, CA, USA) and used a Kernel Density analysis that together produced an accurate spatial distribution of our gathered data. We then applied the Kernel Density analysis again, this time performed by matching each point with its nearest neighbour using hierarchical clustering. Our active variables became those "hotspots" on our spatial pattern showing these participant data-points, revealing which areas attracted the most attention on our Kernel Density heat maps. Using this heat map, our team was thus able to gather a quantitative overview of the development preferences of our local participants.

4.2.3. Overlapping the Spatial Patterns of Local Resident Views with the Current Plan

The current corridor plan also contains a spatial outline that indicates the objectives and development directions of the Silk Roads area based on decision-makers' preferences. By overlapping the spatial patterns we had documented from local resident preferences with those of the current corridor plan, we could discover and further explore potential conflict areas between the two sets of development preferences. The layout of development projects on the current plan (e.g., vineyard corridor, desert sports park, rural tourism greenbelt) reveal the development preferences and patterns of the decision-makers involved in initial corridor planning. As such, overlapping each group's preferences also helped us analyze which areas had similar data stratification, showing us where the development preferences of each party are compatible, or similar. From there, we selected those villages where local residents had few or no development preferences, indicating the most profound conflict areas are between the two groups.

4.2.4. Alternative Futures Assessment Through Scenario Planning for the Conflict Area

Scenario planning is a method used for making long-term plans about the future, with the aim of fair decision and policymaking. This practice is widely used in geography, business, and politics [30] where scenarios are identified as a description of the possible future state of something, including event sequences that could lead from the current state of affairs to (or toward) a preferred future state [31]. We completed several alternative futures assessments through scenario planning for the conflict areas as mentioned in the above overlapping step. Based on the different development preferences of local residents and actual decision-maker, we constructed different scenarios to describe plausible alternative futures of our focus corridor planning area. While there is no single approach to scenario planning, and related literature shows various methodologies for building such scenarios [34,35], scenario planning generally emphasizes the following steps:

(1) Identify and establish the driving factors of change;
(2) Facilitate scenario-building and visualization;
(3) Elicit issues and qualify the impact of each scenario [36]

In this context, the existing corridor plan—with updated development directions—would be considered to represent the driving force behind changes to the local land. As a consequence, alternative futures based on either resident or decision-maker preferences were used as our scenario guidelines. It turned out that photorealistic visualizations based on land photos comprised an efficient tool for presenting planning metrics [37] (less-effective tools included GIS-based modelled landform surfaces, drawing, figures, etc.) [38,39].

Further exploration of scenario-planning itself showed that this method is especially useful for visualizing smaller-scale projects because it enables the presentation of details that are important for non-expert stakeholders [36,40]. With these considerations in mind, our team selected photorealistic visualizations as our primary tool for explaining and contextualizing our scenario discussions with non-expert, resident stakeholders. Following this visualization process, we then discussed our findings with these stakeholders during group meetings, using these mapped tools to best present the planning, construction, and function issues of each scenario, as well as their impact on local development.

Our study used a single, bird's eye view photograph rendered using Google Earth's satellite imagery, adding to it our data relating to those villages with development planning conflicts between local residents and decision-makers. Using Photoshop CS 6.0 (Adobe, San Jose, CA, USA), we further visualized our given scenarios by adding different layers, each containing different elements of our alternative future assessments to the original bird's-eye photograph.

During the process producing these photographs, significant new landscape elements of each alternative future were turned "on" or "off" depending on responses from the stakeholders' development preferences. Gradually these photographs were supplemented with detailed illustrations describing the full spectrum of activities and infrastructures each scenario represented. All of these materials helped the stakeholders to better understand the areas. We then used the two bird's eye photographs to discuss the scenario planning with different stakeholder groups, which includes three steps:

(1) Verification;
(2) Combination;
(3) Assessment.

First of all, we performed a verification process: each of the two bird's eye photographs presented an alternative future based on separate expressions of the two sets of development preferences (residents and decision-makers; scenarios 1 and 2, respectively). The alternative futures visualized in two additional scenarios were subsequently discussed with decision-makers as well as small groups of residents. Based on feedback from resident groups, we modified the then-current scenario 1 into one that was even more consistent with the residents' development preference. We made a new scenario 2 by performing the same process with decision-makers (Figure 2). All of this is a part of verifying that our constructed scenarios corresponded to the preferences of the two groups.

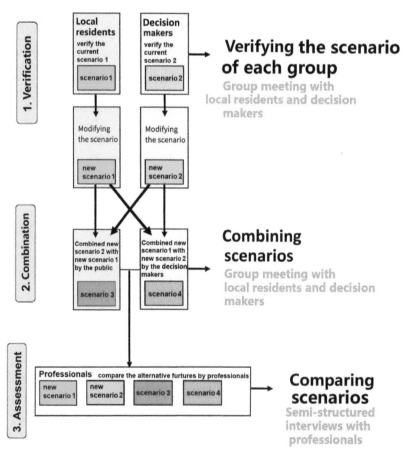

Figure 2. Framework of alternative futures assessment through scenario planning.

For step two, respondents were asked to discuss the differences that arose when combining the development preferences of both residents and decision-makers. We then modified and improved the new scenarios 1 and 2 to become scenarios 3 and 4 by combining these two different group opinions (Figure 2).

For step three, we invited five professionals who participated in the original Silk Roads corridor project to examine our four scenarios, as well as share their views as to the impact of each scenario. These conversations took the form of semi-structured interviews. The five professionals had already worked as a team together for one month on this project. All of them participated during the whole process, from fieldwork to final overall proposal (Figure 1C). However, they were not aware of details within the proposal, or specific impacts on specific villages. In this respect, they could be regarded as the outsiders (third parties) who were familiar enough to the case study are to assess the scenarios. We sent the photos of these scenarios to them one day before the interview and invited them for individual assessment interviews. Each interview lasted from 30–60 min and concerned how the individual professional valued each scenario, including how he or she ranked them in order of suitability degree for local development, and the reasons behind the ranking.

5. Results

5.1. Areas With Conflicting Development Preferences

5.1.1. Characteristics of Respondents

Our survey covered full- or part-time local residents from ten villages. In total, 200 respondents participated in the survey (53% male and 47% female). Sixty-nine percent were farmers, with the remaining residents being either technology staff (11%), administrators (6%), tourism service (5%), students (9%) or jobless (1%). Thirty-eight percent of respondents had been to high-school and 32 percent were under 30 years old, while 47 percent were between 30 and 60, and two percent were above 60 years old. The majority of our respondents had been settled in the local area of study for more than 20 years (73%).

The questions we asked directly concerned which land use developments local residents preferred. The most frequently supported development preferences were rural tourism and homestays (27%), followed by fruit cultivation (23%). The desert sports park idea (6%) is less attractive for local residents when compared with other development opportunities.

5.1.2. Spatial Patterns of Development Preferences

A Kernel density analysis of the development preferences of specific sites based on local resident perspectives is shown in Figure 3. Similar preferences for rural tourism and homestays were present across all villages within the corridor proposal coverage. Our heat map shows that the highest degree of clustering for given development preferences pertaining to the vineyard and wine cultural resort were common to the southeast part of the region (village 1, village 2). Support for the Danxia landform and Silk Roads cultural heritage tourist sites also clustered in these villages, along with village 3. For the Chinese medicine health resort and the desert sports park, support was relatively limited, and present only in villages 8 and 9.

A summary of the compatibility between the development preferences of local residents and decision-makers is as follows: Concerning rural tourism greenbelt planning, no major conflicts were found in nine of the ten included villages (1–9). Decision-makers and local residents shared a common preference for building a Chinese medicine-focused health resort and medicinal garden in villages 8 and 9, and showed a similar interest in developing a vineyard corridor and wine cultural resort in villages 2, 3, and 4. Based on the consensus of local residents and decision-makers, villages 8 and 9 were also considered as potential areas for the cultivation of fruit.

At the same time, there were a few significant differences between the development preferences of decision-makers and local residents. Most importantly, we found that residents frequently pointed out village 1 as the best area for developing a Danxia Landform and Silk Roads cultural tourism area, as it already had red cliff landforms and some Great Wall historical importance. At the same time, decision-makers in the real plan had considered a bare land south of village 10, because of its existing National Danxia Landform Geopark along the Silk Roads corridor.

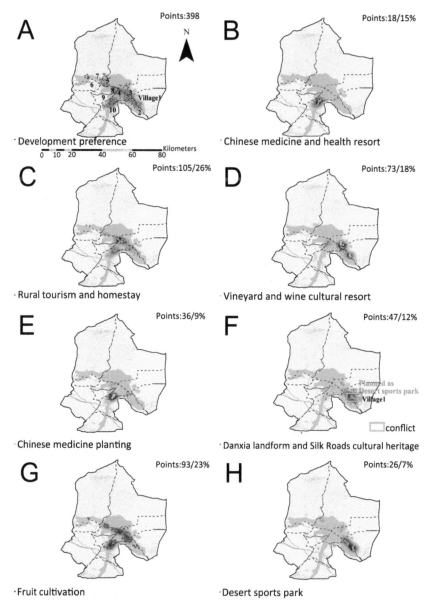

Figure 3. Kernel density heat maps of development preferences for local residents. (**A**) All kinds of development preferences; (**B**) Chinese medicine health resort; (**C**) Rural tourism; (**D**) Vineyard and wine cultural resort; (**E**) Chinese medicine planting; (**F**) Danxia landform and Silk Roads cultural tourism area (as the location of the planned desert sports park); (**G**) Fruit cultivation; and (**H**) Desert sports park. A higher density of points is visualized in dark purple, with lower densities in light purple.

To summarize, we narrowed down the major conflict areas to be within the development preferences of village 1—which included differing local resident preferences for the Danxia landform and Silk Roads cultural tourism areas, and different decision-maker preferences for the Desert sports park—and of village 5, which included differences in local preferences for fruit cultivation as well as in decision-makers' preferences limiting the development to a vineyard corridor. In this case, there are no 'hotspots' of development preferences for local residents in village 10, even though decision-makers have in fact planned the Danxia landform and Silk Roads cultural tourism areas there.

When we started this research, both the municipality and all local village committees had confirmed the proposed village 1 desert park. Meanwhile, village 5 merged with another village and a new village committee had been elected. To date, the vineyard corridor proposal has not yet been confirmed at the village level, and the new committee is considering changes. On this background we selected village 1, which has an unambiguous conflict between the preferences of local residents and current proposal (Figure 3F), for further analysis and assessment. Our goal was to work toward a better and more focused understanding of possible alternatives (or combinations of alternatives) during the next phase of our study.

5.2. Comparing the Differences between Alternative Futures through Scenario Planning

Within our assigned conflict area, we then made our detail bird's eye view photo of the planned sample area for scenario planning in village1, following different development guidelines (Figure 4A–C). The local original landscape includes cultivated land along the river, artificial surfaces, and bare land with red Danxia landforms (Figure 4B,C). The guidelines for scenario planning included (1) Local resident development preferences: Danxia landform and Silk Roads cultural tourism areas, and (2) Decision-makers' development preferences (the current plan): Desert sports park.

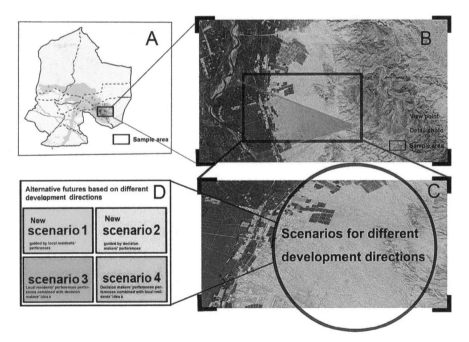

Figure 4. The four conflicting scenarios for village 1. (**A**) Conflict area in village 1; (**B**) Detailed view of the project area: all circular images in subsequent figures are visualizations of this original landscape; (**C**) The Enlarged area for scenario planning (**D**) Four scenarios based on different development directions.

After a thorough discussion with local residents and decision-makers during our scenario verification and combination process, the following four sections with subtitles describe the scenarios that we finalized as Figure 4. The scenario descriptions are followed by the results we collected based on local resident and decision-maker comments during group meetings. In addition, these scenarios reveal the presence of the planned sample area in relation to the current plan, in the event that one of the four development directions dominates after the Silk Roads cultural landscape corridor plan has been completed (Figure 4).

Finally, we created four scenarios of the future, each guided by a different set of development preferences (see Figures 5–8). For better presenting the differences of different scenarios; we enlarged the landscape change area as Figure 4C. These assumptions are of crucial importance toward an accurate presentation of the complex processes and alternative futures that shape landscapes in general, and this narrow corridor in the specific.

Figure 5. Scenario 1, guided by local resident development preferences. (The figure is a detail from the planning area of the scenario 1 poster shown to the local residents.).

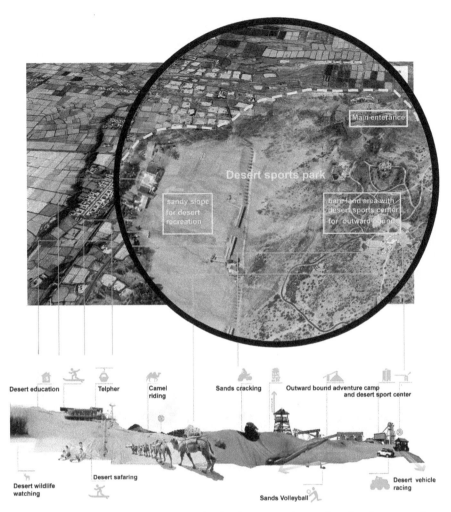

Figure 6. Scenario 2, guided by decision-maker development preferences based on the current plan (The figure enlarged the planning area of the poster of scenario 2 shown to the decision-makers).

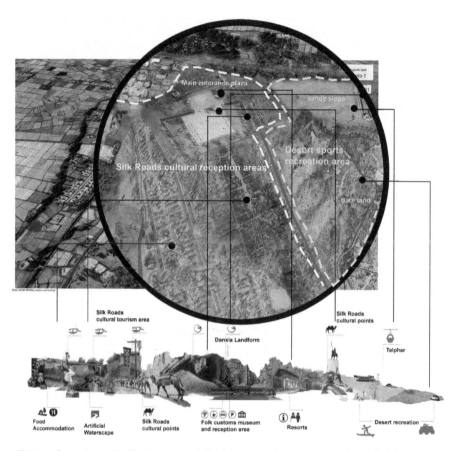

Figure 7. Scenario 3, guided by local resident development preferences combined with decision maker ideas (The figure enlarged the planning area of the poster of scenario 3 shown to the local residents).

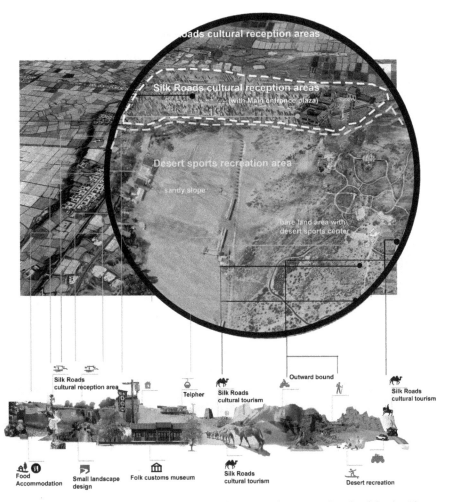

Figure 8. Scenario 4, guided by decision-maker development preferences combined with local resident ideas (The figure enlarged the planning area of the poster of scenario 2 shown to the decision-makers).

5.2.1. Scenario 1—A Scenario Guided by Local Residents' Development Preferences

In our first scenario, we assume that the area has been developed based exclusively on the preferences of local residents, meaning it has become a tourist area focusing on the Danxia landform and Silk Roads cultural histories. The plan mainly consists of a large central plaza, a recreational and residential district built in the historical style, and a temple with adequate open space. The entrance plaza includes a folk customs museum, a sculpture of a local legend and hero, and other supporting amenities like parking areas. The folk recreational and residential district includes various recreational and reception facilities such as shopping streets, homestay venues, and distinctive small hotels, restaurants, and bars; the recreational and residential districts are separated by a landscaped artificial water channel. At this future moment, planning continues for a Loong Temple at the end of the channel, complete with open space for local market days and religious festivals. In this scenario, this cultural landscape corridor project provides cultural education, recreation and tourism functions relating to the Silk Roads culture, experiences, and folk customs, all with local Danxia landform scenery as the background.

5.2.2. Scenario 2—The Scenario of Decision-Makers' Development Preferences

In the second scenario, which visualizes decision-makers' development preferences as reflected in the current plan, we see an area that has developed as a desert sports park attractive to tourists. According to the requirements outlined by decision-makers, the desert sports park consists of a main gate (with supporting facilities such as a parking area), a large-scale sandy slope for desert recreation, a large-scale bare land area as the desert sports center, an outward-bound camp, and a small garden to commemorate the ruins of a Silk Roads fire tower. The sandy slope area contains facilities for sandboarding, desert safaris, sand cracking, camel riding and a telpher. The bare land area includes hiking trails that facilitate the exploration of the Danxia landforms (with the roads and spaces also built for cross-country training and other events like mountain biking and off-road vehicle racing), a reception centre for desert sporting events, and an outward-bound camping area for outward-bound and desert adventurers. In this scenario, the desert park has been established as a unique tourist attraction along the regional cultural landscape corridor, and the developed area provides all the functions expected for cultural heritage conservation, desert recreation, hosting sports training and events, and professional outward-bound activities.

5.2.3. Scenario 3—A Scenario Guided By Local Residents' Development Preferences Combined with Decision-Makers' Ideas

In our third scenario, local resident development preferences are combined with existing decision-maker ideas, and we visualize a Danxia landform and Silk Roads cultural tourism area, complete with desert recreation facilities. The layout of the large center plaza, recreational and residential district is built in the historical style, and includes a temple with open space. For this alternative future, facilities have been added to the hillside for desert recreation, such as sandboarding on the sandy slopes, and a telpher connects tourist populations to the Danxia landform. In addition to the main function of cultural education, this scenario focuses on Silk Roads cultural recreation experiences and folk custom tourism, while also providing some desert recreation amenities for more adventurous travellers.

5.2.4. Scenario 4: A Scenario Guided By Decision Makers' Development Preferences Combined with Local Resident Ideas

In the fourth scenario, we assume that identified conflict areas have been developed according to decision-makers' development preferences paired with ideas from local residents. The area has therefore become a desert park with Silk Roads cultural reception areas near the main entrance. This scenario consists of a Silk Roads cultural reception area, an entrance plaza, a large-scale sandy slope area for desert recreation, a large-scale bare land area with an accompanying desert sports centre, an outward-bound camp, and a small garden to commemorate the ruins of a Silk Roads fire tower. The sandy slope and bare land areas provide similar functions as desert recreation amenities, as well as supporting facilities for desert hiking trails and outward-bound centres. The entrance plaza includes a folk customs museum and a main gate as historical features, while the reception area includes specialty restaurants and homestays that are located close to existing villages. In this scenario, this area acts as a functional source for cultural heritage conservation, cultural and folk experience-based tourism, desert recreation, sports training and recreation, and professional outward-bound activities.

5.3. Professionals' Responses to Possible Future Scenarios with and without Local Resident Participation

After scenario planning, the five professionals involved in the current corridor plan proposal included in our study were interviewed to provide an additional professional assessment of the four scenarios. The five professionals ranged ranging in age from 26 to 41, and two were female while three were male. Each professional came from a different (but relevant) educational background (geography, landscape architecture, urban planning, and architecture), and thus played a different role in developing the current corridor plan.

5.3.1. Differences between Scenarios

When the professionals compared each of the four scenarios, they discovered that any differences between them could be generalized as going in two main directions: on the one hand, scenarios 1 and 3 represented a direction toward cultural tourism, whereas scenarios 2 and 4 focused more on a recreational desert sports park. As a result, the professionals found that scenario 1 was the only of the four to have cultural characteristics mainly consisting of historical buildings. By contrast, these professionals felt scenario 3 included more desert-oriented recreation facilities than scenario 1, owing to the presence of the small-scale desert recreation area planned for that scenario.

Scenario 2, on the other hand, focused on the building of a desert sports park that would contain professional recreation facilities. The reception areas in scenario 2 were planned mainly for cross-country training and camping functions. Out of the four, scenario 4 shows a desert park that integrates characteristics of the Silk Roads culture into supporting facilities that can then be used for the leisurely reception of regular tourists. The professionals also noted differences between scenarios 3 and 4, where each combined the ideas of local residents and related decision-makers:

"When compared with scenario 3, scenario 4 did not include many of the 'unreasonable' design elements supported by residents. A large-scale shopping street feature, or a canal for water scenes would, for instance, be very difficult to maintain, and would affect the local environment. As these are both 'unreasonable' design elements for this area, scenario 4 instead developed an industrial output centered on the main desert resource." (Professional 2)

"Scenario 4 has better functional zoning than scenario 3. The functions in scenario 4 have a clearer focus on desert recreation, complemented by the supporting facilities containing local features, while the functional zoning in scenario 3 looks more piecemeal." (Professional 1)

5.3.2. The Most Suitable Scenario for Local Sustainable Development

The professionals were asked to rank the assessed potential alternative futures for local development in the four scenarios, as shown in Table 1 below. Scenario 4, guided by decision-maker development preferences, and combined with the desires and ideas stemming from local residents, was evaluated as the most positive alternative future for local development.

Table 1. Outcome of rankings and selection of the most suitable scenario.

Interviewee	Most Suitable Scenario	Scenario of Second Rank	Scenario of Third Rank	Scenario of Fourth Rank
Professional 1	Scenario 4	Scenario 2	Scenario 3	Scenario 1
Professional 2	Scenario 2	Scenario 4	Scenario 3	Scenario 1
Professional 3	Scenario 4	Scenario 3	Scenario 1	Scenario 2
Professional 4	Scenario 4	Scenario 1	Scenario 2	Scenario 3
Professional 5	Scenario 4	Scenario 2	Scenario 3	Scenario 1

Four of the five listed professionals found scenario 4 to be the most suitable development scenario, while professional 2 preferred scenario 2 and listed scenario 4 as the second-most suitable, in their opinion. From our report, these professionals also thought scenario 4 has the potential to combine local cultural features, folklore characteristics, and desert recreation, all unified by characteristic tourism features that support the main service areas and ultimately enrich the local tourism production chain. It should be noted that this scenario makes full use of local natural resources within a reasonable investment. As one of the professionals explained during the interview, for example:

"Scenario 4 has the most diverse facilities for regional development. I really like the idea of combining desert sports recreation with local Silk Roads cultural elements for the supporting facilities. For example, I like the idea of involving the 'Journey to the West' myth in the design of local resident

spaces as well as trails in the desert sports park. This scenario also highlighted the natural features of the desert and Danxia landforms. It could become a unique attraction in this corridor region among other places referred to by the current proposal. Compared with scenario 1 and 3, scenario 4 demands less investment. And I also like that its functional zoning design fully used local resources." (Professional 3)

Another professional found that:

"Scenario 4 is the most suitable scenario for me. Firstly, it has the most fully developed functionality for both the desert and the local Danxia landscape features. The desert recreation elements could create a unique tourist attraction for the benefit of the corridor region as a whole. In addition, the desert sports area could attract a strong, long-term source of visitors attending sports events such as hiking or cross-country vehicle racing. Moreover, there is a tourism-supporting service area that bears local cultural characteristics, combined with the desert recreation and folklore experience in this area. Tourists might stay longer where the amenities are best, which inflates local consumption growth and thus benefits residents. Last but not the least, this scenario has the most comprehensive functionality for its related economic investment. Compared with scenario 1 and 3, this scenario requires less construction work volume, while also fully covering the recreation amenities for both the cultural tourism and desert sports features at the same time. It also makes full use of local natural resources, for example the desert sports area in the desert landscape and the hiking trails located in the Danxia landform area. This scenario minimizes environmental consequences due to its more economical plan for construction work volume". (Professional 5)

One professional regarded scenario 2—"guided by decision-maker preferences in the current proposal"—as the most suitable direction for development. He states:

A desert park with a desert recreation theme as a priority service function could highlight the particular organic qualities of this area. This area has special natural resources, as well as Silk Roads cultural resources that can be found along the whole corridor region. This scenario has an intense purpose. (Professional 2)

Nevertheless, this professional evaluated scenario 4 positively:

"Scenario 4 requires the smallest construction work volume when compared with other scenarios, and provides space for potential changes in planning, all at the lowest cost."

5.3.3. The Scenarios with Negative Comments for Local Development

Overall, none of our professionals found scenario 1 (guided by local resident development preferences) or scenario 3 (guided by local resident development preferences combined with decision-maker ideas) to be very engaging. Each of the professionals mentioned that the high construction investment involved—not to mention the problems of undesirable competition between this proposed Silk Roads tourism town and other villages along the corridor—was the main reason that they ranked these scenarios poorly. As shown from the testimonies below, most felt that these scenarios are not realistic for creating sustainable development. One professional pointed out that a large initial investment in these historical buildings is not wise, as it could decrease much-needed room for future adjustments and thereby affect the flexibility the project has for future improvements.

"The scenario focusing on the Silk Roads cultural tourism area looks tedious. The scenario that combines elements of the Silk Roads cultural tourism area with some of the desert recreation facilities looks better; however, people would not be able to effectively use these facilities as the desert recreation area is quite small compared with other scenarios. In addition, due to the large scale of historical style buildings such as the folklore expo museum and the shopping and recreation streets, there is limited space for future adjustments if new projects demand development in this area. (Professional 1)

Another stated that

"The scenarios involving the Danxia landform and the Silk Roads cultural tourism area require a large amount of investment for building each block in a historical style. And there will be a Danxia landform-Silk Roads characteristic tourism area planned along the peripheral zones of the national geopark area of the Danxia landform in the current corridor plan anyway. The focus of developing a cultural tourism area in this village has no advantages in relation to the homogenous competition." (Professional 2)

A third professional summarized his view of scenarios 1 and 3 in this way:

"These scenarios will tend toward development in the direction of commercialization, and is likely because of other tourism-estate projects. There are already similar projects along the Silk Roads. Thus, these developments would not highlight the local features of the area. Besides, the 'cultural tourism area developing direction' requires more investment for building and maintaining properties and attractions, which places a high risk on the future of the whole operation." (Professional 5)

The interviews showed that four of the five professionals recognize that scenarios with different degrees of local resident involvement indeed facilitated new and innovative thinking into each applicable area's regional development when compared with the scenario 2 based on the current plan from important decision-makers alone. Each scenario that included the opinions and requests of local residents often brought in new and creative ideas not considered by current decision-maker plans—such as expanding the tourism market and connecting local residents. Our professionals explained this phenomenon as follows:

"The new scenarios enriched the planning content based on the development directions of the current plan. The current desert park provides recreational opportunities for desert sports and related adventures. It provides professional Outward-Bound facilities and has a clear targeted customer market. However, it does limit the average number of visitors that will explore this region.". (Professional 1)

"Compared with the current plan from the decision-makers alone, the other scenarios reminded us to add a theme to the desert sports park that is based on local, cultural resources. I like the idea of this region in village 1 mainly being focused on desert sports and recreation, with an integration of local Silk Roads cultural elements. Then we could make this desert park stand out from other desert recreation areas, using a unique theme. Additionally, the new ideas in other scenarios reminded us to improve the connections between the plan and local village representatives. The supporting area combined with local culture could provide board opportunities involving residents that benefit their life (such as the spaces for village markets and snack streets). This region could not only be a desert recreation destination or an outdoor training camp, but also a rustic luxury resort combining special Silk Roads cultural resources and desert scenery. " (Professional 4)

In sum, our professional respondents pointed out some general limitations of decision-makers and, as a result, these new scenarios can help related planners incorporate knowledge from all available resources during planning to understand the existing environmental local, and social expectations for land use. As one professional put it:

"The newer scenarios bring in some elements that might inspire us about the expectations of local residents. A decision maker's knowledge can sometimes be limited in this area, and even an expert's skills can vary. Thus, local resident views can provide a localized, contextual knowledge aiding in long-term development." (Professional 5)

6. Discussion

6.1. Scenarios Based on Development Preferences and Professional Planner Assessments of Local Sustainable Development

Our results indicate that the professional assessments of the different scenarios were largely in favour of combining local resident and decision-maker development preferences. In general, professionals found that many of the suggestions proposed by local residents were unrealistic because of cost—both in construction and in maintenance. As a result, scenario 1 (the scenario primarily guided by local resident preferences) received the lowest score, as ranked by professionals who offered more negative comments about this scenario in general. A major reason was that related facilities such as the artificial canal and the larger-scale historical buildings required correspondingly large investments. That said, four of the five professionals nevertheless found that local resident participation did add new and good ideas to each scenario. Consequently, professional respondents marked scenario 4 (the scenario where decision-makers' development preferences are combined with local ideas) as the most preferred alternative.

The arguments for giving scenario 4 the highest rank can be summarized in three points: (1) Scenario 4 has a more comprehensive functionality that combines both the cultural tourism and desert sports recreation aspects of the possible development plans; (2) Scenario 4 requires moderate investments as well as only limited construction volume; (3) Scenario 4 has better functional zoning solutions and more sustainable use of local natural resources and environmental opportunities.

Together, the four professionals asserted that scenario 1 (the scenario-based exclusively on local resident preferences) required the highest level of investment for building, and would also be the most expensive to maintain. There was an additional worry among the group regarding competition with other similar historical tourism centres already planned along the Silk Roads corridor region. The professionals found that scenario 2 (the scenario-based exclusively on decision-makers' development preferences) had some inherent limitations; in this case, the mono-functional land uses for the desert sports park in scenario 2 would mean more special desert sports functions than general recreation, which would appeal to a more limited group of tourists. For scenario 3, professionals expressed concern that, even though the planning of this scenario (local preferences combined with decision-makers' ideas) did, in fact, consider both the function of cultural tourism and desert recreation, the functional zoning plans in scenario 3 were unfortunately piecemeal and required higher construction investments. In short, the smaller scale of the desert recreation zones compared with scenario 4 would limit desert recreation activities, reduce service functions, and ultimately compromise tourists' overall experience of the town.

In general, scenarios 2 and 4 generally have higher ranks than scenarios 1 and 3. Professionals stated that their rankings were not influenced by their prior knowledge of the direction of desert sport park. In their collective view, a well-designed desert sports park indeed could be an attractive tourism feature. Further, in comparison with cultural tourism, they viewed the desert sport park as a novelty in the local area.

The fourth professional preferred scenario 2 to scenario 4, as he felt resident suggestions did not add value to the development and planning strategies. In the review of the professionals, professional number four also stated that their reason for ranking scenario 4 as the most suitable for sustainable development was because it has a clear focus on desert sports recreation, naturally highlighting the particular organic qualities of unique, local, and natural resources.

To sum up, even though our professional respondents thought that many of the suggestions from local residents were unrealistic, the majority of them nevertheless asserted that local resident participation contributed to a better planning solution when combined with current decision-maker plans (shown in scenario 1). For that reason, local resident participation has the potential to cement the success of an improved corridor-planning proposal for local sustainable development, benefitting both the local residents and the decision-makers involved.

6.2. The Difficulties of Involving Local Residents during Development and Planning Processes in China

In China, local urban and rural communities have already developed some participatory institutions [41], where currently a range of experiences with local residents' participation has been gained [29,42]. This shift could ultimately develop into an era of increased local resident participation in China.

While researching solutions, scholars have thus concluded that barriers to local residents participation and collaborative planning in China lie not only in its weak framework for environmental legislation, but also in part because of Chinese culture [30], the weakness of its planning system, and limited tradition of incorporating local resident values into planning practice [43,44]. Over the course of our study, we found that the existing gap between the differing views and education levels of our professional respondents, local decision-makers, and local residents also contributed to the difficulties of involving the general population during landscape corridor planning in China. We therefore discuss this condition from both a local resident and planner perspective as follows:

Our study used face-to-face interviews combined with local resident participatory GIS and group meetings for the purpose of communicating with local residents and collecting ideas and preferences concerning the future development of this part of the Silk Roads corridor. When local residents were asked about their preferences, we found that they more often had higher expectations of construction works, and that their development preferences were highly influenced by their general views of modernization and economic development on the whole (i.e., as seen via social media). For instance, the idea of developing a canal to provide artificial water scenery in the historical district from scenario one was an idea local resident participants mentioned they got from another cultural tourism town they saw on television. Clearly, local resident views of future developments cannot replace more trained, professional inputs on development opportunities and limitations.

During our communications process, the professionals saw themselves as key players, essential to the planning process and with the capacity to link existing resources with development opportunities and limitations. They were the experts with the technological skills necessary for the plan-making process and did not really recognize the need to explore, much less incorporate, local preferences and livelihoods during the process. As a result, their spatial and planning processes were viewed more as an exercise in technical engineering, than as a social activity that included collaboration with local stakeholders. This technical view of planning may explain why planners in China currently have such a low motivation to share their values and experiences with local residents.

Our data collection process revealed the current practical difficulties inherent in coordinating the steps it takes to consider local resident opinions during planning processes in China. These difficulties relate to the existing gaps between educated planners and "uneducated" residents, which can reduce a planners' motivation to incorporate local resident participation in their practices. To further develop sustainable participation practices, decision-makers must therefore address this gap, and local residents must on their side accept that not all their ideas and preferences are feasible from economic and technical points of view.

Educating planners, authorities, and residents is a prerequisite to more collaborative planning practices. Local resident participation in China may not only represent a better approach to formally allowing people to express their views, but it is also a process that promotes cooperation between planners and motivating them to share their experiences and knowledge with the local residents. This should also include raising local awareness about sustainable development options, strengths, and limitations. The planning process could then combine local resident opinion and high-level planning to better analyze the conditions of planning areas; better ascertain geological and socioeconomic strengths and weaknesses; and ultimately further explore development factors (i.e., opportunities and threats). The value of local resident participation is thus demonstrated as an enrichment mechanism for the current plan.

7. Conclusions

Landscape corridor planning is a complex process which requires close collaboration between various stakeholders to be efficient and innovative, concerning both the place-making and conflict-management dimensions of planning practice. In the current state of participatory landscape corridor planning, our study illustrates a practical way to involve public participation in LCP solutions that may also be applicable to other forms of spatial planning. Our study demonstrates the concrete differences that exist between planning solutions with and without local resident participation, thereby revealing how spatial conflicts might be reduced through collaborative processes. In addition, our results highlight the extent to which local resident participation indeed contributes to innovation, and professionals acknowledged that an enriched current corridor plan could contribute to local sustainable development in China. Further research is required to understand—and remove—the barriers to promoting local resident participation in Chinese spatial planning practice, including research on various forms of involvement, combined planning and educational processes, and ways to mobilize local knowledge and ideas during the planning process.

Author Contributions: Conceptualization, H.X.; Methodology, J.P., and T.P.; Software, H.X.; G.Z.; Validation, H.X. and J.P.; Investigation, H.X.; Data Curation, H.X.; Writing Original Draft Preparation, H.X. and J.P.; Writing Review & Editing, J.P. and T.P.; Visualization, H.X.; Supervision, J.P. and T.P.; Project Administration, H.X.

Funding: This paper is sponsored by the Chinese Scholar Council.

Acknowledgments: We are grateful to the great help and support of Prof. Xuesong Xi from China Agricultural University for providing the master plan and supporting the cooperation with the Municipality. The landscape architects and urban planners, Bingbing Zhang, Leng Gang, Xinya Bei, Yanli Tao also provide their input in this paper.

Conflicts of Interest: The authors declare no conflict of interest.

References

1. Xu, H.; Plieninger, T.; Primdahl, J. A Systematic Comparison of Cultural and Ecological Landscape Corridors in Europe. *Land* **2019**, *8*, 41. [CrossRef]
2. Fábos, J.G.; Ryan, R.L. International greenway planning: An introduction. *Landsc. Urban Plan.* **2004**, *2*, 143–146. [CrossRef]
3. Jongman, R.H.; Pungetti, G. *Ecological Networks and Greenways: Concept, Design, Implementation*; Cambridge University Press: Cambridge, UK, 2004.
4. Fábos, J.G. Greenway planning in the United States: Its origins and recent case studies. *Landsc. Urban Plan.* **2004**, *68*, 321–342. [CrossRef]
5. Ian McHarg, L. *Design with nature*; University of Pennsylvania: New York, NY, USA, 1969.
6. Lewis, P.H.J. Quality corridors for Wisconsin. *Landsc. Archit.* **1964**, *1*, 100–108.
7. Davidoff, P. Advocacy and pluralism in planning. *J. Am. Inst. Plan.* **1965**, *31*, 331–338. [CrossRef]
8. Rowe, G.; Frewer, L.J. Public participation methods: A framework for evaluation. *Sci. Technol. Hum. Values* **2000**, *25*, 3–29. [CrossRef]
9. West, S.E. Understanding participant and practitioner outcomes of environmental education. *Environ. Educ. Res.* **2015**, *21*, 45–60. [CrossRef]
10. Lafont, C. Deliberation, participation, and democratic legitimacy: Should deliberative mini-publics shape public policy? *J. Political Philos.* **2015**, *23*, 40–63. [CrossRef]
11. McLain, R.J.; Banis, D.; Todd, A.; Cerveny, L. Multiple methods of public engagement: Disaggregating socio-spatial data for environmental planning in western Washington, USA. *J. Environ. Manag.* **2017**, 61–74. [CrossRef]
12. Marzuki, A.; Hay, I.; James, J. Public participation shortcomings in tourism planning: The case of the Langkawi Islands, Malaysia. *J. Sust. Tour.* **2012**, *20*, 585–602. [CrossRef]
13. Irvin, R.A.; Stansbury, J. Citizen participation in decision making: Is it worth the effort? *Public Adm. Rev.* **2004**, *64*, 55–65. [CrossRef]
14. Rourke, F.E. *Bureaucracy, Politics, and Public Policy*; Little, Brown: Boston, MA, USA, 1984.

15. Echeverria, J.D. No success like failure: The Platte River collaborative watershed planning process. *Wm. Mary Envtl. L. Pol'y Rev.* **2000**, *25*, 559.

16. UNESCO. Silk Roads: The Routes Network of Chang'an-Tianshan Corridor. Available online: https://whc.unesco.org/en/list/1442 (accessed on 17 November 2019).

17. UNESCO. The Grand Canal. Available online: https://whc.unesco.org/en/list/1443 (accessed on 17 November 2019).

18. Lundquist, L. 'Some views on the concept of political planning', Political science. *Political Sci. J.* **1976**, *79*, 121–139.

19. Healey, P. Collaborative planning in a stakeholder society. *Town Plan. Rev.* **1998**, *69*, 1–21. [CrossRef]

20. Hall, P.; Tewdwr-Jones, M. *Urban and Regional Planning*; Routledge: London, UK, 2010.

21. Healey, P. The treatment of space and place in the new strategic spatial planning in Europe. In *Steuerung und Planung im Wandel*; Springer: Berlin/Heidelberg, Germany, 2004; p. 46.

22. Selman, P. Community participation in the planning and management of cultural landscapes. *J. Environ. Plan. Manag.* **2004**, *47*, 365–392. [CrossRef]

23. Stahlschmidt, P.; Swaffield, S.; Primdahl, J.; Nellemann, V. *Landscape Analysis: Investigating the Potentials of Space and Place*; Routledge: New York, NY, USA, 2017.

24. Turner, T. Pattern analysis. *Landsc. Des.* **1991**, *10*, 39–41.

25. Turner, T. *City as Landscape: A Post Post-Modern View of Design and Planning*; Taylor & Francis: London, UK, 2014.

26. Stiles, R. The limits of patterns analysis. *Landsc. Des.* **1992**, *11*, 51–52.

27. Shuwei, W.; Xiping, C.; Zhaowu, L. Reflection and Reconstruction of the Role of Planners from the Perspective of Urban Inclusive Development. *Planners* **2012**, *28*, 91–95.

28. Sager, T.Ø. *Reviving Critical Planning Theory: Dealing with Pressure, Neo-Liberalism, and Responsibility in Communicative Planning*; Routledge: London, UK, 2012.

29. Enserink, B.; Koppenjan, J. Public participation in China: Sustainable urbanization and governance. *Manag. Envir. Qual. Inter. J.* **2007**, *18*, 459–474. [CrossRef]

30. Tang, B.-s.; Wong, S.-w.; Lau, M.C.-h. Social impact assessment and public participation in China: A case study of land requisition in Guangzhou. *Environ. Impact Assess. Rev.* **2008**, *28*, 57–72. [CrossRef]

31. Li, W.; Liu, J.; Li, D. Getting their voices heard: Three cases of public participation in environmental protection in China. *J. Environ. Manag.* **2012**, *98*, 65–72. [CrossRef]

32. Van der Ploeg, J.D.; Jingzhong, Y.; Schneider, S. Rural development through the construction of new, nested, markets: Comparative perspectives from China, Brazil and the European Union. *J. Peasant Stud.* **2012**, *39*, 133–173. [CrossRef]

33. Jianrong, Y. Social conflict in rural China. *China Secur.* **2007**, *3*, 2–17.

34. Veeneklaas, F.; Van den Berg, L. Scenario building: Art, craft or just a fashionable whim? In Proceedings of the Scenario studies for the rural environment; selected and edited proceedings of the symposium Scenario Studies for the Rural Environment, Wageningen, The Netherlands, 12–15 September 1994; pp. 11–13.

35. Bradfield, R.; Wright, G.; Burt, G.; Cairns, G.; Van Der Heijden, K. The origins and evolution of scenario techniques in long range business planning. *Futures* **2005**, *37*, 795–812. [CrossRef]

36. Amer, M.; Daim, T.U.; Jetter, A. A review of scenario planning. *Futures* **2013**, *46*, 23–40. [CrossRef]

37. Tress, B.; Tress, G. Scenario visualisation for participatory landscape planning—A study from Denmark. *Landsc. Urban Plan.* **2003**, *64*, 161–178. [CrossRef]

38. Orland, B. Visualization techniques for incorporation in forest planning geographic information systems. *Landsc. Urban Plan.* **1994**, *30*, 83–97. [CrossRef]

39. Chermack, T.J.; Lynham, S.A.; Ruona, W.E. A review of scenario planning literature. *Futures Res. Quart.* **2001**, *17*, 7–32.

40. Jansson, A. *Investing in Natural Capital: The Ecological Economics Approach to Sustainability*; Island Press: Washington, DC, USA, 1994.

41. He, B. Participatory and deliberative institutions in China. In *The Search for Deliberative Democracy in China*; Springer: Berlin/Heidelberg, Germany, 2006; pp. 175–196.

42. Qiao, W. A study on public participation system in China' urban and rural planning. *Financ. Manag.* **2017**, *8*, 156–157.

43. Li, D.; Han, G. The historical reason for the lackage of public participation in urban planning in China. *Planners* **2005**, *21*, 12–15.

44. Zhao, Y.; Gao, S. Change and Prospect of Guiding Ideology and Policy Regime of China's Urban Planning in 60 Years. *Int. Urban Plan.* **2016**, *1*, 53–57.

Article

The Spatial Analysis and Sustainability of Rural Cultural Landscapes: Linpan Settlements in China's Chengdu Plain

Qiushan Li [1], Kabilijiang Wumaier [1,*] and Mikiko Ishikawa [2]

1   Institute for Disaster Management and Reconstruction, Sichuan University, Chengdu 610000, China
2   Department of Integrated Science and Engineering for Sustainable Society, Faculty of Science and
    Engineering, Chuo University, Tokyo 112-8551, Japan
*   Correspondence: kabil@scu.edu.cn; Tel.: +86-158-824-61371

Received: 15 July 2019; Accepted: 12 August 2019; Published: 16 August 2019

Abstract: Amid rapid urbanization and globalization, rural zones in many countries have undergone dramatic shifts. Although the future development of cultural landscapes is clear, their planning and management are uncertain. The Chengdu Plain is one of the most prosperous in China, which is home to well-developed irrigation and drainage systems, with the earliest history of planting found in Sichuan Province. The Chengdu Plain's unique farming landscape is an important human resource that represents the natural integration of the material and spiritual forms in the farming era. This study takes the unique farming settlements in Dujiangyan Irrigation District as the research object and analyzes the culture, human environment, and spatial order of the Linpan settlement based on the system theory. From the hierarchical structure of each individual Linpan settlement to the spatial layout of the Linpan community, the changes in the relationship between humans and land in the farming area are explored to explain the sustainability of the rural cultural landscape. With long-term field research, the rural geographic information database is built as a basis for the identification and classification of Linpan Cultural landscape types. The results show that between 2005–2018, the Linpan of Juyuan Town illustrated a decreasing trend, and about six Linpan settlements disappeared per square kilometer. The change in the type of Linpan landscape is spatially unbalanced, which is mainly due to the difference between regional development and residents' needs. This study introduces the concept of "demand" and "restriction" in sustainable development to explore a future strategy of maintaining the cultural landscape, which is expected to provide a basis for future policy formulation to protect the traditional rural landscape.

Keywords: sustainability; cultural landscape; Linpan; traditional settlement; spatial analysis

1. Introduction

Cultural landscapes serve as important material testimony to the interactions and sustainable development between humans and nature, as well as between history and the present; they comprise one of five major research themes in the field of cultural geography. The identification, description, and interpretation of landscapes have long been the primary work of geography [1,2]. The core concept of the cultural landscape is to emphasize the interaction between humans and nature. It considers the landscape to be a visible material phenomenon, emphasizing that the landscape styles created by different cultural groups are different [3]. In his 1927 article, "Recent Developments in Cultural Geography," the American geographer, Sauer, re-examines the research content on landscapes and the application of the morphological method. He defines cultural landscapes as "various forms of human activity attached to the natural landscape" [4]. Although it has been a well-established viewpoint that cultural landscapes be appreciated and inherited, people only began to pay attention to the loss of their cultural value at the start of the 20th century [5]. In the past few decades, throughout rapid urbanization and globalization, the changing trends of many landscapes have been considered negative because they have broken prevailing evolutionary laws, resulting in a loss of landscape diversity, continuity, integrity, and identity; sadly, these features of ancient cultural landscapes are rapidly disappearing [6]. The influence of large-scale urban construction on rural cultural land has drawn attention in numerous countries—not only in developing ones, but also in developed nations where rural cultural landscapes face serious threats. These include (1) wastelands, intensive agriculture, and population; (2) the loss of historical culture, local knowledge and traditional features; (3) the disordered expansion of landscape spaces and the fragmentation of rural areas; (4) the depletion of natural resources with a variety of demands and uncoordinated economic growth; and (5) the disturbance of natural ecological processes and the loss of biodiversity. These problems have raised global concerns regarding the development of rural areas [7–15].

The above issues have been outstanding in most rural parts of China's Chengdu Plain since the Wenchuan earthquake struck in 2008. In particular, due to the urgent need for repairing and rebuilding, the majority of affected rural zones need to be rehabilitated in a short span of time, which undoubtedly leads to the neglect of cultural values [16]. In the process of large-scale rural reconstruction, various challenges have surfaced, which has generated research on the part of government departments and scientific scholars. As a result, policymakers have gradually become aware of the deep significance of exploiting rural resources in a prudent manner, strengthening the protection and inheritance of cultural landscapes, and balancing residents' material and spiritual wealth in relation to rural construction [17].

Linpan settlements represent a distinct lifestyle in the farming area of the Sichuan Plain, where ancestral wisdom and adaptation to nature have formed a clever human-land relationship, which has been passed down to the present. In recent years the settlements' ecological, cultural and aesthetic values have been slowly discovered and attracted the attention of the government, which has issued many protection and restoration policies. Examples include "Promoting the Protection of Western Sichuan Linpan Settlements in the City," the "Conservative Construction Plan for Western Sichuan Rural Landscapes (Linpan settlements) in the City of Chengdu" and the "Implementation Plan for the Protection and Restoration of Western Sichuan Linpan Settlements in the City of Chengdu." All these projects emphasize that the rural panoramas of western Sichuan's Linpan settlements contain not only an ancient form of rural settlements, but also a special cultural heritage with extremely high ecological, economic, and historical value. However, while China's policies focus on the importance of the Linpan settlements and culture, advocating for the protection, they fail to specify rules. In 1992, the 16th session of the World Heritage Committee decided to incorporate cultural landscape heritage into the World Heritage List. Compared with the definition of other types of patrimony, cultural landscape heritage highlights the harmony between long-term production, human life, and nature, stressing a sustainable development ideology in which people coexist with the environment [18]. The sustainability explored in this study not only refers to the continuous development and increase of the material form of the cultural landscape, but also emphasizes the development of the connotation of the landscape cultural awareness based on human needs. Cultural landscapes are a product of the interaction between people and nature. They result from an initial social, economic, administrative, and/or religious imperative and have developed to its present form by association with and in response to its natural environment. Such landscapes reflect the process of evolution in terms of their form and component features [19]. In his book Conservation and Sustainability in Historic Cities, Rodwell asserts that it is not enough to comprehend the historical environment by relying on a limited number of cultural terminologies. He believes that the understanding of regional culture should not be constrained to archaeological research, which focuses on building age and associated historical events; rather, it should also involve the connections between physical entities and intangible cultural aspects [20]. Therefore, cultural landscapes provide a new perspective for exploring social evolution and the human-land relationship, and serve as the best example for understanding the latter. This is also why the protection of traditional Linpan settlements in the Chengdu Plain should begin with a detailed analysis.

2. Materials and Methods

2.1. Research Area

At more than 2300 years old, Western Sichuan's Linpan settlements date back to the ancient civilization era and took shape over a long period of immigration [21]. They are widely distributed across southwestern China, especially in the fan-shaped alluvial Chengdu Plain in western Sichuan Province. Starting from the Dujiangyan Irrigation Project, the Chengdu Plain's irrigation areas are flat, except for a northwest to southeast slope with an average gradient of 3–10%, where the relative surface relief is less than 20 m, and annual rainfall is about 900–1240 mm. The superior topography and abundant water sources form a highly networked gravity irrigation system. This study examines Juyuan Town in Dujiangyan City to scrutinize the sustainable development of cultural landscapes in rural parts of the Chengdu Plain (Figure 1). Juyuan Town covers a total area of 34 km^2 and has 10 administrative villages; the land cover is mainly composed of farmland, with rice and rapeseed as the chief agricultural products. Compared to 2005, by 2018, there were approximately 515 Linpan settlements in the region under the town's jurisdiction, with a decrease of 273 in the total number and 19.87 ha in terms of area. Juyuan Town is selected for the following reasons:

1. Juyuan Town is in the core zone of Dujiangyan City's elite irrigation district—it has a rich cultural history, which is of great significance for understanding the evolution of Linpan settlements. Moreover, the town is the first through which gravity irrigation water flows, making it a good model for examining the relationship between human activity and the natural environment;
2. In China, most rural areas are underdeveloped, and relevant information or data is complicated to obtain. However, since the Wenchuan earthquake in 2008, we have conducted a continuous follow-up study on Juyuan Town for nearly ten years. The previous research findings are very beneficial to the in-depth study of the area and similar rural areas;
3. It is a key municipality in Dujiangyan City (Key towns are geographically big with high populations, well-developed economies, and complete supportive facilities. Their construction effectively balances the distribution of relocated rural residents, and reduces the population burden on large cities), located in an urban-rural intersection. The town is facing the pressure of urbanization, as well as the threat of the destruction of traditional cultural landscapes. Thus, it is urgent that the sustainability of local cultural landscapes be investigated.

2.2. Defining Linpan Cultural Landscape

Based on the definition of CO Sauer, the cultural landscape has both time and space characteristics, which is the result of the continuous influence of human culture on a specific regional environment. Each traditional cultural landscape expresses a singular local spirit that helps shape its identity [22]. Historically, the Linpan settlement system in the western Sichuan Plain has changed with society's prosperity. During the pre-Qin period, the indigenous people of Laos began to establish settlements in the Chengdu Plain. The construction of the Dujiangyan Water Conservancy Project during the Qin and Han Dynasties stimulated the improvement of the agricultural and forestry management structure, and Linpan's living pattern gradually formed. During the Tang and Song Dynasties, with the prosperity of the economy, the Linpan model continued to improve and experienced large-scale expansion. In the late Ming and early Qing dynasties, due to war, natural disasters, and social and economic turmoil, the population of the western Sichuan Plain sharply reduced, and most of the Linpan settlements were damaged or uninhabited. In the late Qing Dynasty, with the gradual recovery of population and social reconstruction, the Linpan settlements cluster became more stable and has sustained itself until now through the process of constant adjustment and adaptation to the environment [21]. In terms of space, the Linpan settlement is geographically unique. The Linpan settlements are spread over the vast plains of western Sichuan with the surname (clan) as the settlement unit. With extensive farmlands, the structure of the Linpan settlement looks like a green island. It serves as a carrier for the harmonious symbiosis of humans, plants, and animals. It is a typical natural village (Figure 2). To date, although the Linpan settlements have not been uniformly defined in the academic community, scholars with different research perspectives have described them in various ways, such as social units with cultural symbols and use value (Duan); composite, rural scattered settlements that integrate human life, production, ecology and land (Fang); and rural living units that combine agricultural production and family relations (Chen) [23–25]. These definitions all focus on social relations, landscape patterns, and ecological structures. In addition, many scholars, such as Zhang, have proposed the concept of the Linpan system, arguing that it is a complete settlement layout representing the semi-natural, semi-artificial wetlands of the Dujiangyan irrigation region [26]. Japanese expert Ishikawa asserts that the Linpan system should be regarded as a remarkable historical and cultural landscape heritage [27]. This study proposes that the Linpan system consists of scattered communities, which are composed of a number of individual (Linpan) settlements. Such settlements are independent living spaces in rural habitats, dispersed throughout the Chengdu Plain. A Linpan community refers to a whole body of spatial continuities comprising river systems, road networks, fields, and settlements of a certain size, which embody a lifestyle with local characteristics (referred to as "the Linpan mode").

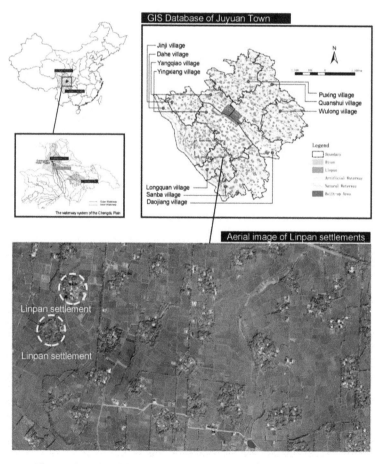

Figure 1. Location of the study region and an aerial view of the landscape.

Figure 2. Linpan landscape during the late Qing Dynasty and the Republic of China. Source: http://www.phoer.net.

2.3. Methodology

The concept of the cultural landscape is rooted in geography and was first put forth by German geographers in classifying fields, villages, towns and roads. Later, it was promoted as the focus of geography research by American geographer Sauer, who also proposed the "morphology of landscape," advocating for the use of actual observation of the surface landscape to study geographical phenomena [28]. The study of cultural landscapes initially focused on the description and observation of performance [29]. With the deepening of people's understanding of the landscape and the mutual influence of various disciplines, quantitative analysis methods such as computer technology and 3S have gradually achieved fruitful results in the study of dynamic landscape characteristics [30]. This study mainly uses the traditional geography description and classification method, combined with modern computer-aided technology, to systematically study the evolution process of Linpan culture landscape. The research methods include typology and comprehensive induction, field research, and case analysis. Based on systematology, this study draws on a large number of theories on cultural landscapes to identify and categorize the landscape features and evolution types of Linpan settlements. Finally, based on the idea of demand in defining sustainable development, the factors affecting landscape change are interpreted. Typology theory is an essential theoretical support for the study of Linpan landscape morphology, which is not just an epistemology that facilitates the understanding of the intrinsic formation mechanism and formal meaning of historical rural settlements; it is also used to guide design based on rational treatment of traditions as a prerequisite and adaptation to the requirements of new lifestyles [31]. A typological approach allows us to extract the structural components of the Linpan system, grounded in an analysis of existing settlements, and to speculate on the morphological transformation of characteristics in different contexts.

With respect to technology, this study establishes a spatial geographic information database of Juyuan Town between 2005–2018, and employs the spatial analysis module of a geographic information system (GIS) to conduct an in-depth examination of the Linpan system's spatial attributes, layout, and structural changes. Moreover, the study uses 1 km × 1 km grid cells as the unit of analysis to statistically count and summarize morphological variation features of the town. The boundaries of the Linpan system are determined by extracting historical remote sensing images from Google

Earth. Correlation analysis of spatial patterns is mainly based on 2005 data, as data after 2008 are largely subject to anthropogenic influence. The data used in the study are derived from relevant information provided by government departments (such as statistical data, planning documents, and local chronicles), as well as details obtained from field surveys (such as mapping, observation, interviews, and transcripts).

3. Case Study

3.1. The Hierarchy of Linpan Settlements

Linpan characteristics are based on two aspects: Spatial structure and human culture [32]. As the lowest level of spatial structure regarding human settlements in the Chengdu Plain, the Linpan system consists of numerous, widely distributed units, which signify the most fundamental building block of life [33]. As shown in Figure 3, each unit consists of five elements (house, courtyard, forest, farmland, and waterway) and is comprised of tens, sometimes dozens, of households. Water channels surround or pass through most Linpan units. The houses are concealed by tall trees (e.g., nanmu and cypress) and low forms (i.e., Omei Mountain and bitter) forms of bamboo, creating a tranquil scene where the inhabitants rest in the forest and work in the fields [34]. The boundaries of forests outside the Linpan settlement are often seen as criteria for defining their scale. Due to the different needs and habits of each family and the limitations of environmental conditions, Linpan settlements present a variety of features with the same hierarchy, but different shapes. For example, in the densely populated areas of the irrigation water system, there are more concentrated Linpans, and the Linpan settlements on both sides of the river are mostly long strips. As for humanistic value, the Linpan system offers a sustainable farming lifestyle that integrates humanity and nature, life and production. It relies and acts on the land, representing a local ideology of "harmony between man and nature" according to Taoist philosophy. Since antiquity, Linpan settlements have been dispersed throughout the fields, with each unit given a surname (based on patriarchal clans); the Linpan system is typical of *naturally formed villages* [35]. Over the long term, the Linpan system has been a vehicle for local farming culture and served as evidence of farming civilization. Embodying ancestral wisdom, this self-contained, composite agriculture and forestry ecosystem, founded on a small-scale peasant economy, has very high ecological, cultural and aesthetic value [36].

3.2. The Spatial Structure of Linpan Settlements

3.2.1. The Distribution of Linpan Settlements

Firstly, based on the GIS database, the area of each Linpan settlement in Juyuan Town was calculated and divided all of them into five grades through natural break classification method. Then the spatial pattern of Linpan settlements in different scales were analyzed and statistics respectively, this step is mainly based on the use of the Average Nearest Neighbor tool in ArcMap software, the calculation results are shown in Figure 4: The settlements in Juyuan Town have an average area of 0.73 ha and are small and medium-sized (less than 1.26 ha), accounting for 87% of the total number of settlements. The density of settlements in the town's jurisdictional region is about 23.62 units per km^2 (settlements larger than 2.16 ha in size are quite rare in most areas), and the average distance between adjacent settlements are about 134.19 m. The smaller the area, the closer the distance between adjacent Linpan settlements, and the spatial distribution is often characterized by agglomerated distribution. Conversely, the larger the area, the more scattered or random the distribution. On the whole, Linpan communities show dispersed distribution patterns in terms of spatial structure, and are characterized by dynamic evolution in the sequence of "aggregation, randomization, and dispersion" corresponding to the serial stages of their "occurrence, development, and stability."

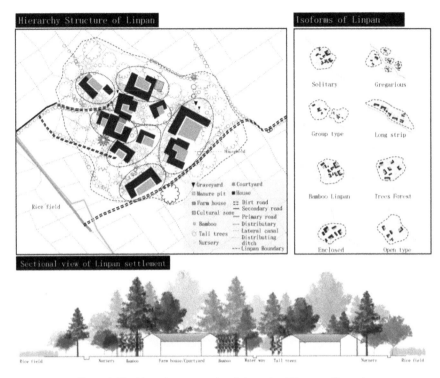

Figure 3. The hierarchy structural and elements of Linpan settlements.

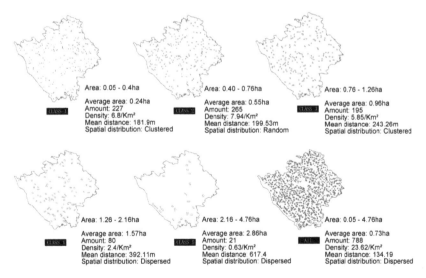

Figure 4. The spatial layout of Linpan settlements.

3.2.2. Relationship between Life and Production Space

Initially, most vernacular landscapes were created by farmers to organize their land better, and ensure their survival [37]. The Linpan system, containing typical farming settlements, provides farmers with *an ideal environment*, wherein production takes place on cultivated land along the periphery. The relationship between living and production spaces is reflected in the tillage radius, which refers to the distance from a settlement to cultivated land. The buffer analysis method is commonly used to analyze the radius of the cultivated land. Theoretically, in a specific area, when the area of the buffer generated by a specific radius is equal to the actual cultivated area, the radius of the buffer of the settlement is equal to the farming radius. At the same time, the ratio of the buffer area to the cultivated area can indicate farming pressure [38]. In this section, the evaluation unit is first determined by the boundary of the administrative division of the village area, and then the peripheral buffer range of different scales is established based on the Linpan boundary in the database (Table 1). The buffer area in each unit is counted using layer superposition in ArcMap software. Finally, the farmland pressure coefficient is calculated using the attribute field of each unit layer to link the corresponding cultivated land area data (Table 2):

$$I_i = H_i / G_i.$$

In the formula, I_i is defined as the cultivated pressure coefficient of the evaluation, which represents the rural labor carried out on the cultivated land in the evaluation unit (village); H_i is defined as the buffer area of the Linpan settlement in the evaluation unit; and G_i refers to the cultivated land within the unit. Tillage radius affects the residential layout in rural zones, serving as an important reference for exploring the human-land connection in farming areas and assessing the layout rationality of future protection planning [39].

The findings demonstrate that 50–60 m is the optimal tillage radius of a Linpan settlement in Juyuan Town (Table 1). Within this range, the tillage pressure coefficient of each village is close to 1, suggesting that this is the best spatial distribution for rural settlements and cultivated land in the process of natural adaptation (Table 2). That is, there is neither excessive labor nor a large amount of wasted arable land, which reflects the environmental adaptation of the Linpan-based residential mode according to different scales of land use.

3.3. Type Conversion

By constructing a spatial geographic information database in rural areas, it is possible to quickly identify changes in the spatial patterns of a given area over a period of time. Cultural landscapes are complex cultural phenomena that generally result from human activities acting on the earth's surface; they are composed of the natural environment, residential architecture, transportation, art, and people. Meanwhile, shifts in spatial patterns reflect long-term interactions between the landscape and the natural environment, the social economy, and historical development [40]. The basis for the classification of Linpan dynamic evolution is the morphological identification and field research (including questionnaires, interview minutes, phenomenon descriptions, etc.) of geospatial databases at two different time points. According to the overall system-to-local rules of the system theory, the dynamic changes of the Linpan settlement are summarized into three first-class categories of internal change, morphological change, and external landscape change, as well as the original site renewal, relocation, primary forest expansion, economic forest planting, municipal infrastructure, and social foundation. The facility has six sub-categories, and the different spatial representations of each sub-category are illustrated based on the typology performance method. The selection of typical samples is shown in Figure 5. In addition, based on interviews with 114 households in Juyuan Town and the results of several rounds of expert consultation, this paper used the demand hierarchy theory to mark the spatial shape change of Juyuan Town. For example, the demand for basic life, agricultural production, or non-agricultural supply is marked as survival demand; the outstanding desire for the development of individuals or local economic development and social development is marked as development demand; and people's spiritual needs, such as cultural heritage, and environmental protection, indicate quality demand. The classification description is shown in Table 3.

Table 1. Statistical table of buffer zones and actual farmland area in each village.

Village Name	Buffer Area (ha)										Actual Cultivated Area (ha)	Village Domain Area (ha)
	10 m	20 m	30 m	40 m	50 m	60 m	70 m	80 m	90 m	100 m		
Dahe Village	41.98	90.87	146.99	209.98	280.17	357.42	441.81	533.26	631.69	737.19	313.55	445.63
Daojiang Village	16.31	35.13	56.62	80.94	107.97	137.86	170.67	206.30	244.90	286.38	129.75	185.26
Jinji Village	18.41	39.93	64.51	92.39	123.29	157.02	193.68	233.26	275.87	321.49	163.61	219.78
Longquan Village	18.69	40.30	64.87	92.40	122.66	155.62	191.13	229.28	270.10	313.49	151.90	211.81
Puxing Village	33.19	71.05	113.37	160.07	211.22	266.88	327.06	391.72	460.55	533.79	284.74	392.81
Quanshui Village	37.53	81.03	130.63	186.46	248.45	316.73	391.25	471.64	558.23	651.23	341.31	484.35
Sanba Village	40.07	85.84	137.27	194.51	257.50	326.38	401.15	481.78	568.15	660.09	333.07	459.39
Wulong Village	31.04	67.60	109.61	156.90	209.59	267.72	331.23	400.17	474.26	553.39	301.53	391.49
Yangqiao Village	13.13	28.64	46.57	66.77	89.28	114.05	141.07	170.33	202.00	235.99	87.68	159.74
Yingxiang Village	35.14	76.11	122.93	175.80	234.74	299.62	370.48	447.30	529.95	618.80	268.31	386.01

Table 2. Statistical table of the farming pressure coefficient of each village.

Village Name	Tillage Pressure Coefficient ($0 \leq I$)									
	10 m	20 m	30 m	40 m	50 m	60 m	70 m	80 m	90 m	100 m
Dahe Village	0.13	0.29	0.47	0.67	0.89	1.14	1.41	1.70	2.01	2.35
Daojiang Village	0.13	0.27	0.44	0.62	0.83	1.06	1.32	1.59	1.89	2.21
Jinji Village	0.11	0.24	0.39	0.56	0.75	0.96	1.18	1.43	1.69	1.96
Longquan Village	0.12	0.27	0.43	0.61	0.81	1.02	1.26	1.51	1.78	2.06
Puxing Village	0.12	0.25	0.40	0.56	0.74	0.94	1.15	1.38	1.62	1.87
Quanshui Village	0.11	0.24	0.38	0.55	0.73	0.93	1.15	1.38	1.64	1.91
Sanba Village	0.12	0.26	0.41	0.58	0.77	0.98	1.20	1.45	1.71	1.98
Wulong Village	0.10	0.22	0.36	0.52	0.70	0.89	1.10	1.33	1.57	1.84
Yangqiao Village	0.15	0.33	0.53	0.76	1.02	1.30	1.61	1.94	2.30	2.69
Yingxiang Village	0.13	0.28	0.46	0.66	0.87	1.12	1.38	1.67	1.98	2.31

Note: I = 1 indicates the optimal spatial distribution state; I > 1 refers to a surplus labor force and high tillage pressure; I < 1 denotes excess arable lands, a labor force deficiency, and wasted resources

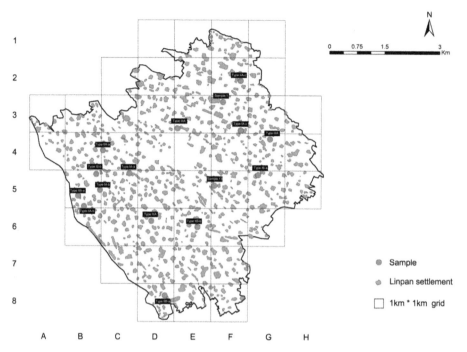

Figure 5. A classification sample of changes in a Linpan community.

Table 3. Classification specifications.

Type			Changes in Spatial Hierarchy		Needs Level
I	Internal changes	IA	on-site update	a. spatial replacement	Survival
				b. functional replacement	Survival
				c. old-for-new house replacement	Survival
		IB	off-site migration	a. concentration in new villages	Development
				b. off-site construction	Survival
II	Changes in shape	IIA	expansion of native forest	a. unconstrained expansion	Survival
				b. joint expansion	Survival
				c. constrained expansion	Survival
		IIB	planting of economic forest	a. non-scale planting	Development
				b. large-scale planting	Development
III	External landscape changes	IIIA	municipal infrastructure construction	a. enhancement and new construction of roads	Development
		IIIB	social infrastructure construction	a. construction of public facilities and business organizations	Quality

3.3.1. Type I: Internal Changes

Internal changes refer to alterations of the residential environment, as well as the structure and function of each individual Linpan settlement over time. Two modes are involved: On-site updates and off-site migration:

1. *Spatial replacement* (Figure 6a). More open space is obtained by reducing the number of trees to meet farmers' daily living needs (such as parking, leisure, and production). This reflects the self-regulation and lifestyle of local residents in relation to the residential environment, based on their customs and behavior when adapting to modern life.
2. *Functional replacement* (Figure 6b). Forests outside Linpan settlements provide shelter from the wind and protect the soil. At present, they are gradually being replaced by cement walls or iron

gauze, demonstrating improvement in residents' security awareness, promotion of land tenure awareness (i.e., boundaries), and greater concern about personal privacy [41].

3. *Old-for-new replacement* (Figure 6c).   With the enhancement of farmers' living standards, many traditional residential buildings in western Sichuan cannot meet the needs of modern living. Numerous old buildings in Linpan settlements have been demolished, with new houses erected on site. After the 2008 earthquake, countless old buildings were damaged to varying degrees. The post-disaster recovery and reconstruction resulted in the construction of multi-story (2–3 stories) single-family buildings with a large number of brick-concrete structures (verified through the field survey). To a large extent, this shows that improved *economic status* leads to changes in rural scenery. Off-site migration consists of two modes: Centralized building (at a different site) versus construction by residents (on their own).

4. *New rural communities*. As presented in Figure 6d, several neighboring Linpan settlements are merged into a group, while uniformly placing the original residents in newly built rural communities in response to the "Three Concentrations" policy (The "Three Concentrations" policy is the fundamental approach for coordinating urban and rural growth, promoting urban-rural integration, and solving the "three rural issues." Industry is concentrated in centralized development zones, farmers are placed in towns and new kinds of communities, and the land is used for operations on a moderate scale) regarding land readjustment in rural areas.

5. *Off-site construction*. As shown in Figure 6e, when Linpan settlements' size and relationship to the land are exploited to the maximum extent, the residents living there will build new homes around the original settlement, mostly 2–3 story, brick-concrete, single-family buildings, similar to those in type IA-c. The selection of the location for the new houses is no longer predicated on whether the new areas have well-developed river systems. Instead, the primary consideration becomes whether the new homes are in close proximity to roads.

3.3.2. Type II: Morphological Changes

As presented in Figure 7a–c, morphological changes consist of three modes: Unconstrained expansion, joint expansion, and constrained expansion:

1. In unconstrained expansion, the extension of the boundary is free from apparent environmental constraints. Afterwards, the morphology is similar to the original one. The expansion can be one of two types: The growth of native trees or the planting of new seedlings;

2. In joint expansion, adjacent Linpan settlements are gradually integrated into an overall body. Linpan settlements established through joint expansion are strip-shaped and contain two or more family clans with different surnames;

3. In constrained expansion, due to geographical obstacles (such as roads, canals and rivers), the boundary undergoes irregular extension. The Linpan settlements are unevenly distributed, and the shape after the expansion is inconsistent with the one from before.

Furthermore, the transformation of land-use types is a direct factor affecting changes in rural landscapes and has clearly identifiable spatial traits. As exhibited in Figure 7d,e, the market economy has triggered industrial structural adjustments, while the planting industry (of economic forests) has brought high economic benefits to farmers, with such market-driven trends becoming increasingly popular [42]. In Juyuan Town, economic forestry planting covers almost every village and can be divided into two types: Large-scale and non-scale. Large-scale planting is primarily carried out by several adjacent Linpan settlements to jointly manage an area and undertake forest planting. Large-scale planting is the uniform planting type, marked by regular borders in the planting area. In contrast, non-scale planting in economic forests mostly indicates spontaneous planting by some Linpan owners in the open space surrounding their own settlements. The forest types are not uniform, with the entire planting area showing a broken, patch-like pattern in the absence of an obvious spatial order.

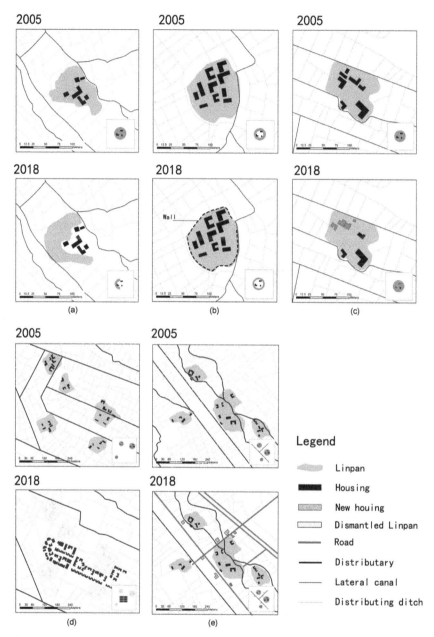

Figure 6. Schematic diagram of Type I changes. (**a**) Spatial replacement; (**b**) Functional replacement; (**c**) Old-for-new replacement; (**d**) New rural communities; (**e**) Off-site construction.

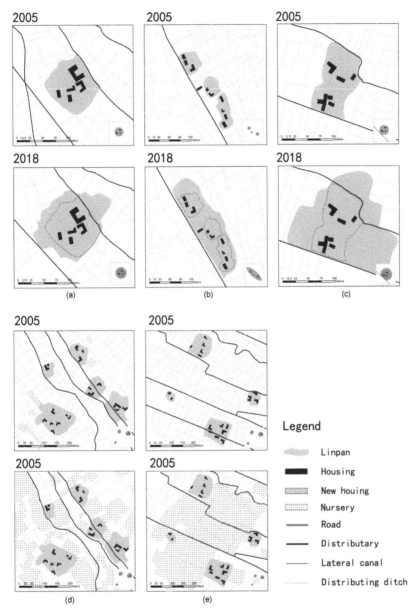

Figure 7. Schematic diagram of Type II changes. (**a**) Unconstrained expansion; (**b**) Joint expansion; (**c**) Constrained expansion; (**d**) Non-scale planting; (**e**) Large-scale planting.

3.3.3. Type III: External Landscape Changes

With the development of Juyuan Town and the surrounding municipalities, residents' demands for resources have become more diverse. The construction of supportive facilities necessary to meet daily needs has broken the spatial integrity and continuity of the original rural landscape. In particular, the layout of the transportation network has divided the continuous landscape matrix, and different kinds of land have been assigned different functions. External landscape changes are mainly reflected in the impact of urbanization-related construction on the landscape matrix, with roads and other facilities being built. There are roads with three purposes (Figure 8): New forms of transportation facilities (for example, two additional highways and one railway have been built in the past decade, as shown in Figure 8a); to enhance internal connections (Figure 8b) (i.e., the accessibility of Linpan settlements); and new village- and town-level roads, as well as upgrades made to them (Figure 8c). The construction of new facilities implies social infrastructure besides the addition and improvement of municipal infrastructure, such as cultural and leisure facilities (e.g., Western-style concert venues, cultural squares, activity centers, and resort hotels), and educational and medical facilities (e.g., Bayi High School, Qiyi Middle School, Sichuan Business and Technology Vocational College, general hospitals). In addition, some commercial organizations (e.g., automotive service companies, horticultural fields, and machine manufacturing companies) are slowly being introduced as the supporting infrastructure continues to be improved.

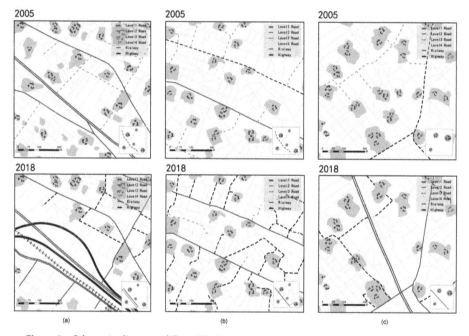

Figure 8. Schematic diagram of Type III changes. (**a**) Enhancement and new construction of roads; (**b**) Enhancement and new construction of roads; (**c**) Construction of public facilities and business organizations.

4. Results and Findings

4.1. Change Statistics

From 2005 to 2018, the total number of Linpan settlements in Juyuan Town decreased by 273, amounting to an average loss of six settlements per square kilometer. As shown in Figure 9, correlation analysis of the various types of Linpan settlements and their reduced areas, with respect to the spatial locations where the reduction occurs, reveals seven statistical grid cells each that have experienced a change in the quantity of Linpan settlements by one or less than one, largely in the border zone within the central urban area, with the city's surrounding green belt designed to control disorderly expansion and to protect the rural cultural landscape in the restoration and reconstruction process following an earthquake [43]. By contrast, areas—where the number of Linpan settlements has decreased by more than ten—are chiefly concentrated in the south of the built-up zone, indicating that the urban construction of Juyuan Town tends to extend southward. In most areas, the number of Linpan settlements was approximately ten, which also reflected the fact that the broad area of Linpan's traditional settlements was reduced. In other areas, the number of Linpan settlements was within ten. According to the change test, the main external material conditions that lead to a decrease in the number of Linpan settlements are the construction of roads (e.g., public roads, highways, and railways), the development of towns and new rural communities, and improvements in living service facilities (e.g., hospitals, schools, factories, and power stations). This new construction requires more land. The expansion of land for such purposes in the town is the most direct cause of the decline in the number of Linpan settlements.

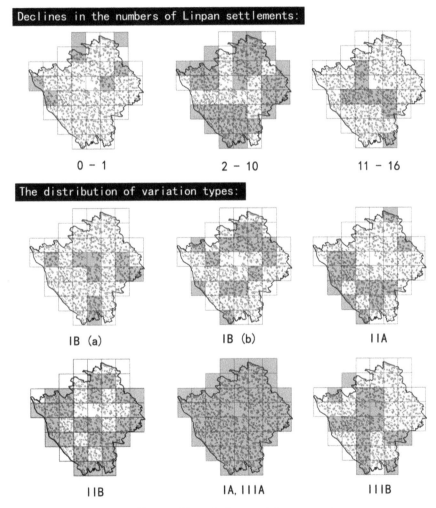

Figure 9. Schematic diagram of changes.

The grid statistics reveal that type IIIA (municipal infrastructure construction) and type IA (on-site renewal) are the most frequent landscape changes in the study region. Second, type IIB (economic forest planting) is a factor causing apparent visual shifts in the overall landscape pattern. A large number of trees planted for their economic value has destroyed the original landscape matrix, making the landscape discontinuous. Such fragmentation leads to a decrease in regional identity and sense of belonging [44]. Changes in settlement morphology resulting from the growth of the Linpan settlements' native and secondary forests (type IIA) are also common. Moreover, with respect to the geographical positions of these transformations (Figure 9), the construction of new rural communities (type IB) is largely concentrated in the center of town, where there are well-developed transportation facilities with convenient living conditions. Farmers' self-built homes are mostly distributed along both sides of the roads not far from the above-mentioned, large Linpan settlements. The survey and interviews reveal that when the number of people living in a Linpan settlement increases, people will build new houses in the adjacent open space (while in the past, a new Linpan settlement would be erected, rather than a new, single-family home). This embodies a randomized, jump-like expansion mode. Moreover, the construction of new and modern homes is preferentially conducted close to roads rather than river systems. This reveals that the traditional functions of the Linpan settlement have changed, and their farming culture is also disintegrating.

4.2. Discussion on Demand and Sustainability

The World Commission on Environment and Development defines sustainable development in the book Our Common Future (Brundtland Commission) as "development that meets the needs of today's people without compromising the ability of future generations to meet their needs." Moreover, it points out that sustainability includes the two concepts of "needs" and "restriction" [45]. The results of this study indicate that the needs of Juyuan Town's residents are transitioning from those of survival to those of development. Such growth, alongside the hierarchy of needs, is engendering changes in the cultural landscape of Juyuan Town's Linpan community. For example, types I (IA,IB-b) and II (IIA) tend to meet survival needs. The residents' new housing locations also indicate that the prerequisites for choosing a home have shifted from the convenience of production to the convenience of living. Meanwhile, types II (IIB-a) and III (IIIA) meet development needs. The layout of a large number of transportation networks and the construction of new rural communities represent a trend toward urbanization. Type III (IIIB) shows that people are striving to obtain more education and entertainment resources, improve living standards, and pursue a higher quality of life.

The differences in the categories of need have led to the spatial differentiation of rural landscape patterns. The more categories of need there are, the higher the requirements for landscape diversity. Meanwhile, the changes in the level of demands have caused the time differentiation of landscape patterns. For instance, the closer to the built-up area of the town center, the richer the various types of land, and the more quickly residents' needs can be met, which leads to a desire for another level of resources. The expansion of residents' needs, along with the hierarchy of needs, is the intrinsic driving force that affects the sustainability of the Linpan cultural landscape. Interviews reveal that population in the settlements is dominated by the elderly, as most young people leave to seek work: This means that the main human resources for protecting Linpan culture are constantly being lost. Although the basic structure of the Linpan settlements under investigation has remained intact, some are gradually showing signs of "empty nests", due to the neglect of management. In addition, changes in transportation patterns have a large impact on the land of the studied region. The arrangement of transportation networks accelerates the connections and energy exchanges between the interior and the outside world, while the modernization process of rural cultural landscapes has a high degree of synergy with the construction of transportation infrastructure.

According to this study, the sustainable development of traditional cultural landscape should have traceable spatial continuity, regional uniqueness, cultural accumulation, and landscape inheritance. The discussion of the sustainability of cultural landscape mainly refers to the quality requirements in the future development process, rather than the pursuit of the maximum number of regional resources or similar landscapes. The cultural landscape is a real reflection of personal lifestyle and an expression of how people meet their own needs at all levels. The cultural value of the traditional cultural landscape lies in the fact that the environment records and reproduces the methods of people's survival under previous material technology and thinking in the form of carriers. Then, with the progress of society, economic development, and the change of value system, traditional culture and lifestyle are gradually detached from modern society. Therefore, to achieve sustainability, traditional cultural landscapes must continue to meet the expectations of people's needs according to different levels of survival, development, and quality of life. For example, according to the case study, at the level of basic survival needs, the impact of people on the landscape is mainly due to the creating of residences, using the most inexpensive choice. In this respect, the maintenance of cultural landscape depends chiefly on the realization of science, that is, rapid advances in technology. From the perspective of development needs, once basic needs are guaranteed, people begin to consider physical renewal and optimization. At this stage, the impact of human activities on the environment essentially comes from the social environment and personal feelings. Here, the cultural landscape can sustain itself if the traditional functions of the landscape are balanced with contemporary needs. Finally, the relationship between the demand for quality of life and cultural landscape sustainability lies in the use of traditional values and aesthetic concepts to combine the material, technology, and culture of the time to establish a new cultural philosophy. To sum up, the sustainability of cultural landscapes does not mean that the traditional landscape is preserved and continued in its entirety; rather, it is adapted to social life, is people-centered, and can be organically inclusive and continuously updated in future development.

5. Conclusions

Linpan settlements are distributed as independent living units in the vast rural area of the Chengdu Plain. During conventional planning, the integrity and continuity of the Linpan community often become fragmented. As a result, traditional Linpan farming settlements have failed to receive effective protection and attention in the planning preparation process. This study establishes a spatial geographic information database of Juyuan Town. Through the application of the hierarchy description of the settlement system and the description of spatial features, the internal logic of the spatial layout of the traditional settlement culture landscape is expounded from three aspects: Hierarchy, geographical layout, and type conversion. Using the typology theory, the Linpan settlement group of Juyuan Town was used as a case study to demonstrate the dynamic evolution process of agricultural settlements in the western Sichuan Plain. Otherwise, the Linpan settlement landscape type conversion process was described in stages based on the demand hierarchy theory in sustainable development. The findings were as follows.

The Linpan settlement in Juyuan Town has a decreasing trend as a whole. Approximately six traditional Linpans per square kilometer have been demolished in the past decade. The new rural settlements in the region do not substantially retain the typical characteristics of the traditional settlements. The degradation of the Linpans' traditional functions and the changes in people's contemporary life concepts are the root causes of the abandonment of Linpan settlements. Historically, although these settlements have been highly self-sustainable and capable of organic self-replication and renewal, in the process of transformation and evolution, residents and settlements often have a unique symbiotic relationship of separation and integration. The differences in the level and type of need affect the changing trends of the landscape, and growth in terms of the hierarchy of needs influences the speed and magnitude of landscape transformations, thus resulting in the continuous reduction of the integrity, continuity, and recognizability of the landscape. Additionally, the development of the social economy will inevitably introduce higher requirements for the construction of rural areas. The construction of rural infrastructure not only promotes internal convenience, but also strengthens energy exchange with surrounding areas; however, it is also a double-edged sword. Due to the lack of awareness of traditional culture in the early stage, and the lack of inheritance, regional culture cannot resist the erosion from a foreign culture. Coupled with the dilemma of the younger generation migrating to urban environments and the aging of the local population, the leading human resources that protect and inherit the Linpan culture are continually being lost with the laying of the transportation network.

Landscapes have always been subject to alterations in order to better adapt to evolving social needs. It is necessary to correctly grasp the harmonious human-nature relationship in the Linpan farming settlements of western Sichuan plain, to explore the potential for the sustainable development of cultural landscapes, and to strengthen the understanding of rural cultural landscapes, which is an inevitable requirement to transform natural environment. The sustainable development of western Sichuan's Linpan communities depends on the strategic measures of spatial planning. According to this study, sustainable development of cultural landscape is an original implementation path that takes protection as the main factor and development as a supplement to maintain the identities, continuity, authenticity, and outstanding cultural value of the cultural landscape. House, courtyard, forest, field, and waterway are the five critical elements of the Linpan settlement hierarchy, which are the essential factors to ensure the sustainability of cultural landscape construction and landscape culture communication. In addition, the Linpan community appears to be in a random, but orderly spatial order, and these universal commonalities within the Linpan settlement are sustainable landscape inheritance characteristics. The hierarchy of needs to examine the spatial evolution of Linpan settlements' landscapes is adopted as a result of an inspiration during the field survey: It is hoped that the ultimate beneficiaries of the sustainable development of cultural landscapes are human beings, particularly the original residents, and that this will be not only the fundamental starting point of scientific development, but also a basic requirement for sustainable growth.

Author Contributions: Conceptualization, Q.L. and M.I.; methodology, Q.L. and M.I.; software, Q.L.; validation, Q.L., K.W. and M.I.; formal analysis, Q.L., K.W. and M.I.; investigation, Q.L., K.W. and M.I.; resources, Q.L., K.W. and M.I.; data curation, Q.L.; writing—original draft preparation, Q.L.; writing—review and editing, Q.L., K.W. and M.I.; visualization, Q.L.; supervision, K.W., M.I.

Funding: This research was funded by Sichuan University, Fundamental Research Funds for the Central Universities, grant number 2018skzx-pt243.

Acknowledgments: We gratefully acknowledge the support provided by the Dujiangyan government and villagers live in Juyuan Town. Thanks for the data and relative sources offered by the government.

Conflicts of Interest: The authors declare no conflict of interest.

References

1. Anderson, K.; Domosh, M.; Pile, S.; Thrift, N. *Handbook of Cultural Geography*; Sage: Thousand Oaks, CA, USA, 2002.
2. Mikesell, M.W. Landscape. *Int. Encycl. Soc. Sci.* **1968**, *8*, 575–580.
3. Fowler, P.J. *World Heritage Paper 6: World Heritage Cultural Landscape 1992–2002*; UNESCO World Heritage Centre: Paris, France, 2003; p. 22.
4. Sauer, C.O. The Morphology of Landscape. *Univ. Calif. Publ. Geogr.* **1925**, *2*, 19–54.
5. Bi, J. Research on the Sustainable Development of Wutai Mountain Cultural Landscape from the Perspective of Ecological Philosophy. Ph.D. Thesis, Shanxi University, Taiyuan, China, 2013.
6. Antrop, M. Why Landscapes of the Past Are Important for the Future. *Landsc. Urban. Plan.* **2005**, *70*, 21–34. [CrossRef]
7. Wei, H. *China's Urbanization: The Road to Harmony and Prosperity*; Social Sciences Academic Press: Beijing, China, 2014.
8. Antrop, M. Changing Patterns in the Urbanized Countryside of Western Europe. *J. Landsc. Ecol.* **2000**, *35*, 257–270. [CrossRef]
9. Ahadnejad, M.; Maruyama, Y.; Yamazaki, F. Evaluation and Forecast of Human Impacts Based on Land Use Changes Using Multi-Temporal Satellite Imagery and GIS: A Case Study on Zanjan, Iran. *J. Indian Soc. Remote* **2009**, *37*, 659–669. [CrossRef]
10. Araya, Y.H.; Cabral, P. Analysis and Modeling of Urban Land Cover Change in Setúbal and Sesimbra, Portugal. *Remote Sens.* **2010**, *2*, 1549–1563. [CrossRef]
11. Bürgi, M.; Straub, A.; Gimmi, U.; Salzmann, D. The Recent Landscape History of Limpach Valley, Switzerland: Considering Three Empirical Hypotheses on Driving Forces of Landscape Change. *Landsc. Ecol.* **2010**, *25*, 287–297. [CrossRef]

12. Jiao, Y. Synthetic Assessment on the Multifunction of Terrace Landscape in Ailao Mountains. *Yunnan Geogr. Environ. Res.* **2008**, *20*, 7–10.

13. Shi, M.; Xie, Y.; Cao, Q. The Landscape Evolution and Mechanism Analysis of Rural Settlements in the Oasis of Arid Region. *Geogr. Res.* **2016**, *35*, 692–702.

14. Stockdale, A.; Barker, A. Sustainability and the Multifunctional Landscape: An Assessment of Approaches to Planning and Management in the Cairngorms National Park. *Land Use Policy* **2009**, *26*, 479–492. [CrossRef]

15. World Rural Landscapes: A Worldwide Initiative for Global Conservation and Management of Rural Landscapes. Available online: http://www.worldrurallandscapes.org/ (accessed on 2 February 2015).

16. Fan, Y.; Wu, Y. The Sustainable Development of Qiang Ethnic Indigenous Villages in the Context of New Urbanization. *Eco-Econ.* **2014**, *30*, 47–51.

17. Liu, F. Study on Creating Rural Landscape Features under New Local View. *Anhui Archit.* **2013**, *19*, 19–20.

18. Fowler, P. World Heritage Cultural Landscapes, 1992–2002: A Review and Prospect. In *Cultural Landscapes: The Challenges of Conservation, Ferrara, Italy*; UNESCO World Heritage Centre: Paris, France, 2002; p. 16.

19. Council of Europe. *The European Landscape Convention*; Council of Europe: Strasbourg, France, 2000.

20. Rodwell, D. *Conservation and Sustainability in Historic Cities*; John Wiley & Sons: Hoboken, NJ, USA, 2008.

21. Fang, Z.; Zhou, J. Population, Cultivated Land and the Self-organization of Traditional Rural Settlement—With the Case of the Linpan Settlement System (1644–1911) in Chuanxi Plain. *Chin. Gard.* **2011**, *27*, 83–87.

22. Antrop, M.; Van, E.V. Holistic Aspects of Suburban Landscapes: Visual Image Interpretation and Landscape Metrics. *Landsc. Urban. Plan.* **2000**, *50*, 43–58. [CrossRef]

23. Duan, P.; Liu, T. *Linpan: Ecological Home of Shu Culture*; Sichuan Science and Technology Press: Chengdu, China, 2004.

24. Fang, Z. Essentials of Linpan Culture in Western Sichuan. Ph.D. Thesis, Chongqing University, Chongqing, China, 2012.

25. Sun, D. Research on the Protection and Development Model of Linpan Landscape Resources in Western Sichuan. Ph.D. Thesis, Sichuan Agricultural University, Ya'an, China, 2011.

26. Zhang, Y. Research on the Protection and Development of Linpan System in Western Sichuan. Ph.D. Thesis, Southwest Jiaotong University, Sichuan, China, 2008.

27. Ishikawa, M. Sustainable Development of Heritage Protection. *China Homes* **2016**, *5*, 44–45.

28. Turne, M.C.; Gardner, R.H. *Quantitative Methods in Landscape Ecology: The Analysis and Interpretation of Landscape Heterogeneity*; Springer: New York, NY, USA, 1990.

29. Tang, M. The Inventory and Progress of Cultural Landscape Study. *Prog. Geogr.* **2000**, *19*, 70–79.

30. Fan, J.; Zhang, W.; Lei, Y. Review and prospect of the development of landscape geography abroad. *World Geogr. Res.* **2007**, *1*, 83–89.

31. Zhang, Z.; Qiu, Z.; Wang, Z. Exploration of Evolution Mechanism and Renewal Strategy of Traditional Rural Settlements Based on Typology. *Archit. J.* **2017**, *S2*, 7–12. [CrossRef]

32. Liu, H. Study on the Rural Landscape Pattern and Settlement (Linpan) Pattern of Xujia Town, Dujiangyan City. Ph.D. Thesis, Sichuan Agricultural University, Sichuan, China, 2012.

33. Yin, L.; Cai, J. Study on the Sustainable Development Path of Linpan Landscape in Chengdu Plain. *Lands. Res.* **2010**, *12*, 11. (In English)

34. Fang, Z.; Li, X. Analysis on Linpan Cultural Value in Western Sichuan. *J. Xihua Univ. Philos. Soc. Sci.* **2011**, *30*, 26–30.

35. Chen, Q.; Yang, J.; Luo, S.; Sun, D. Identification and Extraction of the Linpan Culture Landscape Gene. *Trop. Geogr.* **2019**, *39*, 254–266.

36. Zheng, J. On Ecological Significance of Linpan Domain in West of Sichuan. *Shanxi Archit.* **2010**, *36*, 50–52.

37. Roberts, B. *Landscapes of Settlement: Prehistory to the Present*; Routledge: Abingdon, UK, 2013.

38. Qiao, W.; Wu, J.; Zhang, X.; Ji, Y.; Li, H.; Wang, Y. Optimization of Spatial Distribution of Rural Settlements at County Scale Based on Analysis of Farming Radius—A Case Study of Yongqiao District in Anhui Province. *Resour. Environ. Yangtze Basin* **2013**, *22*, 1557–1563.

39. Jiao, Y.; Hu, W.; Su, S.; Fan, T.; Yang, Y. Spatial Pattern and Farming Radius of Hani's Settlements in Ailao Mountain Using GIS. *Resour. Sci.* **2006**, *28*, 66–72.

40. Birks, H.H.; Birks, H.J.B.; Kaland, P.E.; Moe, D. (Eds.) *The Cultural Landscape: Past, Present and Future*; Cambridge University Press: Cambridge, UK, 1988.

41. Cai, X. Analysis of the Value of Traditional Rural Dwellings–Taking the Linpan Settlement in Western Sichuan as an Example. *Theory Reform* **2009**, *4*, 151–153.
42. Zheng, J.; Zhu, H. Research Report on the Protection, Development and Utilization of Linpan Culture in Western Sichuan. *J. Chengdu Munic. Party Coll. C.P.C.* **2008**, *2*, 71–74.
43. Ding, S.; Dianhong, Z. Research on the Zoning Methods in Rural Areas from the Perspective of Cultural Landscape: A Case Study of the Fan-Shape Area in Dujiangyan. *Urban. Rural Plan.* **2018**, *6*, 88–97.
44. Whelan, Y. *Heritage, Memory and the Politics of Identity: New Perspectives on the Cultural Landscape*; Routledge: Abingdon, UK, 2016.
45. Holden, E.; Linnerud, K.; Banister, D. Sustainable Development: Our Common Future Revisited. *Glob. Environ. Chang.* **2014**, *26*, 130–139. [CrossRef]

MDPI

St. Alban-Anlage 66

4052 Basel

Switzerland

Tel. +41 61 683 77 34

Fax +41 61 302 89 18

www.mdpi.com

Sustainability Editorial Office

E-mail: sustainability@mdpi.com

www.mdpi.com/journal/sustainability

Printed in the USA
CPSIA information can be obtained
at www.ICGtesting.com
LVHW061932231123
764772LV00010B/49